# APPLIED ASYMPTOTIC METHODS IN NONLINEAR OSCILLATIONS

# SOLID MECHANICS AND ITS APPLICATIONS
Volume 55



*Aims and Scope of the Series*

The fundamental questions arising in mechanics are: *Why?, How?,* and *How much?* The aim of this series is to provide lucid accounts written by authoritative researchers giving vision and insight in answering these questions on the subject of mechanics as it relates to solids.

The scope of the series covers the entire spectrum of solid mechanics. Thus it includes the foundation of mechanics; variational formulations; computational mechanics; statics, kinematics and dynamics of rigid and elastic bodies; vibrations of solids and structures; dynamical systems and chaos; the theories of elasticity, plasticity and viscoelasticity; composite materials; rods, beams, shells and membranes; structural control and stability; soils, rocks and geomechanics; fracture; tribology; experimental mechanics; biomechanics and machine design.

The median level of presentation is the first year graduate student. Some texts are monographs defining the current state of the field; others are accessible to final year undergraduates; but essentially the emphasis is on readability and clarity.

*For a list of related mechanics titles, see final pages.*

# Applied Asymptotic Methods in Nonlinear Oscillations

by

YU. A. MITROPOLSKII
*Institute of Mathematics,*
*Kiev, Ukraine*

and

NGUYEN VAN DAO
*Vietnam National University,*
*Hanoi, Vietnam*

SPRINGER-SCIENCE+BUSINESS MEDIA, B.V.

A C.I.P. Catalogue record for this book is available from the Library of Congress.

ISBN 978-90-481-4865-3 ISBN 978-94-015-8847-8 (eBook)
DOI 10.1007/978-94-015-8847-8

---

*Printed on acid-free paper*

## ABOUT THE AUTHORS

Professor Mitropolskii Yu. A. is internationally well known to mathematicians, physicists, scientists and researchers. He has to his credit more than 40 monographs translated in the many languages outside of his country and a large number of research papers. Professor Mitropolskii Yu. A. has been variously a member of the Russian and Ukrainian Academies of Sciences and a member of the Academy of Sciences of Bologne. At present he is actively engaged as a professor at Kiev State University and Academician - Secretary of the branch of mathematics, mechanics and cybernetics of the Ukrainian Academy of sciences.

Professor Nguyen Van Dao has earned a great reputation for his specialized work in the theory of nonlinear oscillations and mechanics. He has published more than one hundred papers and monographs. He has served as Vice - President of NCSR Vietnam and President of Vietnamese Association of Mechanics. He has been Editor in Chief of the Journal of Mechanics of Vietnam and a member of the Editorial Board of International Journals: Advances of Mechanics and Engineering and Automation. He has been elected as a foreign member of Academy of Sciences of Czechoslovakia (1988) and is an honorary member of Association "Znania" of Russia (1992).

CONTENTS

# PREFACE

Many dynamical systems are described by differential equations that can be separated into one part, containing linear terms with constant coefficients, and a second part, relatively small compared with the first, containing nonlinear terms. Such a system is said to be weakly nonlinear. The small terms rendering the system nonlinear are referred to as perturbations. A weakly nonlinear system is called quasi-linear and is governed by quasi-linear differential equations. We will be interested in systems that reduce to harmonic oscillators in the absence of perturbations.

This book is devoted primarily to applied asymptotic methods in nonlinear oscillations which are associated with the names of N. M. Krylov, N. N. Bogoliubov and Yu. A. Mitropolskii. The advantages of the present methods are their simplicity, especially for computing higher approximations, and their applicability to a large class of quasi-linear problems. In this book, we confine ourselves basically to the scheme proposed by Krylov, Bogoliubov as stated in the monographs [6, 21]. We use these methods, and also develop and improve them for solving new problems and new classes of nonlinear differential equations.

Although these methods have many applications in Mechanics, Physics and Technique, we will illustrate them only with examples which clearly show their strength and which are themselves of great interest.

A certain amount of more advanced material has also been included, making the book suitable for a senior elective or a beginning graduate course on nonlinear oscillations.

Usually, dynamic systems are classified and examined according to their number of "degrees of freedom". The number of degrees of freedom that a system possesses is the number of co-ordinates which must be specified in order to define its configuration. This book is constructed from another point of view: the originating mechanism of the oscillations. We divide oscillations into the following kinds:

1. Free Oscillations (Chapter 1)
2. Self-excited Oscillations (Chapter 2)
3. Forced Oscillations (Chapter 3)
4. Parametrically-excited Oscillations (Chapter 4)
5. Interactions between nonlinear Oscillations (Chapter 5)

The first two belong to autonomous systems described by differential equations in which time does not appear explicitly. The third and fourth are non-autonomous, governed by equations which contain time explicitly.

*Free or Natural Oscillations* - are motions occurring in an oscillatory system without external excitation; no energy is supplied from an external source during the oscillation. Usually, free oscillations arise when an oscillatory system is displaced from its equilibrium state.

*Self-Excited Oscillation* is one induced without any external force acting upon the oscillatory system. As in the case of free oscillation, the frequency of self-excited oscillation is determined by the oscillatory system itself. Self-excited oscillation is maintained by an energy source of non-oscillatory character.

*Forced Oscillation* is characterized by external excitation. The differential equations for forced oscillation always have a time-dependent term. The frequency of forced oscillation and of parametrically-excited oscillation is determined by the external excitation.

*Parametrically-Excited* Oscillation occurs in an oscillatory system when one or more of its parameters change, usually periodically, in time. In contrast to forced oscillation, the excitation is not effective if the oscillator remains in its position of equilibrium.

Nonlinear oscillations always exhibit some kinds of N. Minorsky said that "Perhaps the whole theory of nonlinear oscillations could be formed on the basis of interaction".

The last chapter - Chapter 6 - is devoted to the mathematical foundation of averaging methods and related problems.

The text has been designed to cover material from the elementary to the more advanced, in increasing order of difficulty. The exposition is largely by means of examples and problems, and these are drawn solely from mechanics. It naturally draws heavily on our own research and our experiences in using the asymptotic methods for many years. The methods discussed will certainly find fruitful application in other branches of Physics and Engineering, and we hope this book will prove useful to workers in those fields.

In the text, the symbols employed are largely conventional and the use of the same symbol for vastly different quantities has been avoided where possible. Definition of symbols are given where necessary. The formulae within each chapter of the text are numbered without stating the chapter number. If a formula of another Chapter is stated, the number of this Chapter is shown after the section number, for example, formula (1.2), Chapter 2.

This book is aimed at engineers, undergraduate and postgraduate students and scientists working in mechanics, physics, engineering and applied mathematics.

The authors are indebted to many colleagues, especially to Prof. G. M. L. Gladwell at University of Waterloo, Canada, for proofreading the manuscript and for making valuable suggestions. Any errors that remain are, of course, our responsibility. Last but not least, the authors wish to thank Mr. Phung Van Tieu for his expert and conscientious typing of the manuscript.

CHAPTER 1

# FREE OSCILLATIONS OF QUASI-LINEAR SYSTEMS

## 1. Free Oscillations of Systems Governed by a General Second Order Differential Equation

Many problems in Engineering, Technology and Physics, notably in electrical circuit theory, lead to the equation :

$$\frac{d^2x}{dt^2} + \omega^2 x = \varepsilon f\left(x, \frac{dx}{dt}\right), \tag{1.1}$$

where $\omega$ is a constant, $\varepsilon$ is a small positive dimensionless parameter characterizing the smallness of the function $f\left(x, \frac{dx}{dt}\right)$ in comparison with the linear term $\omega^2 x$. Sometimes, $\varepsilon$ is introduced artificially and used as a book-keeping device and will be set equal to unity in the final solution.

In general, the function $f\left(x, \frac{dx}{dt}\right)$ is a nonlinear function of the variables $x$, $\frac{dx}{dt}$. For the sake of simplicity, this function is frequently replaced by a linear function. In some cases, this is sufficient, but there are phenomena which cannot be explained by analysis of the linear approximation.

In this book, we concentrate on some aspects of differential equations which depend very strongly upon the fact that (1.1) is nonlinear.

If $\varepsilon = 0$, one has the harmonic oscillator

$$\frac{d^2x}{dt^2} + \omega^2 x = 0,$$

whose general solution is

$$x = a\cos\psi, \qquad \psi = \omega t + \theta.$$

The amplitude $a$ is constant and the total phase $\psi$ increases monotonically with $t$:

$$\frac{da}{dt} = 0, \qquad \frac{d\psi}{dt} = \omega.$$

The term $\varepsilon f(x, \dot{x})$ perturbs this simple situation. The presence of the nonlinear excitation ($\varepsilon \neq 0$) leads to the dependence of the momentary frequency $\frac{d\psi}{dt}$ on the amplitude and may give rise to a systematic increase or decrease of the amplitude of oscillations, depending respectively on the generation or absorption of energy by the exciting forces.

Taking into consideration all these facts, we seek the approximate solution of differential equation (1.1) by using the asymptotic expansion [6]

$$x = a\cos\psi + \varepsilon u_1(a,\psi) + \varepsilon^2 u_2(a,\psi) + \dots, \tag{1.2}$$

where $u_1, u_2, \dots$ are certain periodic functions of $\psi$ with period $2\pi$. As to $a$ and $\psi$ themselves, we shall try to determine them from the equations

$$\begin{aligned} \frac{da}{dt} &= \varepsilon A_1(a) + \varepsilon^2 A_2(a) + \dots, \\ \frac{d\psi}{dt} &= \omega + \varepsilon B_1(a) + \varepsilon^2 B_2(a) + \dots \end{aligned} \tag{1.3}$$

The problem is to determine the functions $u_1(a,\psi)$, $u_2(a,\psi)$, …, $A_1(a)$, $A_2(a)$, …, $B_1(a)$, $B_2(a)$ … in such a manner that their substitution into (1.1) satisfies this equation with prescribed accuracy.

Thus, instead of the single differential equation of the second order (1.1) in the unknown $x$, we have two separate differential equations (1.3) of the first order in the two unknowns $a$, $\psi$. As will be seen later, the determination of the coefficients of expansions (1.2), (1.3) is not difficult, but in view of the great complexity of the formulae, practically, if $\varepsilon$ is small enough, only two or three terms of the series are sufficient.

Truncating the expansions of these terms, i.e. putting

$$x = a\cos\psi + \varepsilon u_1(a,\psi) + \varepsilon^2 u_2(a,\psi) + \dots + \varepsilon^m u_m(a,\psi), \tag{1.4}$$

$$\begin{aligned} \frac{da}{dt} &= \varepsilon A_1(a) + \varepsilon^2 A_2(a,\psi) + \dots + \varepsilon^m A_m(a), \\ \frac{d\psi}{dt} &= \omega + \varepsilon B_1(a) + \varepsilon^2 B_2(a) + \dots + \varepsilon^m B_m(a), \\ &(m = 1, 2, \dots), \end{aligned} \tag{1.5}$$

we can obtain approximations of the first, second ... and subsequent orders. Therefore, the practical applicability of this method is defined not by the properties of convergence of the sums (1.4), (1.5) when $m \to \infty$, but by their asymptotic properties for given fixed $m$ and $\varepsilon \to 0$.

For this reason, we will not study the problem of convergence when $m \to \infty$, but we will consider the expansions (1.2), (1.3) as the formal developments which are necessary for construction of the asymptotic approximations (1.4).

In other words, we formulate the problem more carefully, as the problem of finding the functions:

$$u_1(a,\psi), u_2(a,\psi), \dots, A_1(a), A_2(a), \dots, B_1(a), B_2(a), \dots, \tag{1.6}$$

so that the asymptotic series (1.4), in which the functions $a$, $\psi$ are determined by equations (1.5), satisfies equation (1.1) to an accuracy of $\varepsilon^{m+1}$ and is the asymptotic expansion of the solution of equation (1.1) i.e.

$$x(t) = \sum_{n=0}^{m} \varepsilon^n u_n\big[a(t), \psi(t)\big] + O(\varepsilon^{m+1}), \text{ when } \varepsilon \to 0, \tag{1.7}$$

where: $u_0\big(a(t), \psi(t)\big) = a(t)\cos\psi(t)$; and $a(t)$, $\psi(t)$ are the solutions of equations (1.5).

One must note that there is a certain arbitrariness in selecting the functions $u_i(a, \psi)$. Suppose that one starts with some arbitrary functions $\alpha_1(a)$, $\alpha_2(a)$, ..., $\beta_1(a)$, $\beta_2(a)$, ..., respectively, for $A_i$ and $B_i$ and one replaces in (1.4) and (1.5) $a$ and $\psi$ by expressions

$$a = b + \varepsilon\alpha_1(b) + \varepsilon^2\alpha_2(b) + \dots,$$
$$\psi = \varphi + \varepsilon\beta_1(b) + \varepsilon^2\beta_2(b) + \dots$$

One obtains then, instead of (1.4), similar equations but with different coefficients:

$$\begin{aligned} x &= b\cos\varphi + \varepsilon\big[\alpha_1(b)\cos\varphi - b\beta_1(b)\sin\varphi + u_1(b,\varphi)\big] + O(\varepsilon^2), \\ \frac{db}{dt} &= \varepsilon A_1(b) + \varepsilon^2\Big[\frac{dA_1(b)}{db}\alpha_1(b) - \frac{d\alpha_1(b)}{db}A_1(b) + A_2(b)\Big] + O(\varepsilon^3), \\ \frac{d\varphi}{dt} &= \omega + \varepsilon B_1(b) + \varepsilon^2\Big[\frac{dB_1(b)}{db}\beta_1(b) - \frac{d\beta_1(b)}{db}A_1(b) + B_2(b)\Big] + O(\varepsilon^3). \end{aligned} \tag{1.8}$$

It is clear that one has to impose some additional conditions in order to remove this arbitrariness at the start. One could require, for instance, that there is no fundamental harmonic in $u_i(a, \psi)$. This assumption leads to the conditions

$$\begin{aligned} &\int_0^{2\pi} u_1(a,\psi)\cos\psi\, d\psi = 0, \quad \int_0^{2\pi} u_2(a,\psi)\cos\psi\, d\psi = 0, \dots \\ &\int_0^{2\pi} u_1(a,\psi)\sin\psi\, d\psi = 0, \quad \int_0^{2\pi} u_2(a,\psi)\sin\psi\, d\psi = 0, \dots \end{aligned} \tag{1.9}$$

The physical significance of this assumption is that $a$ will be the full amplitude of the fundamental harmonic of oscillation.

From (1.2), one has:

$$
\begin{aligned}
\frac{dx}{dt} = &\frac{da}{dt}\Big\{\cos\psi + \varepsilon\frac{\partial u_1}{\partial a} + \varepsilon^2\frac{\partial u_2}{\partial a} + \dots\Big\} + \\
&+ \frac{d\psi}{dt}\Big\{-a\sin\psi + \varepsilon\frac{\partial u_1}{\partial\psi} + \varepsilon^2\frac{\partial u_2}{\partial\psi} + \dots\Big\},\\
\frac{d^2x}{dt^2} = &\frac{d^2a}{dt^2}\Big\{\cos\psi + \varepsilon\frac{\partial u_1}{\partial a} + \varepsilon^2\frac{\partial u_2}{\partial a} + \dots\Big\} + \frac{d^2\psi}{\partial t^2}\Big\{-a\sin\psi + \\
&+ \varepsilon\frac{\partial u_1}{\partial\psi} + \varepsilon^2\frac{\partial u_2}{\partial\psi} + \dots\Big\} + \Big(\frac{da}{dt}\Big)^2\Big\{\varepsilon\frac{\partial^2 u_1}{\partial a^2} + \varepsilon^2\frac{\partial^2 u_2}{\partial a^2} + \dots\Big\} + \\
&+ 2\frac{da}{dt}\cdot\frac{d\psi}{dt}\Big\{-\sin\psi + \varepsilon\frac{\partial^2 u_1}{\partial a\partial\psi} + \varepsilon^2\frac{\partial^2 u_2}{\partial a\partial\psi} + \dots\Big\} + \\
&+ \Big(\frac{d\psi}{dt}\Big)^2\Big\{-a\cos\psi + \varepsilon\frac{\partial^2 u_1}{\partial\psi^2} + \varepsilon^2\frac{\partial^2 u_2}{\partial\psi^2} + \dots\Big\}.
\end{aligned} \tag{1.10}
$$

From (1.3), one obtains also expressions:

$$
\begin{aligned}
\frac{d^2a}{dt^2} &= \Big(\varepsilon\frac{dA_1}{da} + \varepsilon^2\frac{dA_2}{da} + \dots\Big)\big(\varepsilon A_1 + \varepsilon^2 A_2 + \dots\big) = \varepsilon^2 A_1\frac{dA_1}{da} + O(\varepsilon^3),\\
\frac{d^2\psi}{dt^2} &= \Big(\varepsilon\frac{dB_1}{da} + \varepsilon^2\frac{dB_2}{da} + \dots\Big)\big(\varepsilon A_1 + \varepsilon^2 A_2 + \dots\big) = \varepsilon^2 A_1\frac{dB_1}{da} + O(\varepsilon^3),\\
\Big(\frac{da}{dt}\Big)^2 &= \big(\varepsilon A_1 + \varepsilon^2 A_2 + \dots\big)^2 = \varepsilon^2 A_1^2 + O(\varepsilon^3),
\end{aligned} \tag{1.11}
$$

$$
\begin{aligned}
\frac{da}{dt}\frac{d\psi}{dt} &= \big(\varepsilon A_1 + \varepsilon^2 A_2 + \dots\big)\big(\omega + \varepsilon B_1 + \varepsilon^2 B_2 + \dots\big) = \\
&= \varepsilon A_1\omega + \varepsilon^2\big(A_2\omega + A_1 B_1\big) + O(\varepsilon^3),\\
\Big(\frac{d\psi}{dt}\Big)^2 &= \big(\omega + \varepsilon B_1 + \varepsilon^2 B_2 + \dots\big)^2 = \\
&= \omega^2 + \varepsilon 2\omega B_1 + \varepsilon^2\big(B_1^2 + 2\omega B_2\big) + O(\varepsilon^3).
\end{aligned}
$$

Substituting these various expressions into (1.10), one has:

$$
\begin{aligned}
\frac{dx}{dt} = &-a\omega\sin\psi + \varepsilon\Big\{A_1\cos\psi - aB_1\sin\psi + \omega\frac{\partial u_1}{\partial\psi}\Big\} + \\
&+ \varepsilon^2\Big\{A_2\cos\psi - aB_2\sin\psi + A_1\frac{\partial u_1}{\partial a} + B_1\frac{\partial u_1}{\partial\psi} + \omega\frac{\partial u_2}{\partial\psi}\Big\} + O(\varepsilon^3),\\
\frac{d^2x}{dt^2} = &-a\omega^2\cos\psi + \varepsilon\Big\{-2\omega A_1\sin\psi - 2\omega aB_1\cos\psi + \omega^2\frac{\partial^2 u_1}{\partial\psi^2}\Big\} + \\
&+ \varepsilon^2\Big\{\Big(A_1\frac{dA_1}{da} - aB_1^2 - 2\omega aB_2\Big)\cos\psi - \Big(2\omega A_2 + 2A_1B_1 + \\
&+ A_1\frac{dB_1}{da}a\Big)\sin\psi + 2\omega A_1\frac{\partial^2 u_1}{\partial a\partial\psi} + 2\omega B_1\frac{\partial^2 u_1}{\partial\psi^2} + \omega^2\frac{\partial^2 u_2}{\partial\psi^2}\Big\} + O(\varepsilon^3).
\end{aligned}
$$

Then the left-hand side of equation (1.1) can be represented in the form:

$$\frac{d^2x}{dt^2}+\omega^2x=\varepsilon\Big\{-2\omega A_1\sin\psi-2\omega aB_1\cos\psi+\omega^2\frac{\partial^2u_1}{\partial\psi^2}+\omega^2u_1\Big\}+$$
$$+\varepsilon^2\Big\{\Big(A_1\frac{dA_1}{da}-aB_1^2-2\omega aB_2\Big)\cos\psi-$$
$$-\Big(2\omega A_2+2A_1B_1+A_1\frac{dB_1}{da}a\Big)\sin\psi+ \tag{1.12}$$
$$+2\omega A_1\frac{\partial^2u_1}{\partial a\partial\psi}+2\omega B_1\frac{\partial^2u_1}{\partial\psi^2}+\omega^2\frac{\partial^2u_2}{\partial\psi^2}+\omega^2u_2\Big\}+O(\varepsilon^3).$$

As to the right-hand side of (1.1), it is:

$$\varepsilon f\Big(x,\frac{dx}{dt}\Big)=\varepsilon f(a\cos\psi,-a\omega\sin\psi)+\varepsilon^2\Big\{u_1f'_x(a\cos\psi,-a\omega\sin\psi)+$$
$$+\Big(A_1\cos\psi-aB_1\sin\psi+\omega\frac{\partial u_1}{\partial\psi}\Big)f'_{\dot{x}}(a\cos\psi,-a\omega\sin\psi)\Big\}+\varepsilon^3\dots \tag{1.13}$$

Equating the like powers of $\varepsilon$ on both sides, we obtain a recursive system of equations:

$$\omega^2\Big(\frac{\partial^2u_1}{\partial\psi^2}+u_1\Big)=f_0(a,\psi)+2\omega A_1\sin\psi+2\omega aB_1\cos\psi,$$
$$\omega^2\Big(\frac{\partial^2u_2}{\partial\psi^2}+u_2\Big)=f_1(a,\psi)+2\omega A_2\sin\psi+2\omega aB_2\cos\psi, \tag{1.14}$$
$$\dots\dots\dots\dots\dots\dots$$
$$\omega^2\Big(\frac{\partial^2u_m}{\partial\psi^2}+u_m\Big)=f_{m-1}(a,\psi)+2\omega A_m\sin\psi+2\omega aB_m\cos\psi,$$

where:

$$f_0(a,\psi)=f(a\cos\psi,-a\omega\sin\psi),$$

$$f_1(a,\psi)=u_1f'_x(a\cos\psi,-a\omega\sin\psi)+\Big[A_1\cos\psi-aB_1\sin\psi+$$
$$+\omega\frac{\partial u_1}{\partial\psi}\Big]f'_{\dot{x}}(a\cos\psi,-a\omega\sin\psi)+\Big(aB_1^2-A_1\frac{dA_1}{da}\Big)\cos\psi+$$
$$+\Big(2A_1B_1+A_1a\frac{dB_1}{da}\Big)\sin\psi-2\omega A_1\frac{\partial^2u_1}{\partial a\partial\psi}-2\omega B_1\frac{\partial^2u_1}{\partial\psi^2}\cdot \tag{1.15}$$

It is clear that $f_k(a,\psi)$ is a periodic function of $\psi$ with period $2\pi$, depending also on the amplitude $a$ its explicit expression is determined as soon as one determines $A_j(a)$, $B_j(a)$, $u_j(a,\psi)$ to the $k$-th order.

In order to determine the functions $A_1(a)$, $B_1(a)$ and $u_1(a,\psi)$ from the first equation of the system (1.14), we consider the Fourier developments for $f_0(a,\psi)$ and $u_1(a,\psi)$:

$$\begin{aligned} f_0(a,\psi) &= g_0(a) + \sum_{n=1}^{\infty} \Big\{ g_n(a)\cos n\psi + h_n(a)\sin n\psi \Big\}, \\ u_1(a,\psi) &= v_0(a) + \sum_{n=1}^{\infty} \Big\{ v_n(a)\cos n\psi + w_n(a)\sin n\psi \Big\}. \end{aligned} \tag{1.16}$$

Substituting these values into the first equation of (1.14), one has:

$$\begin{aligned} &\omega^2 v_0(a) + \sum_{n=1}^{\infty} \omega^2(1-n^2)\{v_n(a)\cos n\psi + w_n(a)\sin n\psi\} = \\ &= g_0(a) + \big\{g_1(a) + 2\omega a B_1\big\}\cos\psi + \big\{h_1(a) + 2\omega A_1\big\}\sin\psi + \\ &+ \sum_{n=2}^{\infty}\{g_n(a)\cos n\psi + h_n(a)\sin n\psi\}. \end{aligned}$$

Equating coefficients of like harmonics, one obtains:

$$\begin{aligned} &g_1(a) + 2\omega a B_1 = 0, \quad h_1(a) + 2\omega A_1 = 0, \quad v_0(a) = \frac{g_0(a)}{\omega^2}, \\ &v_n(a) = \frac{g_n(a)}{\omega^2(1-n^2)}, \quad w_n(a) = \frac{h_n(a)}{\omega^2(1-n^2)} \quad (n = 2,3,\ldots). \end{aligned} \tag{1.17}$$

One has thus $A_1(a)$, $B_1(a)$ and all harmonics of the development of $u_1(a,\psi)$ except the first two $v_1(a)$ and $w_1(a)$. However, in view of the requirement that none of the $u_1(a,\psi)$ should have a fundamental harmonic, one has $v_1(a) = 0$, $w_1(a) = 0$, so that

$$u_1(a,\psi) = \frac{g_0(a)}{\omega^2} + \frac{1}{\omega^2}\sum_{n=2}^{\infty} \frac{g_n(a)\cos n\psi + h_n(a)\sin n\psi}{1-n^2}. \tag{1.18}$$

Since $u_1(a,\psi)$, $A_1(a)$ and $B_1(a)$ are determined, one knows also $f_1(a,\psi)$; its Fourier development is:

$$f_1(a,\psi) = g_0^{(1)}(a) + \sum_{n=1}^{\infty}\{g_n^{(1)}(a)\cos n\psi + h_n^{(1)}(a)\sin n\psi\}. \tag{1.19}$$

Using the second equation of (1.14) and conditions (1.9), one has also:

$$\begin{aligned} &g_1^{(1)}(a) + 2\omega a B_2 = 0, \qquad h_1^{(1)}(a) + 2\omega A_2 = 0, \\ &u_2(a,\psi) = \frac{g_0^{(1)}(a)}{\omega^2} + \frac{1}{\omega^2}\sum_{n=2}^{\infty} \frac{g_n^{(1)}(a)\cos n\psi + h_n^{(1)}(a)\sin n\psi}{1-n^2}. \end{aligned} \tag{1.20}$$

The determination of the higher order approximations is traightforward. The conditions (1.9) eliminate the fundamental harmonic in the functions $u_i(a,\psi)$ and this, in turn, guarantees the absence of secular terms in all successive approximations.

Consider the first approximation:

$$x = a\cos\psi + \varepsilon u_1(a,\psi), \tag{1.21}$$

$$\begin{aligned} \frac{da}{dt} &= \varepsilon A_1(a), \\ \frac{d\psi}{dt} &= \omega + \varepsilon B_1(a). \end{aligned} \tag{1.22}$$

The system of equations (1.22) will be called the equations of the first approximation.

Let $\widetilde{A}_1$ and $\widetilde{B}_1$ be average values of $A_1(a)$ and $B_1(a)$ in an interval $(0,t)$, that is from (1.22):

$$\Delta a = a(t) - a(0) \sim \varepsilon t \cdot \widetilde{A}_1,$$
$$\Delta(\psi - \omega t) = [\psi(t) - \omega t] - \psi(0) \sim \varepsilon t \cdot \widetilde{B}_1.$$

It is seen that the time $t$, during which $a$ and $\psi-\omega t$ may acquire finite changes, is of the order $1/\varepsilon$. On the other hand, the equations (1.22) of the first approximation arise from neglecting terms with $\varepsilon^2$ in equation (1.3). An error of this order in $\frac{da}{dt}$ and $\frac{d\psi}{dt}$ during $t$ leads to an error of order $\varepsilon^2 t$ in the functions $a$ and $\psi$. Hence, in the interval, during which $a$ and $\psi - \omega t$ undergo finite changes, errors in these quantities are of the order $\varepsilon$. Thus, there is no necessity for keeping the term $\varepsilon u_1(a,\psi)$ in (1.21) inasmuch as the error of using $x = a\cos\psi$ is of the same order. It is simpler to take:

$$x = a\cos\psi, \tag{1.23}$$

as **the first approximation**, where $a$ and $\psi$ are determined from (1.22).

In a similar manner, for **the second approximation**, one has the expression (1.21), in which, $a$ and $\psi$ are given by the differential equations

$$\begin{aligned} \frac{da}{dt} &= \varepsilon A_1(a) + \varepsilon^2 A_2(a), \\ \frac{d\psi}{dt} &= \omega + \varepsilon B_1(a) + \varepsilon^2 B_2(a). \end{aligned} \tag{1.24}$$

From the general formulae (1.18) - (1.20), one has the following explicit expressions for $A_1(a)$, $B_1(a)$, $A_2(a)$, $B_2(a)$ and $u_1(a,\psi)$:

$$
\begin{aligned}
A_1(a) &= -\frac{1}{2\pi\omega}\int_0^{2\pi} f(a\cos\psi, -a\omega\sin\psi)\sin\psi d\psi,\\
B_1(a) &= -\frac{1}{2\pi\omega a}\int_0^{2\pi} f(a\cos\psi, -a\omega\sin\psi)\cos\psi d\psi,\\
u_1(a,\psi) &= \frac{g_0(a)}{\omega^2} - \frac{1}{\omega^2}\sum_{n=2}^{\infty}\frac{g_n(a)\cos n\psi + h_n(a)\sin n\psi}{n^2-1},
\end{aligned}
\tag{1.25}
$$

where $g_n(a)$, $h_n(a)$ are defined by the formulae:

$$
\begin{aligned}
g_n(a) &= \frac{1}{\pi}\int_0^{2\pi} f(a\cos\psi, -a\omega\sin\psi)\cos n\psi d\psi,\\
h_n(a) &= \frac{1}{\pi}\int_0^{2\pi} f(a\cos\psi, -a\omega\sin\psi)\sin n\psi d\psi,
\end{aligned}
\tag{1.26}
$$

$$
\begin{aligned}
A_2(a) = &-\frac{1}{2\omega}\left\{2A_1B_1 + A_1\frac{dB_1}{da}a\right\} - \frac{1}{2\pi\omega}\int_0^{2\pi}\Big[u_1(a,\psi)f'_x(a\cos\psi, -a\omega\sin\psi)\\
&+ \left(A_1\cos\psi - aB_1\sin\psi + \omega\frac{\partial u_1}{\partial\psi}\right)f'_{\dot{x}}(a\cos\psi, -a\omega\sin\psi)\Big]\sin\psi d\psi,\\
B_2(a) = &-\frac{1}{2\omega}\left\{B_1^2 - \frac{A_1}{a}\frac{dA_1}{da}\right\} - \frac{1}{2\pi\omega a}\int_0^{2\pi}\Big[u_1(a,\psi)f'_x(a\cos\psi, -a\omega\sin\psi) +\\
&+ \left(A_1\cos\psi - aB_1\sin\psi + \omega\frac{\partial u_1}{\partial\psi}\right)f'_{\dot{x}}(a\cos\psi, -a\omega\sin\psi)\Big]\cos\psi d\psi.
\end{aligned}
\tag{1.27}
$$

It is noted that the equations of the second approximation (1.24), from which $A_2(a)$ and $B_2(a)$ are determined by (1.27), are complicated because they are written in the general form. As will be seen later, for concrete oscillatory systems, those equations become simpler.

Sometimes, instead of (1.23), it is useful to consider the expression (1.21) with $a$ and $\psi$, determined by equations (1.22), i.e. by the equations of the first approximation. Such an approximate solution is called **the refinement of the first approximation**, inasmuch as it incorporates the high harmonics contained in the functions $u_1(a,\psi)$, while the degree of the error remains the same as for formula (1.23).

In many applied problems, the equations of the first approximation are sufficient. We return to a more detailed study of these equations:

$$\begin{aligned} \frac{da}{dt} &= \frac{-\varepsilon}{2\pi\omega}\int_0^{2\pi} f(a\cos\psi, -a\omega\sin\psi)\sin\psi d\psi, \\ \frac{d\psi}{dt} &= \omega - \frac{\varepsilon}{2\pi\omega a}\int_0^{2\pi} f(a\cos\psi, -a\omega\sin\psi)\cos\psi d\psi. \end{aligned} \tag{1.28}$$

This form of the equations of the first approximation is generally used in applied problems.

Van der Pol derived equations (1.28) in a different way, by varying phase and amplitude. We first observe that for $\varepsilon = 0$, (1.1) has the solution:

$$\begin{aligned} x &= a\cos(\omega t + \theta), \\ \frac{dx}{dt} &= -a\omega\sin(\omega t + \theta), \end{aligned} \tag{1.29}$$

where the amplitude $a$ and the phase $\theta$ are constant. When $\varepsilon \neq 0$, we consider $a$, $\theta$, as new unknown functions of the time, which are to be determined so that (1.29) becomes a solution of (1.1). We must have first:

$$\frac{dx}{dt} = \frac{da}{dt}\cos(\omega t + \theta) - a\frac{d\theta}{dt}\sin(\omega t + \theta) - a\omega\sin(\omega t + \theta). \tag{1.30}$$

Hence, if we impose (1.29), then:

$$\frac{da}{dt}\cos(\omega t + \theta) - a\frac{d\theta}{dt}\sin(\omega t + \theta) = 0. \tag{1.31}$$

From the second equation of (1.29), we deduce:

$$\frac{d^2x}{dt^2} = -\omega\cdot\frac{da}{dt}\sin(\omega t + \theta) - \omega a\frac{d\theta}{dt}\cos(\omega t + \theta) - a\omega^2\cos(\omega t + \theta),$$

and so finally from (1.1)

$$-\omega\frac{da}{dt}\sin\psi - \omega a\frac{d\theta}{dt}\cos\psi = \varepsilon f(a\cos\psi, -a\omega\sin\psi), \quad \psi = \omega t + \theta. \tag{1.32}$$

By combining with (1.31), we arrive at

$$\begin{aligned} \frac{da}{dt} &= -\frac{\varepsilon}{\omega} f(a\cos\psi, -a\omega\sin\psi)\sin\psi, \\ \frac{d\theta}{dt} &= -\frac{\varepsilon}{\omega a} f(a\cos\psi, -a\omega\sin\psi)\cos\psi. \end{aligned} \tag{1.33}$$

Up to now no more than a coordinate transformation has been carried out, and the equations (1.33) are still exactly equivalent to (1.1). The derivatives $\frac{da}{dt}, \frac{d\theta}{dt}$ are proportional to the small parameter $\varepsilon$, so that $a$, and $\theta$ will be slowly varying functions of the time during the period $T = \frac{2\pi}{\omega}$, and as a first approximation we may, therefore, consider them as constant and replace the right-hand sides of (1.33) by their time-averages over $(t, t+T)$. If this is done, one has the equations of form (1.28).

If we compare the approximate equations (1.28) with the exact equations (1.33), we find that **the former are obtained from the latter by averaging the right-hand sides with respect to time, assuming $a$ and $\theta$ to be constant.** This process, duly generalized, will be described as the **Averaging Principle**, the mathematical foundation of which will be given in Chapter 6.

We conclude that the first approximation to the solution of (1.1) is of the form (1.23), where the amplitude $a$ and the full phase $\psi$ are to be determined from (1.28).

## 2. Conservative Systems

Let us consider a particular case of equation (1.1) when the function $f$ does not contain $\frac{dx}{dt}$ (for example, the equation for the frictionless pendulum)

$$\frac{d^2x}{dt^2} + p(x) = 0. \tag{2.1}$$

Here, one has a conservative system.

It is assumed that the function p(x) is nearly linear so that

$$p(x) = \omega^2 x + \varepsilon\Phi(x), \tag{2.2}$$

and hence, instead of (1.28):

$$\frac{da}{dt} = \frac{\varepsilon}{2\pi\omega}\int_0^{2\pi} \Phi(a\cos\psi)\sin\psi d\psi, \tag{2.3}$$

$$\frac{d\psi}{dt} = \omega + \frac{\varepsilon}{2\pi\omega a}\int_0^{2\pi} \Phi(a\cos\psi)\cos\psi d\psi = \Omega_I(a). \tag{2.4}$$

It is clear that

$$\int_0^{2\pi} \Phi(a\cos\psi)\sin\psi d\psi = 0$$

and hence

$$\frac{da}{dt} = 0.$$

Thus, the amplitude of the oscillations is now constant $a = a_0$, and so, from (2.4), we have:

$$\psi = \Omega_I(a)t + \psi_0,$$

where the phase $\psi_0$ is constant and equal to the initial value of $\psi$.

An approximate solution of (2.1) is then

$$x = a\cos[\Omega_I(a)t + \psi_0]. \tag{2.5}$$

We may say that, here the nonlinear character of the equation has no other effect in the first approximation, than to make the frequency depend upon the amplitude.

If we square equation (2.4) and retain only terms in $\varepsilon$, we obtain:

$$\Omega_I^2(a) = \omega^2 + \frac{\varepsilon}{\pi a}\int_0^{2\pi} \Phi(a\cos\psi)\cos\psi d\psi \tag{2.6}$$

or in view of (2.2), we have

$$\Omega_I^2(a) = \frac{1}{a\pi}\int_0^{2\pi} p(a\cos\psi)\cos\psi d\psi. \tag{2.7}$$

Formula (2.7) has the considerable advantage that the function $p(x)$ enters into it directly and not merely through its nonlinear part as it does in (2.6).

One can improve the accuracy of the first approximation by taking into account the small oscillatory terms. The function $\Phi(a\cos\phi)$ can be expanded in a trigonometric series:

$$\Phi(a\cos\psi) = \sum_{n=0}^{\infty} C_n(a)\cos n\psi. \tag{2.8}$$

Using the formulae (1.21) - (1.24), we have for the second approximation:

$$x = a\cos\psi + \varepsilon\sum_{\substack{n=0\\ n\neq 1}}^{\infty} \frac{C_n(a)\cos n\psi}{n^2 - 1}, \tag{2.9}$$

where $a$ and $\psi$ are defined by the equations of the second approximation:

$$\begin{aligned} &\frac{da}{dt} = 0, \qquad \frac{d\psi}{dt} = \Omega_{II}(a), \\ &\Omega_{II}(a) = \omega + \frac{\varepsilon C_1(a)}{2\omega a} - \frac{\varepsilon^2}{2\omega}\left[\frac{C_1(a)}{2\omega a}\right]^2 + \\ &\quad + \frac{\varepsilon^2}{2\omega k a}\left\{\sum_{n=2}^{\infty} \frac{C_n(a)\dfrac{dC_n(a)}{da}}{n^2 - 1} - 2C_0(a)\cdot\frac{dC_0(a)}{da}\right\}. \end{aligned} \tag{2.10}$$

It is easy to show that for the conservative systems described by equation (2.1), all quantities $A_n(a)$ are equal to zero, so that the equation for the amplitude of the fundamental harmonic to within an accuracy of $\varepsilon$, is $\frac{da}{dt} = 0$, and hence, $a = a_0 = \text{const}$.

**Example 1. Duffing equation**

Let us consider the equation

$$\frac{d^2x}{dt^2} + x + \varepsilon x^3 = 0. \tag{2.11}$$

Using the formulae (1.17), (1.18), we obtain

$$u_1(a, \psi) = \frac{a^3}{32} \cos 3\psi,$$
$$A_1(a) = 0, \qquad B_1(a) = \frac{3a^2}{8}.$$

Following formulae (1.20), we have

$$u_2(a, \psi) = \frac{-a^5}{1024}(21 \cos 3\psi - \cos 5\psi),$$
$$A_2(a) = 0, \qquad B_2(a) = -\frac{15a^4}{256}.$$

Then, for the approximate solution of equation (2.11), we get the following asymptotic expansion

$$x(t, \varepsilon) = a \cos \psi + \frac{\varepsilon a^3}{32} \cos 3\psi - \frac{\varepsilon^2 a^5}{1024}(21 \cos 3\psi - \cos 5\psi) + O(\varepsilon^3),$$

where $a$ and $\psi$ are determined from the equations

$$\frac{da}{dt} = 0, \qquad \text{or} \quad a = a_0 = \text{const},$$
$$\frac{d\psi}{dt} = 1 + \frac{3\varepsilon a^2}{8} - \frac{15\varepsilon^2 a^4}{256} + O(\varepsilon^3),$$
$$\psi = \left[1 + \frac{3\varepsilon a_0^2}{8} - \frac{15\varepsilon^2 a_0^4}{256}\right] t + \psi_0 + O(\varepsilon^3),$$

$a_0$, $\psi_0$ are constants.

**Example 2. The mathematical pendulum**

When the damping is neglected, the differential equation, governing the free oscillation of the mathematical pendulum, is given by

$$m\ell\ddot{x} + mg\sin x = 0,$$

or

$$\ddot{x} + \omega^2 \sin x = 0. \tag{2.12}$$

Here, $m$ is the mass, $\ell$ is the length of the pendulum, $g$ is the gravitational acceleration, and $\omega^2 = \frac{g}{\ell}$. The angle $x$ designates the deviation from the vertical equilibrium position (Fig. 1).

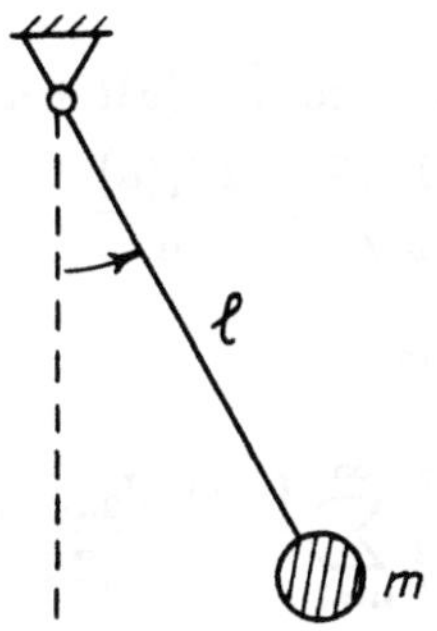


*Fig. 1.* Mathematical pendulum.

For small oscillations (say not exceeding 30°), the equation (2.12) is reduced to the form:

$$\ddot{x} + \omega^2\left(x - \frac{x^3}{6}\right) = 0.$$

From (2.1) and (2.7), we have at once

$$p(x) = \omega^2\left(x - \frac{x^3}{6}\right),$$

$$\Omega_I^2(a) = \frac{\omega^2}{\pi a}\int_0^{2\pi}\left(a\cos\psi - \frac{a^3}{6}\cos^3\psi\right)\cos\psi d\psi,$$

and so

$$\Omega_I^2(a) = \omega^2\left(1 - \frac{a^2}{8}\right),$$

$$\frac{\Omega_I(a)}{\omega} = \sqrt{1 - \frac{a^2}{8}} \approx 1 - \frac{a^2}{16}.$$

As the amplitude increases, the frequency decreases and hence, the period $T$ of free oscillation increases also:

$$T = \frac{2\pi}{\Omega_I(a)} \approx \frac{2\pi}{\omega}\left(1 + \frac{a^2}{16}\right).$$

We return to the exact equation (2.12). By comparing equations (2.12) and (2.1), we have

$$p(x) = \omega^2 \sin x,$$

and hence

$$p(a\cos\psi) = 2\omega^2 \sum_{n=0}^{\infty} (-1)^n J_{2n+1}(a)\cos(2n+1)\psi,$$

where $J_k(a)$ is the Bessel function.

According to equations (2.5) - (2.7), we have in the first approximation:

$$x = a\cos\left[\Omega_I(a)t + \psi_0\right],$$
$$\left(\frac{\Omega_I}{\omega}\right)^2 = \frac{2J_1(a)}{a},$$

and in the second approximation:

$$\begin{aligned} x &= a\cos\psi + 2\frac{\omega^2}{\Omega_I^2}\sum_{n=1}^{\infty}\frac{(-1)^n J_{2n+1}(a)\cos(2n+1)\psi}{(2n+1)^2-1}, \\ \left(\frac{\Omega_{II}}{\omega}\right)^2 &= \left(\frac{\Omega_I}{\omega}\right)^2 + 4\frac{\omega^2}{\Omega_I^2 a}\sum_{n=1}^{\infty}\frac{J_{2n+1}(a)\,J'_{2n+1}(a)}{(2n+1)^2-1}. \end{aligned} \tag{2.13}$$

In view of the very rapid convergence of the series on the right-hand side of equation (2.13), it is sufficient to take into account only the first two terms.

**Example 3. Oscillatory shaft**

Consider a shaft, composed of two revolving masses joined by a nonlinear elastic connection (see Fig. 2). Let $J_1$ and $J_2$ be the moments of inertia of the revolving masses, and $\theta_1$, $\theta_2$ their angles of rotation. Further, let $M = c(\theta_1 - \theta_2)$ be the angular momentum of the elastic connection, represented as a function of the angle of rotation $\theta = \theta_1 - \theta_2$. The equations of motion for each of the two masses are

$$J_1\frac{d^2\theta_1}{dt^2} + c(\theta_1 - \theta_2) = 0,$$
$$J_2\frac{d^2\theta_2}{dt^2} - c(\theta_1 - \theta_2) = 0.$$

Hence, the equation governing the oscillations is

$$J_1 J_2\frac{d^2\theta}{dt^2} + (J_1 + J_2)c(\theta) = 0. \tag{2.14}$$

Suppose that $M = c(\theta)$ is represented by the graph of Fig. 3, or more explicitly that

$$c(\theta) = \begin{cases} h + k\theta & \text{for} \quad \theta > 0, \\ -h + k\theta & \text{for} \quad \theta < 0. \end{cases}$$

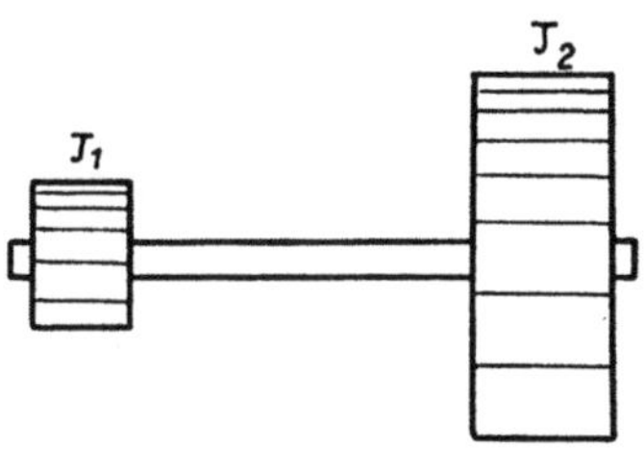


*Fig. 2.* Oscillatory shaft.

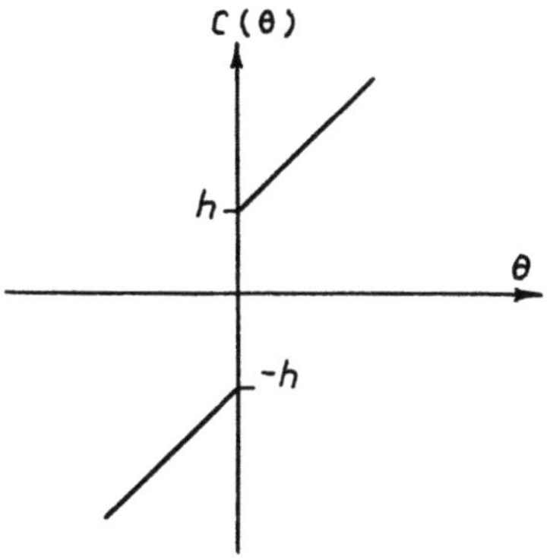


*Fig. 3.* Angular momentum of the elastic connection.

Comparing equations (2.14) and (2.1), we find that

$$p(\theta) = \frac{J_1 + J_2}{J_1 \cdot J_2} c(\theta).$$

Hence, we have from (2.7):

$$\Omega_I^2(a) = \frac{1}{\pi \cdot a} \cdot \frac{J_1 + J_2}{J_1 J_2} \int_0^{2\pi} c(a \cos\psi) \cos\psi d\psi.$$

Since

$$\int_0^{2\pi} c(a\cos\psi)\cos\psi d\psi = \int_{-\pi/2}^{\pi/2} (h + ka\cos\psi)\cos\psi d\psi +$$

$$+ \int_{\pi/2}^{3\pi/2} (-h + ka\cos\psi)\cos\psi d\psi = 4h + \pi ka,$$

then

$$\Omega_I^2(a) = \frac{J_1 + J_2}{J_1 \, J_2} k\Big(1 + \frac{4h}{\pi ak}\Big).$$

In order that this formula be applicable, it will clearly be necessary that $\frac{h}{ak}$ be small, in a sense, this quantity measures the deviation of the angular momentum $M$ from linearity.

**Example 4. Electrical circuit without resistance**

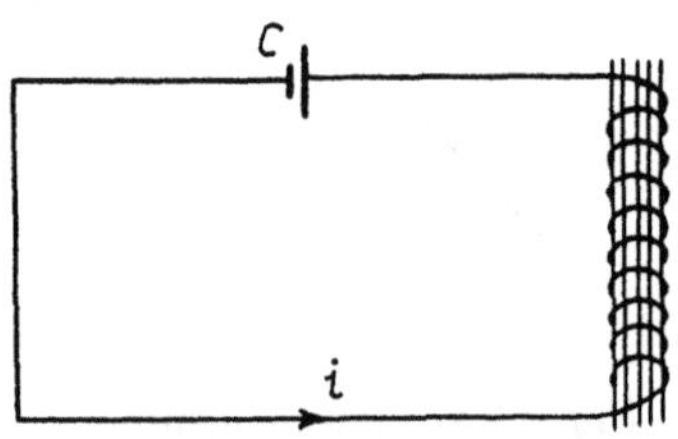


*Fig. 4.* Electrical circuit.

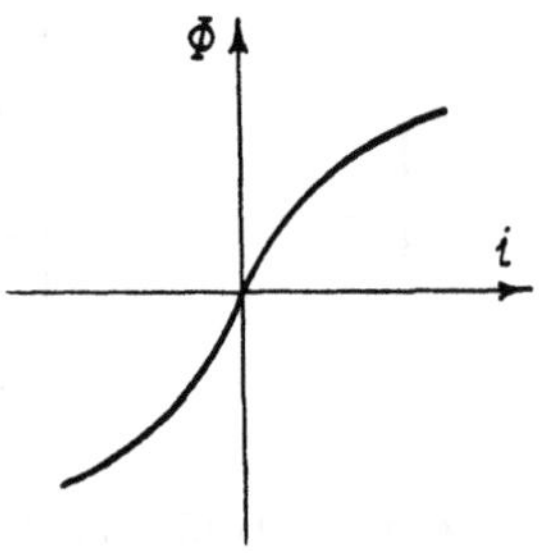


*Fig. 5.* Relation between magnetic flux and the current.

Consider an electrical oscillatory circuit (Fig. 4), containing an iron core. Let $\Phi$ denote the magnetic flux, $i$ the line current, $C$ the capacity. We then have

$$\frac{d^2\Phi}{dt^2} + \frac{i}{C} = 0.$$

The relation between $\Phi$ and $i$ is shown in Fig. 5:

$$i = \alpha\Phi - \beta\Phi^3,$$

where $\alpha$ and $\beta$ are positive constants. For $\Phi$, we have the differential equation:

$$\frac{d^2\Phi}{dt^2} + \frac{1}{C}\left(\alpha\Phi - \beta\Phi^3\right) = 0.$$

Assuming $\frac{\beta}{\alpha}\Phi^2$ small, we find by (2.7)

$$\Omega_1^2(a) = \frac{\alpha}{C}\left(1 - \frac{3\beta a^2}{4\alpha}\right),$$

from which it follows approximately that

$$\Omega_1(a) = \sqrt{\frac{\alpha}{C}}\left(1 - \frac{3\beta a^2}{8\alpha}\right).$$

## 3. Dissipative Systems

As a second particular case of equation (1.1), we consider the equation of oscillation of a mass $m$ under the action of the elastic force $kx$ and nonlinear weak friction $\varepsilon F(\frac{dx}{dt})$ depending on the velocity

$$m\frac{d^2x}{dt^2} + kx = \varepsilon F\left(\frac{dx}{dt}\right). \tag{3.1}$$

This equation belongs to the type (1.1), where

$$f\left(x, \frac{dx}{dt}\right) = \frac{1}{m}F\left(\frac{dx}{dt}\right).$$

We use the general formulae (1.16) - (1.20). We have

$$\frac{1}{m}F(-a\omega \sin\psi) = \sum_{n=0}^{\infty} F_n(a\omega)\cos n\left(\psi + \frac{\pi}{2}\right).$$

Comparing this with the general form of the expansions (1.16), one has:

$$g_n(a) = F_n(a\omega)\cos\left(\frac{n\pi}{2}\right), \quad h_n(a) = -F_n(a\omega)\sin\frac{n\pi}{2},$$

hence

$$A_1(a) = \frac{1}{2\omega}F_1(a\omega), \quad B_1(a) = 0,$$

and we have the first approximation:

$$\begin{aligned} x &= a\cos\psi, \\ \frac{da}{dt} &= \frac{\varepsilon}{2\omega}F_1(a\omega), \\ \frac{d\psi}{dt} &= \omega, \quad \omega = \sqrt{\frac{k}{m}}. \end{aligned} \tag{3.2}$$

It is seen that the frequency remains the same as in the linear case. In the first approximation, the nonlinear friction has influence only on the amplitude of oscillation.

For the second approximation, one has:

$$x = a\cos\psi + \varepsilon u_1(a, \psi),$$

where

$$u_1(a, \psi) = -\frac{1}{\omega^2}\sum_{\substack{n\neq 1\\ n=0}}^{\infty} \frac{F_n(a\omega)\cos n\left(\psi + \frac{\pi}{2}\right)}{n^2 - 1},$$

and for $A_2(a)$ and $B_2(a)$, one has the following expressions:

$$A_2(a) = 0, \tag{3.3}$$

$$B_2(a) = \frac{F_1(a\omega)}{8\omega^3 a} \cdot \frac{dF_1(a\omega)}{da} - \frac{F_1^2(a\omega)}{4\omega^3 a^2} - \frac{1}{2\omega^3 a^2} \sum_{n=2}^{\infty} \frac{n^2 F_n^2(a\omega)}{n^2 - 1}.$$

The second approximation then has the form

$$x = a \cos \psi - \frac{\varepsilon}{\omega^2} \sum_{\substack{n=0 \\ n \neq 1}}^{\infty} \frac{F_n(a\omega) \cos n(\psi + \frac{\pi}{2})}{n^2 - 1}, \tag{3.4}$$

$$\frac{da}{dt} = \frac{\varepsilon F_1(a\omega)}{2\omega}, \quad \frac{d\psi}{dt} = \omega + \varepsilon^2 B_2(a).$$

**Example 5. Oscillator with linear damping**

Consider the equation:

$$\frac{d^2 x}{dt^2} + \lambda \frac{dx}{dt} + \omega^2 x = 0, \quad \lambda < 2\omega.$$

One has

$$m = 1, \quad F\left(\frac{dx}{dt}\right) = -h \frac{dx}{dt}, \quad \lambda = \varepsilon h,$$

$$F_1(a\omega) = -h\omega a, \quad F_n(a\omega) = 0, \quad n \neq 1,$$

so that, the second approximation is

$$x = a \cos \psi, \quad \frac{da}{dt} = -\frac{\lambda}{2} a, \quad \frac{d\psi}{dt} = \omega \left[1 - \frac{1}{8} \left(\frac{\lambda}{\omega}\right)^2\right].$$

For the amplitude, this approximation gives the exact value $a = a_0 \exp\left(-\frac{\lambda}{2} t\right)$, and for the frequency, it gives a correction corresponding to the first two terms of the expansion of the exact formula:

$$\omega \sqrt{1 - \frac{1}{4} \left(\frac{\lambda}{\omega}\right)^2} = \omega \left[1 - \frac{1}{8} \left(\frac{\lambda}{\omega}\right)^2 - \frac{1}{128} \left(\frac{\lambda}{\omega}\right)^4 + \ldots\right].$$

This is to be expected, since we do not take any terms beyond the second order.

**Example 6. Oscillator with nonlinear damping**

Consider a differential equation of the form

$$\frac{d^2x}{dt^2} + \alpha\left(\frac{dx}{dt}\right)^2 \mathrm{sign}\dot{x} + \omega^2 x = 0, \quad \alpha = \varepsilon h, \dot{x} = \frac{dx}{dt},$$

where the sign$\dot{x}$ means $+1$ for $\dot{x} > 0$ and $-1$ for $\dot{x} < 0$.

In this case

$$f(x, \dot{x}) = -h\left(\frac{dx}{dt}\right)^2 \mathrm{sign}\dot{x}.$$

The formula (1.28) gives:

$$\frac{da}{dt} = -\frac{\varepsilon h\omega a^2}{2\pi}\int_0^\pi \sin^3\psi d\psi = -\frac{4\alpha\omega}{3\pi}a^2.$$

Integrating this equation with the initial condition $a(t_0) = a_0$, we have:

$$t(a) = -\frac{3\pi}{4\alpha\omega}\int_{a_0}^{a}\frac{da}{a^2} = \frac{3\pi}{4\alpha\omega}\left(\frac{1}{a} - \frac{1}{a_0}\right).$$

Hence

$$a = \frac{a_0}{1 + \dfrac{4\alpha\omega a_0}{3\pi}t}.$$

Thus, after half an oscillation from $t = 0$ to $t = \dfrac{T}{2} = \dfrac{\pi}{\omega}$, the amplitude $a_0$ decreases to the value

$$a_1 \approx \frac{a_0}{1 + \dfrac{4\alpha}{3}a_0}.$$

As the third particular case of the equation (1.1), we consider a system governed by the nearly linear differential equation of the form

$$\frac{d^2x}{dt^2} + \omega^2 x = \varepsilon f(x)\cdot\frac{dx}{dt}, \tag{3.5}$$

which will be treated in Chapter 2. In order to use the formulae (1.23) - (1.25), it is necessary to consider the Fourier expansion of

$$f(a\cos\psi)a\omega\sin\psi.$$

We define the function

$$F^*(x) = \int_0^x f(x)dx \tag{3.6}$$

and consider its development

$$F^*(a\cos\psi) = \sum_{n=0}^{\infty} F_n^*(a)\cos n\psi.$$

Differentiating this relation with respect to $\psi$, and using (3.6), we get

$$f(a\cos\psi)a\omega\sin\psi = \sum_{n=0}^{\infty} \omega n F_n^*(a)\sin n\psi.$$

Comparing with (1.17), we have

$$A_1(a) = \frac{1}{2}F_1^*(a), \quad B_1(a) = 0.$$

Hence, in the first approximation

$$\begin{aligned} x &= a\cos\psi, \\ \frac{da}{dt} &= \frac{\varepsilon}{2}F_1^*(a), \quad \frac{d\psi}{dt} = \omega. \end{aligned} \tag{3.7}$$

For the second approximation, we have

$$x = a\cos\psi + \frac{\varepsilon}{\omega}\sum_{n=2}^{\infty}\frac{nF_n^*(a)\sin n\psi}{n^2-1}, \tag{3.8}$$

where $a$ and $\psi$ are determined by the differential equations

$$\begin{aligned} \frac{da}{dt} &= \frac{\varepsilon}{2}F_1^*(a), \\ \frac{d\psi}{dt} &= \omega + \varepsilon^2 B_2(a), \end{aligned} \tag{3.9}$$

and

$$B_2(a) = -\frac{1}{8a\omega}F_1^*(a)\frac{dF_1^*(a)}{da} - \frac{1}{2\omega a^2}\sum_{n=2}^{\infty}\frac{n^2F_n^{*2}(a)}{n^2-1}. \tag{3.10}$$

## 4. Stationary Amplitudes and their Stability

In this chapter, we have always encountered the amplitude equation in the form

$$\frac{da}{dt} = \Phi(a), \tag{4.1}$$

where

$$\Phi(a) = \varepsilon A_1(a) + \varepsilon^2 A_2(a) + \ldots$$

This differential equation may be reduced to quadratures, but it is useful to study the behaviour of the solution $a = a(t)$ of equation (4.1) without integration. Observe that there cannot exist $a^* > 0$, such that

$$\Phi(a) > 0 \quad \text{for} \quad a > a^*,$$

for if such an $a^*$ existed then, equation (4.1) with the initial condition $a(0) = a_0 > a^*$ would yield $a(t) \to +\infty$ when $t \to +\infty$, which is physically impossible.

Referring to (4.1), we see that if the initial amplitude $a_0$ is not stationary, i.e., does not satisfy:

$$\Phi(a) = 0, \tag{4.2}$$

then with increasing $t$, the amplitude $a(t)$ will steadily tend to a stationary value.

The tendency of every oscillation to approach a steady oscillation points to the special role of steady oscillations. In such systems, the intermediate regime tends very rapidly to a stationary regime and, therefore, stationary oscillations play a great role, especially, for high-frequency oscillatory processes.

A noteworthy special case exists, when the function $\Phi(a)$ is identically equal to zero, for example, for a conservative system. In this case, there are no transitional regimes and every oscillation is stationary. However, this conservative case never occurs and there is always dissipation, hence loss of energy.

We will now consider the stability of the stationary oscillations. Let $a_0$ be any root of the equation (4.2): $\Phi(a_0) = 0$. Obviously, $a_0$ is also a solution of (4.1). To study the stability of this solution, we introduce a perturbation $\delta a$ of $a$, namely

$$a = a_0 + \delta a.$$

Substituting this expression into equation (4.1) and neglecting the second and higher powers of $\delta a$, one obtains the following variational equation:

$$\frac{d}{dt}(\delta a) = \Phi'(a_0) \cdot \delta a. \tag{4.3}$$

Hence,

$$\delta a = (\delta a)_0 \exp\left[\Phi'(a_0)t\right].$$

This shows that $a_0$ is stable, that is to say, corresponds to a stable stationary oscillation if

$$\Phi'(a_0) < 0, \tag{4.4}$$

and is unstable if

$$\Phi'(a_0) > 0. \tag{4.5}$$

In particular, the static regime ($a = 0$) will be unstable whenever

$$\Phi'(0) > 0. \tag{4.6}$$

This is the condition for self-excitation (see Chapter 2).

**Example 7**

Suppose that

$$\Phi(a) = \frac{\varepsilon a}{2}\left(1 - \frac{a^2}{4}\right), \quad \varepsilon > 0.$$

The stationary value $a_0$ is determined from the equation

$$\frac{\varepsilon a}{2}\left(1 - \frac{a^2}{4}\right) = 0.$$

Hence, $\quad a = 0$ or $a = 2$.

The stationary solution $a = 0$ is unstable because $\Phi'(0) = \frac{\varepsilon}{2} > 0$, and the solution $a = 2$ is stable because $\Phi'(2) = -\varepsilon < 0$.

## 5. Equivalent Linearization of Nonlinear Oscillatory Systems

Equations of the first approximation play an important role in applied problems. We will endeavor to obtain suitable interpretations for these equations. For this purpose, we write the basic differential equation (1.1) in the form

$$m\frac{d^2x}{dt^2} + kx = \varepsilon f\left(x, \frac{dx}{dt}\right), \tag{5.1}$$

where $m$, $k$ are positive constants.

As a first approximation, we have obtained a solution

$$x = a\cos\psi, \tag{5.2}$$

where $a$ and $\psi$ are given by the differential equations of the first approximation:

$$\begin{aligned} \frac{da}{dt} &= \frac{-\varepsilon}{2\pi m\omega}\int_0^{2\pi} f(a\cos\psi, -a\omega\sin\psi)\sin\psi d\psi, \\ \frac{d\psi}{dt} &= \omega_e(a), \end{aligned} \tag{5.3}$$

where

$$\omega^2 = \frac{k}{m},$$
$$\omega_e^2(a) = \omega^2 - \frac{\varepsilon}{\pi m a}\int_0^{2\pi} f(a\cos\psi, -a\omega\sin\psi)\cos\psi d\psi, \tag{5.4}$$

is the natural frequency for $\varepsilon = 0$.

Note that the first approximation (5.2) is the fundamental harmonic of the true nonlinear series solution

$$x = a\cos\psi + \varepsilon u_1(a,\psi) + \varepsilon^2 u_2(a,\psi) + \ldots,$$

which can be calculated by approximations, as was shown previously.

Let us introduce the functions of the amplitude $\lambda_e(a)$, $k_e(a)$ defined by

$$\lambda_e(a) = \frac{\varepsilon}{\pi a\omega}\int_0^{2\pi} f(a\cos\psi, -a\omega\sin\psi)\sin\psi d\psi,$$
$$k_e(a) = k - \frac{\varepsilon}{\pi a}\int_0^{2\pi} f(a\cos\psi, -a\omega\sin\psi)\cos\psi d\psi. \tag{5.5}$$

These equations define the so-called linearized parameters $\lambda_e(a)$ and $k_e(a)$. In terms of the quantities (5.5), the equations (5.3) for the first approximation become

$$\frac{da}{dt} = -\frac{\lambda_e(a)\cdot a}{2m},$$
$$\frac{d\psi}{dt} = \omega_e(a) \quad \text{with} \quad \omega_e^2(a) = \frac{k_e(a)}{m}. \tag{5.6}$$

In order to see that the linearized differential equations (5.6) have a solution differing from the solution of the nonlinear differential equation by a remainder $O(\varepsilon^2)$, we differentiate (5.2), taking into account (5.6); this gives

$$\frac{dx}{dt} = -a\omega_e(a)\sin\psi - \frac{\lambda_e(a)}{2m}a\cos\psi \tag{5.7}$$

and, differentiating once more:

$$\frac{d^2x}{dt^2} = -a\omega_e^2(a)\cos\psi + \frac{\lambda_e(a)}{m}a\omega_e(a)\sin\psi + \frac{\lambda_e^2(a)}{4m^2}a\cos\psi +$$
$$+\frac{\lambda_e(a)}{2m}a^2\cdot\frac{d\omega_e(a)}{da}\sin\psi + \frac{d\lambda_e(a)}{da}\cdot\frac{a}{2m}\frac{\lambda_e(a)}{2m}a\cos\psi.$$

This can be rewritten by taking into account (5.7) and $x = a\cos\psi$ as:

$$\frac{d^2x}{dt^2} = -\frac{k_e(a)}{m}x - \frac{\lambda_e(a)}{m}\frac{dx}{dt} + \frac{\lambda_e^2(a)}{4m^2}x + \\ + \frac{\lambda_e(a)}{2m}a^2\frac{d\omega_e(a)}{da}\sin\psi + \frac{1}{2m}\frac{d\lambda_e(a)}{da}a \cdot \frac{\lambda_e(a)}{2m}x.$$

Substituting this value into (5.1), and using (5.5), one has:

$$m\frac{d^2x}{dt^2} + \lambda_e(a)\frac{dx}{dt} + k_e(a)x = O(\varepsilon^2). \tag{5.8}$$

Thus, the first approximation $x = a\cos\psi$ of the quasi-linear differential equation (5.1) also satisfies the linear differential equation (5.8) with accuracy of the order $\varepsilon^2$. In short, in the first approximation, the oscillations of the quasi-linear system are equivalent to those of a linear system

$$m\frac{d^2x}{dt^2} + \lambda_e(a)\frac{dx}{dt} + k_e(a)x = 0, \tag{5.9}$$

with a coefficient of damping $\lambda_e(a)$ and a coefficient of restoring force $k_e(a)$. The accuracy is to the order $\varepsilon^2$, that is to say neglecting quantities of the same order as when we formed the first approximation. For this reason, $\lambda_e(a)$ is called the equivalent coefficient of damping, and $k_e(a)$ - the equivalent stiffness. The linear system (5.9) will also be said to be equivalent to the assigned system. It is noteworthy that inasmuch as both $\lambda_e(a)$ and $k_e(a)$ are functions of the amplitude $a$, the direct solution of the linearized equation is likely to be more complicated than the original quasi-linear differential equation.

By comparing (5.9) with the given equation (5.1), we see that the equation (5.9) arises from the equation (5.1) by replacing the nonlinear term

$$F = \varepsilon f(x, \frac{dx}{dt}) \tag{5.10}$$

by the linear term

$$F_e = -\left[k_1(a)x + \lambda_e(a)\frac{dx}{dt}\right], \tag{5.11}$$

where $k_1(a) = k_e(a) - k$.

Note that

$$\delta_e(a) = \frac{\lambda_e(a)}{2m}$$

is the dissipation decrement in the equivalent linear system, and

$$\omega_e(a) = \sqrt{\frac{k_e(a)}{m}}$$

is the natural frequency of its oscillations.

Thus, we may conclude that the equation (5.6) of the first approximation may be derived as follows. Linearize the oscillatory system by substituting the linear term (5.11) in the basic equation (5.1) for the nonlinear term (5.10). The functions $\lambda_e(a)$, $k_1(a)$ are defined by

$$\begin{aligned}\lambda_e(a) &= \frac{\varepsilon}{\pi a\omega}\int_0^{2\pi} f(a\cos\psi, -a\omega\sin\psi)\sin\psi d\psi,\\ k_1(a) &= \frac{-\varepsilon}{\pi a}\int_0^{2\pi} f(a\cos\psi, -a\omega\sin\psi)\cos\psi d\psi.\end{aligned} \tag{5.12}$$

In order to obtain the equivalent linear system (5.9) with mass $m$, coefficient of damping $\lambda_e(a)$ and coefficient of restoring force $k_e(a) = k_1(a) + k$ in the usual way, we find the dissipation decrement $\delta_e(a)$ and frequency of the free oscillations $\omega_e(a)$, neglecting quantities of order $\varepsilon^2$. We have

$$\delta_e(a) = \frac{\lambda_e(a)}{2m}, \quad \omega_e^2(a) = \frac{k_e(a)}{m}. \tag{5.13}$$

The differential equations are now linear

$$\frac{da}{dt} = -\delta a, \quad \frac{d\psi}{dt} = \omega. \tag{5.14}$$

and if in equation (5.14), we replace $\delta$ by $\delta_e(a)$ and $\omega$ by $\omega_e(a)$ from equations (5.13) and (5.12), we obtain the solution (5.3) of equation (5.1) with accuracy $\varepsilon^2$.

The formal process, just described, will be referred to as the principle of linearization.

## 6. Nonlinear Oscillatory Systems with Slowly Varying Parameters. Adiabatic Invariants.

The asymptotic method has been applied by Mitropolskii [29] to the investigation of non-stationary phenomena, in which, the parameters of the dynamical systems vary slowly in time. In this case, "slowly" means that the rate of variation of parameters is much slower than the frequency of the motion considered at a fixed value of these parameters. The phenomena of this nature are not necessarily periodic.

We consider a general problem, specified by a differential equation of the form

$$\frac{d}{dt}\left[m(\tau)\frac{dx}{dt}\right] + k(\tau)x = \varepsilon f\left(\tau, x, \frac{dx}{dt}\right), \tag{6.1}$$

where $t$ is the normal time, $\tau = \varepsilon t$ is *slow* time, $m(\tau)$ is a mass (or moment of inertia, or coefficient of inductance), and $k(\tau)$ is the coefficient of restoring force (that is *stiffness* in a mechanical system, or $1/c$ in an electrical system, etc), but here we make the assumption that $m$ and $k$ depend on the slow time $\tau$ and are, therefore, $m(\tau)$ and $k(\tau)$.

The problem is now specified in terms of two independent variables $t$ and $\tau$ with an additional relation $\frac{d\tau}{dt} = \varepsilon$, when differentiations are to be carried out with respect to $t$, as usual.

It is obvious that the functions $m(\tau)$ and $k(\tau)$ in (6.1) are always positive; otherwise the problem would have no physical meaning. Moreover, in differentiations, we have:

$$\frac{dm(\tau)}{dt} = \frac{dm(\tau)}{d\tau} \cdot \frac{d\tau}{dt} = \varepsilon \cdot \frac{dm(\tau)}{d\tau},$$

likewise for $k(\tau)$. A differentiation of a function of $\tau$ with respect to $t$ always introduces a small factor $\varepsilon$ and this is to be taken into account in the approximation procedure where the terms are arranged as coefficients of like powers of $\varepsilon$.

The construction of approximate solutions of equation (6.1) does not give rise to significant supplementary difficulties and may be realized by means of the asymptotic method presented in the previous sections. For construction of an asymptotic series, it is necessary that the coefficients of equation (6.1) $m(\tau)$, $k(\tau)$ and also $f(\tau, x, \frac{dx}{dt})$ are continuously differentiable an indefinite number of times, for all finite values of their arguments, in a certain interval $[0 \leq t \leq L/\varepsilon]$ (or for $0 \leq \tau \leq L$).

If $\tau$ is constant in (6.1), we have the differential equation, which has been studied in the previous sections. When $\tau = \varepsilon t$, i.e. if the parameters vary slowly, the situation is more complicated in view of the presence of the additional variable $\tau$.

We are interested here only in formal solutions of equation (6.1) by approximations, and concentrate on the derivation of successive approximations disregarding the physical significance of the problem for the moment.

Taking all these matters into account, we look for a solution of equation (6.1) of the form:

$$x = a\cos\psi + \varepsilon u_1(\tau, a, \psi) + \varepsilon^2 u_2(\tau, a, \psi) + \ldots, \tag{6.2}$$

where $u_1(\tau, a, \psi)$, $u_2(\tau, a, \psi)$, ... are periodic functions of $\psi$ with period $2\pi$. As previously, we assume that the amplitude $a$ and the phase $\psi$ are given by the d.e.

$$\begin{aligned} \frac{da}{dt} &= \varepsilon A_1(\tau, a) + \varepsilon^2 A_2(\tau, a) + \ldots, \\ \frac{d\psi}{dt} &= \omega(\tau) + \varepsilon B_1(\tau, a) + \varepsilon^2 B_2(\tau, a) + \ldots, \end{aligned} \tag{6.3}$$

where $\omega(\tau) = \sqrt{k(\tau)/m(\tau)}$ does not now remain constant. One can retain the concept of frequency, but it will now be a function of the slow time $\tau$.

The problem is to find functions

$$u_1(\tau, a, \psi), u_2(\tau, a, \psi) \dots, A_1(\tau, a),$$
$$A_2(\tau, a), \dots, B_1(\tau, a), B_2(\tau, a), \dots, \tag{6.4}$$

such that the series (6.4), after the substitution of $a$ and $\psi$, as given by the series (6.3), should be the actual solution of (6.1). The problem of determination of the functions (6.4) has a certain degree of arbitrariness. One can impose, however, an additional condition, as we did in the previous sections, which consists in the requirement that the fundamental harmonics $\sin\psi$, $\cos\psi$ should be absent in all functions $u_1, u_2, \dots$. This requirement is expressed by the conditions:

$$\int_0^{2\pi} u_i(\tau, a, \psi)\cos\psi d\psi = 0, \quad \int_0^{2\pi} u_i(\tau, a, \psi)\sin\psi d\psi = 0, \tag{6.5}$$

$i = 1, 2, \dots$

Physically, this means that we take as amplitude $a$ the full amplitude of the fundamental harmonic.

In order to find the functions (6.4) under the conditions (6.5), one has to determine $\frac{d^2a}{dt^2}, \frac{d^2\psi}{dt^2}, \frac{da}{dt}\cdot\frac{d\psi}{dt}, \left(\frac{d\psi}{dt}\right)^2$. One has

$$\begin{aligned}
\frac{d^2a}{dt^2} &= \varepsilon\frac{\partial A_1}{\partial a}\cdot\frac{da}{dt} + \varepsilon^2\frac{\partial A_1}{\partial\tau} + \dots = \varepsilon^2\Big(\frac{\partial A_1}{\partial a}\cdot A_1 + \frac{\partial A_1}{\partial\tau}\Big) + O(\varepsilon^3),\\
\frac{d^2\psi}{dt^2} &= \varepsilon\cdot\frac{d\omega(\tau)}{d\tau} + \varepsilon^2\Big(\frac{\partial B_1}{\partial a}A_1 + \frac{\partial B_1}{\partial\tau}\Big) + O(\varepsilon^3),\\
\frac{da}{dt}\cdot\frac{d\psi}{dt} &= \varepsilon A_1\cdot\omega(\tau) + \varepsilon^2\big[A_2\omega(\tau) + A_1B_1\big] + O(\varepsilon^3),\\
\Big(\frac{d\psi}{dt}\Big)^2 &= \omega^2(\tau) + 2\varepsilon\omega(\tau)B_1 + \varepsilon^2\big[B_1^2 + 2\omega(\tau)B_2\big] + O(\varepsilon^3).
\end{aligned} \tag{6.6}$$

One obtains for $\frac{dx}{dt}$ and $\frac{d^2x}{dt^2}$ the following expressions:

$$\begin{aligned}
\frac{dx}{dt} = \frac{da}{dt}\Big\{\cos\psi + \varepsilon\frac{\partial u_1}{\partial a} + \varepsilon^2\frac{\partial u_2}{\partial a} + \dots\Big\} +\\
+ \frac{d\psi}{dt}\Big\{-a\sin\psi + \varepsilon\frac{\partial u_1}{\partial\psi} + \varepsilon^2\frac{\partial u_2}{\partial\psi} + \dots\Big\} + \varepsilon^2\frac{\partial u_1}{\partial\tau} + \varepsilon^3\frac{\partial u_2}{\partial\tau} + \dots,
\end{aligned}$$

$$\frac{d^2x}{dt^2} = \frac{d^2a}{dt^2}\Big\{\cos\psi + \varepsilon\frac{\partial u_1}{\partial a} + \varepsilon^2\frac{\partial u_2}{\partial a} + \ldots\Big\} + \frac{d^2\psi}{dt^2}\Big\{-a\sin\psi + \tag{6.7}$$
$$+\varepsilon\frac{\partial u_1}{\partial \psi} + \varepsilon^2\frac{\partial u_2}{\partial \psi} + \ldots\Big\} + \Big(\frac{da}{dt}\Big)^2\Big\{\varepsilon\frac{\partial^2 u_1}{\partial a^2} + \varepsilon^2\frac{\partial^2 u_2}{\partial a^2} + \ldots\Big\} +$$
$$+2\frac{da}{dt}\cdot\frac{d\psi}{dt}\Big\{-\sin\psi + \varepsilon\frac{\partial^2 u_1}{\partial a\partial\psi} + \varepsilon^2\frac{\partial^2 u_2}{\partial a\partial\psi} + \ldots\Big\} +$$
$$+\Big(\frac{d\psi}{dt}\Big)^2\Big\{-a\cos\psi + \varepsilon\frac{\partial^2 u_1}{\partial\psi^2} + \varepsilon^2\frac{\partial^2 u_2}{\partial\psi^2} + \ldots\Big\} +$$
$$+2\frac{da}{dt}\Big\{\varepsilon^2\frac{\partial^2 u_1}{\partial\tau\partial a}\Big\} + 2\frac{d\psi}{dt}\Big\{\varepsilon^2\frac{\partial^2 u_1}{\partial\tau\partial\psi}\Big\} + O(\varepsilon^3).$$

If one substitutes $x$ and $\frac{dx}{dt}$ into the right-hand side of (6.1), and develops it into a Taylor's series, one obtains

$$\varepsilon f\Big(\tau, x, \frac{dx}{dt}\Big) = \varepsilon f(\tau, a\cos\psi, -a\omega\sin\psi) + \varepsilon^2\Big\{f'_x(a\cos\psi, -a\omega\sin\psi)u_1 +$$
$$+\Big(A_1\cos\psi - aB_1\sin\psi + \omega(\tau)\frac{\partial u_1}{\partial\psi}\Big)f'_{\dot{x}}(\tau, a\cos\psi, -a\omega\sin\psi)\Big\} + O(\varepsilon^3). \tag{6.8}$$

The next step, as usual, is the identification of coefficients with like powers of $\varepsilon$ in (6.1). One has:

$$k(\tau)\Big[\frac{\partial^2 u_1}{\partial\psi^2} + u_1\Big] = f_0(\tau, a, \psi) + 2m(\tau)\omega(\tau)A_1\sin\psi +$$
$$+2m(\tau)\omega(\tau)aB_1\cos\psi + \frac{d\big[m(\tau)\omega(\tau)\big]}{d\tau}a\sin\psi, \tag{6.9}$$

$$k(\tau)\Big[\frac{\partial^2 u_2}{\partial\psi^2} + u_2\Big] = f_1(\tau, a, \psi) + m(\tau)\Big[2\omega(\tau)aB_2 - \frac{\partial A_1}{\partial a}A_1 +$$
$$+aB_1^2 - \frac{\partial A_1}{\partial\tau} - \frac{dm(\tau)}{d\tau}\frac{A_1}{m(\tau)}\Big]\cos\psi + m(\tau)\Big[2\omega(\tau)A_2 +$$
$$+2A_1B_1 + aA_1\frac{\partial B_1}{\partial a} + a\frac{\partial B_1}{\partial\tau} + \frac{dm(\tau)}{d\tau}\frac{aB_1}{m(\tau)}\Big]\sin\psi,$$

........................

where

$$
\begin{aligned}
f_0(\tau, a, \psi) &= f(\tau, a\cos\psi, -a\omega\sin\psi),\\
f_1(\tau, a, \psi) &= u_1 f'_x(\tau, a\cos\psi, -a\omega\sin\psi) + \\
&+ f'_{\dot{x}}(\tau, a\cos\psi, -a\omega\sin\psi)\big[A_1\cos\psi - aB_1\sin\psi + \\
&+ \frac{\partial u_1}{\partial\psi}\omega(\tau)\big] - m(\tau)\Big[2\frac{\partial^2 u_1}{\partial\tau\partial\psi}\omega(\tau) + 2\frac{\partial^2 u_1}{\partial a\partial\psi}A_1\omega(\tau) + \\
&+ 2\frac{\partial^2 u_1}{\partial\psi^2}\omega(\tau)B_1 + \frac{\partial u_1}{\partial\psi}\frac{d\omega(\tau)}{d\tau} + \frac{\partial u_1}{\partial\psi}\frac{\omega(\tau)}{m(\tau)}\frac{dm(\tau)}{d\tau}\Big].
\end{aligned}
\tag{6.10}
$$

In order to determine the functions (6.4), a Fourier series is used. Thus, for instance,

$$
\begin{aligned}
f_0(\tau, a, \psi) &= \sum_{n=-\infty}^{\infty} f_{0n}(\tau, a)e^{in\psi},\\
u_1(\tau, a, \psi) &= \sum_{\substack{n=-\infty\\(n\neq\pm1)}}^{\infty} g_n(\tau, a)e^{in\psi},
\end{aligned}
\tag{6.11}
$$

where

$$
f_{0n}(\tau, a) = \frac{1}{2\pi}\int_0^{2\pi} f_0(\tau, a, \psi)e^{-in\psi}\,d\psi.
$$

Substituting $f_0$ and $u_1$ from (6.11) into the differential equation (6.9), and equating the coefficients of the like harmonics, one determines $g_n(\tau, a)$, which, finally determines the function $u_1$.

$$
g_n(\tau, a) = \frac{1}{2\pi m(\tau)\omega^2(\tau)(1-n^2)}\int_0^{2\pi} f_0(\tau, a, \psi)e^{-in\psi}\,d\psi, \quad (n \neq \pm1),
\tag{6.12}
$$

hence,

$$
u_1(\tau, a, \psi) = \frac{1}{2\pi k(\tau)}\sum_{n\neq\pm1}\frac{e^{in\psi}}{1-n^2}\int_0^{2\pi} f_0(\tau, a, \psi)e^{-in\psi}\,d\psi.
\tag{6.13}
$$

Comparing the coefficients of the fundamental harmonics of $\psi$ in (6.9), one obtains the coefficients $A_1(\tau, a)$ and $B_1(\tau, a)$:

$$
\begin{aligned}
A_1(\tau, a) &= \frac{-a}{2m(\tau)\omega(\tau)}\frac{d[m(\tau)\omega(\tau)]}{d\tau} - \frac{1}{2\pi\omega(\tau)m(\tau)}\int_0^{2\pi} f_0(\tau, a, \psi)\sin\psi\,d\psi,\\
B_1(\tau, a) &= \frac{-1}{2\pi m(\tau)\omega(\tau)a}\int_0^{2\pi} f_0(\tau, a, \psi)\cos\psi\,d\psi.
\end{aligned}
\tag{6.14}
$$

As the first approximation, one takes

$$x = a\cos\psi, \tag{6.15}$$

where $a$ and $\psi$ are determined from the equations of the first approximation

$$\begin{aligned}\frac{da}{dt} &= \varepsilon A_1(\tau, a),\\ \frac{d\psi}{dt} &= \omega(\tau) + \varepsilon B_1(\tau, a).\end{aligned} \tag{6.16}$$

One proceeds thus to the second approximation, but we shall not reproduce these long calculations.

**Example 8. Oscillation of a pendulum with a variable length**

Denoting by $\theta$ - the angle of deviation of the pendulum from the vertical, by $g$ - the acceleration of gravity, by $m$ - the mass of pendulum, by $\ell(\tau)$ - the length of the pendulum, varying slowly as a function of the slow time $\tau$ and by $2n$ - the coefficient of damping, one has the differential equation of motion of the pendulum as follows:

$$\frac{d}{dt}\left[m\ell^2(\tau)\frac{d\theta}{dt}\right] + 2n\frac{d}{dt}\left[\ell(\tau)\theta\right] + mg\ell(\tau)\sin\theta = 0. \tag{6.17}$$

For small oscillations and small damping, one can use the first two terms of the development of $\sin\theta$, then (6.17) can be written

$$\frac{d}{dt}\left[m\ell^2(\tau)\frac{d\theta_1}{dt}\right] + mg\ell(\tau)\theta_1 = \varepsilon f\left(\tau, \theta_1, \frac{d\theta_1}{dt}\right),$$

where $\theta = \sqrt{\varepsilon}\,\theta_1, \quad n = \varepsilon n_1$

$$\varepsilon f\left(\tau, \theta_1, \frac{d\theta_1}{dt}\right) = \frac{mg\ell(\tau)}{6}\theta_1^3 - 2n_1\ell(\tau)\frac{d\theta_1}{dt} - 2n_1\frac{d\ell(\tau)}{d\tau}\theta_1.$$

In the first approximation, according to (6.15), (6.16), one has

$$\theta_1 = a\cos\psi,$$

where $a$ and $\psi$ are given by the system of differential equations

$$\begin{aligned}\frac{da}{dt} &= -\frac{na}{m\ell(\tau)} - \frac{3\varepsilon\dfrac{d\ell(\tau)}{d\tau}\cdot a}{4\ell(\tau)},\\ \frac{d\psi}{dt} &= \omega(\tau) - \frac{\varepsilon\omega(\tau)a^2}{16},\end{aligned} \tag{6.18}$$

with $\omega(\tau) = \sqrt{g/\ell(\tau)}$.

Integrating the first differential equation with the initial condition $a(0) = a_0$, one has

$$a = a_0 \left(\frac{\ell(0)}{\ell(\tau)}\right)^{3/4} \exp\left\{-\frac{n}{m}\int_0^t \frac{dt}{\ell(\tau)}\right\}.$$

Replacing this value into the second equation of (6.18), one obtains

$$\psi = \int_0^t \omega(\tau)\left\{1 - \frac{\varepsilon a_0^2}{16}\left(\frac{\ell(0)}{\ell(\tau)}\right)^{3/2} \exp\left[-\frac{2n}{m}\int_0^t \frac{dt}{\ell(\tau)}\right]\right\}dt.$$

If one sets $\ell = \text{const}$ in these expressions, one has the usual formulae

$$a = a_0 e^{-\frac{\lambda}{2}t}, \quad \psi = \omega\left[t + \frac{\varepsilon a_0^2}{16\lambda}\left(e^{-\lambda t} - 1\right)\right] + \varphi,$$

and, therefore

$$\theta_1 = a_0 e^{-\frac{\lambda}{2}t} \cos\left\{\sqrt{\frac{g}{\ell}}\left[t + \frac{\varepsilon a_0^2}{16\lambda}(e^{-\lambda t} - 1)\right] + \varphi\right\},$$

where $\lambda = 2n/m\ell$ and $\varphi$ is the initial phase.

From these expressions, one can form an idea of the effect of the parameter variation in the differential equations. For instance, if the length $\ell$ of the pendulum varies linearly with time, that is

$$\ell(\tau) = \ell_0 + \ell_1\tau = \ell_0 + \frac{d\ell(\tau)}{d\tau}t,$$

then the amplitude $a$ and the phase $\psi$ are given by the expressions:

$$a = a_0\left(\frac{\ell_0}{\ell_0 + \ell_1\tau}\right)^{(3/4 + n/\varepsilon m\ell_1)},$$

$$\psi = \int_0^t \sqrt{\frac{g}{\ell_0 + \ell_1\tau}}\left\{1 - \frac{\varepsilon a_0^2}{16}\left(\frac{\ell_0}{\ell_0 + \ell_1\tau}\right)^{(3/2 + 2n/\varepsilon m\ell_1)}\right\}dt.$$

The amplitude of oscillation will not vary exponentially, but is inversely proportional to a power function of time. If $n < 0$, $\ell_1 > 0$, with $\left|\frac{n}{\varepsilon m\ell_1}\right| < \frac{3}{4}$ and also for $n > 0$, $\ell_1 > 0$ the oscillation will be damped.

For the second approximation, we have

$$\theta_1 = a\cos\psi - \frac{\varepsilon a^3}{192}\cos 3\psi, \tag{6.19}$$

where $a$ and $\psi$ are determined from equations:

$$\begin{aligned}
\frac{da}{dt} &= -\Big[3\varepsilon\frac{\ell'(\tau)}{4\ell(\tau)}+\frac{n}{m\ell(\tau)}\Big]\Big(a+\frac{\varepsilon a^3}{16}\Big),\\
\frac{d\psi}{dt} &= \omega(\tau)-\frac{\varepsilon\omega(\tau)a^2}{16}+\frac{1}{2\omega(\tau)}\Big[\frac{n^2}{m^2\ell^2(\tau)}+\\
&+\frac{\varepsilon n\ell'(\tau)}{m\ell^2(\tau)}+\frac{5\varepsilon^2\ell''(\tau)}{4\ell(\tau)}+\frac{5\varepsilon^2\omega^2(\tau)a^4}{3.2^9}-\frac{3\varepsilon^2\ell'^2(\tau)}{16\ell^2(\tau)}\Big],
\end{aligned} \tag{6.20}$$

which may be integrable. Thus, from the first equation, one obtains

$$\frac{a}{\sqrt{16+\varepsilon a^2}}=\frac{a_0}{\sqrt{16+\varepsilon a_0^2}}\Big[\frac{\ell(0)}{\ell(\tau)}\Big]^{3/4}\exp\Big[-\frac{n}{m}\int_0^t\frac{dt}{\ell(\tau)}\Big],$$

after that, one can find the phase $\psi$ by integrating the second equation of (6.20).

**Adiabatic invariants**

In systems with slowly varying parameters, there exist quantities called adiabatic invariants. These quantities remain invariant for slow changes of the parameters.

As an example, let us consider a mathematical pendulum of mass $m$ and length $\ell$ which is slowly varying (either shortening or lengthening). Change in the length of the pendulum leads to changes in the energy $W$, and the frequency $\omega$ of oscillation of the pendulum. However, it can be shown that for small oscillations, the quantity $W/\omega$ is invariant.

The cord of the pendulum is strained under the action of centrifugal force $m\ell\dot\varphi^2$, and the component of gravity $mg\cos\varphi$, where $\varphi$ is the inclination of the pendulum to the downward vertical. The work $A$, done by these forces when the cord is shortened, is:

$$A=-\int mg\cos\varphi d\ell-\int m\ell\dot\varphi^2 d\ell,$$

(lengthening of the cord corresponds to the opposite sign).

If this shortening of the cord happens slowly, then one can write

$$dA=-mg\overline{\cos\varphi}d\ell-m\ell\overline{\dot\varphi}^2 d\ell,$$

where the bar on the variable denotes the averaging over one period.

For small oscillations, one can set $\cos\varphi\simeq 1-\dfrac{\varphi^2}{2}$ and dA consists of the first component - $mgd\ell$ - the work of lifting of pendulum and the second component $dW$:

$$dW=\Big(\frac{mg}{2}\overline{\varphi}^2-m\ell\overline{\dot\varphi}^2\Big)d\ell, \tag{6.21}$$

the acquired energy of oscillation. The mean values of kinetic and potential energy of the oscillation of the pendulum are equal to each other and they are equal to half of the total energy, so that

$$\frac{W}{2} = \frac{m}{2}\ell^2\overline{\dot{\varphi}}^2 = \frac{m}{2}g\ell\overline{\varphi}^2. \tag{6.22}$$

Comparing the right-hand sides of the expressions (6.21) and (6.22) gives

$$dW = -\frac{W}{2\ell}d\ell.$$

From the relation for the frequency $\omega$ of oscillations $\omega = \sqrt{\frac{g}{\ell}}$, we get $\frac{d\omega}{\omega} = -\frac{d\ell}{2\ell}$. Hence, we have $\frac{dW}{W} = \frac{d\omega}{\omega}$ , and therefore

$$\frac{W}{\omega} = \text{const.} \tag{6.23}$$

Thus, for slow variations of the length of the pendulum, the relation $\frac{W}{\omega}$ is an adiabatic invariant.

We will now examine the oscillation of the pendulum with a slowly variable lengths given by the law $\ell = \ell(\tau)$, and we will show that, in the absence of friction, the adiabatic invariant (6.23) also holds in the first approximation.

In fact, in this case, the equation of motion of the pendulum is

$$\frac{d}{dt}\left[m\ell^2(\tau)\frac{dx}{dt}\right] + mg\ell(\tau)\sin x = 0. \tag{6.24}$$

The solution of this equation in the first approximation has been found in the form

$$x = a\cos\psi, \tag{6.25}$$

where $a$ and $\psi$ are determined from the system of equations (6.18):

$$\begin{aligned} \frac{da}{dt} &= -\varepsilon\frac{3\ell'(\tau)}{4\ell(\tau)}a, \\ \frac{d\psi}{dt} &= \omega(\tau)\left(1 - \frac{\varepsilon a^2}{16}\right), \end{aligned} \tag{6.26}$$

where

$$\omega(\tau) = \sqrt{\frac{g}{\ell(\tau)}}\,.$$

From (6.26), we have

$$a = a_0 \left[\frac{\ell(0)}{\ell(\tau)}\right]^{3/4}. \tag{6.27}$$

We now find the value of the total energy following the formula (6.22), in which $\varphi$ is replaced by $x$ (6.25), and $a$ by (6.27). We have

$$W(\tau) = mg\ell(\tau)\overline{\varphi}^2 = mg\ell(\tau)a^2\overline{\cos^2\varphi} = \frac{1}{2}mg\ell(\tau)a_0^2\left[\frac{\ell(0)}{\ell(\tau)}\right]^{3/2}.$$

Hence

$$\frac{W(\tau)}{\omega(\tau)} = \frac{mg\ell(\tau)a_0^2\left[\frac{\ell(0)}{\ell(\tau)}\right]^{3/2}}{2\sqrt{\frac{g}{\ell(\tau)}}} = \frac{m\sqrt{g}a_0^2}{2}\ell^{3/2}(0) = \text{const},$$

so that, the relation $\dfrac{W(\tau)}{\omega(\tau)}$ is an adiabatic invariant.

We will now consider the problem of the existence of an adiabatic invariant for a Hamilton system, depending on a slowly varying parameter $\lambda$. Let the system be characterized by a Hamiltonian function $H(p, q, \lambda)$. As is known, for this system, the time rate of change of energy is

$$\frac{dE}{dt} = \frac{\partial H}{\partial t} = \frac{\partial H}{\partial \lambda}\cdot\frac{d\lambda}{dt}. \tag{6.28}$$

The right-hand side of (6.28) depends on the slowly varying parameter $\lambda$ and on the fastly varying coordinates $q$ and $p$.

Averaging (6.28) over the period of motion, considering $\lambda$ as a constant, we have

$$\overline{\frac{dE}{dt}} = \overline{\frac{\partial H}{\partial \lambda}}\frac{d\lambda}{dt},$$

or

$$\overline{\frac{dE}{dt}} = \frac{d\lambda}{dt}\cdot\frac{1}{T}\int_0^T \frac{\partial H}{\partial \lambda}dt, \tag{6.29}$$

where $T$ is the period of the variables $p(t)$ and $q(t)$.

Further, according to the Hamilton's equation $\dfrac{dq}{dt} = \dfrac{\partial H}{\partial p}$, we find

$$dt = \frac{dq}{\partial H/\partial p}.$$

Taking into consideration

$$T = \int_0^T dt = \oint \frac{dq}{\partial H/\partial p},$$

one can write the equation (6.29) in the form

$$\overline{\frac{dE}{dt}} = \frac{d\lambda}{dt} \frac{\oint \frac{\partial H/\partial \lambda}{\partial H/\partial p} dq}{\oint \frac{dq}{\partial H/\partial p}}, \tag{6.30}$$

where $\oint$ denotes the integration on the full change of coordinates during a period. Along the trajectory of the integration, the Hamiltonian function $H$ keeps $E$ constant, and the impulse $p$ is a function of $q$, $\lambda$ and $E$: $p = p(q, E, \lambda)$.

By differentiating the equality $H(p, q, \lambda) = E$ with respect to $\lambda$, we find

$$\frac{\partial H}{\partial \lambda} + \frac{\partial H}{\partial p} \cdot \frac{\partial p}{\partial \lambda} = 0,$$

or

$$\frac{\partial H/\partial \lambda}{\partial H/\partial p} = -\frac{\partial p}{\partial \lambda} \cdot \tag{6.31}$$

Setting this expression into (6.30) and substituting $H$ in the denominator by $E$, we get

$$\overline{\frac{dE}{dt}} = -\frac{d\lambda}{dt} \frac{\oint \frac{\partial p}{\partial \lambda} dq}{\oint \frac{\partial p}{\partial E} dq},$$

or

$$\oint \left( \frac{\partial p}{\partial E} \cdot \overline{\frac{dE}{dt}} + \frac{\partial p}{\partial \lambda} \frac{d\lambda}{dt} \right) dq = 0,$$

or

$$\frac{d\overline{J}}{dt} = 0,$$

where

$$J = \frac{1}{2\pi} \oint p dq$$

is the action integral taken over the trajectory of the motion for constant values of $E$ and $\lambda$. Thus, $\overline{J}$ is an adiabatic invariant.

We now examine the problem of the existence of the adiabatic invariant for an oscillatory system with slowly varying parameters, described by the equation

$$\frac{d^2x}{dt^2} + f(\tau, x) = 0, \tag{6.32}$$

which is supposed to have a periodic solution

$$x = z(\tau, a, \psi), \quad \left(z(\tau, a, \psi + 2\pi) = z(\tau, a, \psi)\right), \tag{6.33}$$

$$\frac{dx}{dt} = \omega(\tau, a)z'_\psi(\tau, \psi, a),$$

where

$$\psi = \omega(\tau, a)t + \psi_0, \qquad \tau = \varepsilon t.$$

Differentiating the formulae (6.33), considering $a$ and $\psi$ as new variables, and taking into account

$$\omega^2(\tau, a)z''_{\psi^2} + f(\tau, z) \equiv 0, \tag{6.34}$$

we find that the equation (6.32) becomes:

$$\begin{aligned} z'_a\frac{da}{dt} + z'_\psi\frac{d\psi}{dt} &= \omega(\tau, a)z'_\psi - \varepsilon z'_\tau, \\ (\omega z'_\psi)'_a\frac{da}{dt} + \omega z''_{\psi^2}\frac{d\psi}{dt} &= -\omega^2(\tau, a)z''_{\psi^2} - \varepsilon(\omega z'_\psi)'_\tau. \end{aligned}$$

Solving this system of equations, we get

$$\begin{aligned} \frac{da}{dt} &= \frac{\varepsilon[(\omega z'_\psi)'_\tau z'_\psi - \omega z''_{\psi^2}z'_\tau]}{\omega z'_a z''_{\psi^2} - z'_\psi(\omega z'_\psi)'_a}, \\ \frac{d\psi}{dt} &= \omega(\tau, a) + \frac{\varepsilon[-(\omega z'_\psi)'_\tau z'_a + (\omega z'_\psi)'_a z'_\tau]}{\omega z'_a z''_{\psi^2} - z'_\psi(\omega z'_\psi)'_a}. \end{aligned} \tag{6.35}$$

The denominator of the terms on the right-hand sides of (6.35) does not depend on $\psi$. In fact, differentiating (6.34) with respect to $\psi$ and $a$, we find

$$\begin{aligned} &\omega^2 z'''_{\psi^3} + f'_z(\tau, z)z'_\psi \equiv 0, \\ &2\omega\omega'_a z''_{\psi^2} + \omega^2 z'''_{\psi^2 a} + f'_z(\tau, z)z'_a \equiv 0. \end{aligned}$$

Eliminating $f'_z(\tau, z)$ from these equations, we find:

$$\omega^2 z'''_{\psi^3} z'_a - \omega^2 z'''_{\psi^2 a} z'_\psi - 2\omega\omega'_a z''_{\psi^2} z'_\psi \equiv 0,$$

or

$$\frac{d}{d\psi}[\omega z'_a z''_{\psi^2} - z'_\psi(\omega z'_\psi)'_a] = 0.$$

Integrating the last equation on $\psi$ from 0 to $2\pi$, we obtain:

$$\omega z'_a z''_{\psi^2} - z'_\psi(\omega z'_\psi)'_a = \left[\omega z'_a z''_{\psi^2} - z'_\psi(\omega z'_\psi)'_a\right]_{\psi=0} = \mathcal{D}_0(\tau, a), \tag{6.36}$$

where $\mathcal{D}_0(\tau, a)$ does not depend on $\psi$.

We will now show that the adiabatic invariant for this equation is the action integral

$$\mathcal{J}(\tau, a) = \frac{1}{2\pi} \int_0^{2\pi} \omega(\tau, a) z_\psi'^2(\tau, \psi, a) d\psi. \tag{6.37}$$

When $\tau = \text{const}$, $a$ is a constant and therefore, $\mathcal{J}(\tau, a)$ is also a constant. Below, it is shown that, for slowly varying $\tau$ and $a$, the quantity $\mathcal{J}(\tau, a)$ will be also a constant.

For this, we differentiate $\mathcal{J}(\tau, a)$ with respect to time:

$$\frac{d\mathcal{J}(\tau, a)}{dt} = \mathcal{J}_a'(\tau, a)\frac{da}{dt} + \mathcal{J}_\tau'(\tau, a)\frac{d\tau}{dt};$$

we have

$$\frac{d\mathcal{J}(\tau, a)}{dt} = \frac{\dot{a}}{2\pi} \int_0^{2\pi} \left(\omega_a' z_\psi'^2 + 2\omega z_\psi' z_{\psi a}''\right) d\psi + \frac{\varepsilon}{2\pi} \int_0^{2\pi} \left(\omega_\tau' z_\psi'^2 + 2\omega z_\psi' z_{\psi\tau}''\right) d\psi. \tag{6.38}$$

Taking into account

$$\int_0^{2\pi} z_\psi' z_{\psi a}'' d\psi = z_\psi' z_a' \Big|_{\psi=0}^{\psi=2\pi} - \int_0^{2\pi} z_{\psi^2}'' \cdot z_a' d\psi,$$

and $z_\psi' z_a' \Big|_{\psi=0}^{\psi=2\pi} = 0$, we have for the first integral in (6.38):

$$\mathcal{J}_1(\tau, a) = \frac{\dot{a}}{2\pi} \int_0^{2\pi} \left(\omega_a' z_\psi'^2 + 2\omega z_\psi' z_{a\psi}''\right) d\psi =$$

$$= \frac{\dot{a}}{2\pi} \int_0^{2\pi} \left[\omega_a' z_\psi'^2 - \omega\left(z_{\psi^2}'' z_a' - z_\psi' \cdot z_{\psi a}''\right)\right] d\psi,$$

or

$$\mathcal{J}_1(\tau, a) = -\frac{\dot{a}}{2\pi} \int_0^{2\pi} \mathcal{D}_0(\tau, a) d\psi = -\dot{a}\mathcal{D}(\tau, a). \tag{6.39}$$

Similarly, we have for the second integral in (6.38):

$$\mathcal{J}_2(\tau, a) = \frac{\varepsilon}{2\pi} \int_0^{2\pi} \left(\omega_\tau' z_\psi'^2 + 2\omega z_\psi' z_{\psi\tau}''\right) d\psi =$$

$$= \frac{\varepsilon}{2\pi} \int_0^{2\pi} \left(\omega_\tau' z_\psi'^2 - \omega z_{\psi^2}'' z_\tau' + \omega z_\psi' z_{\psi\tau}''\right) d\psi. \tag{6.40}$$

Taking into account the first equation of (6.33) and (6.39), we can write

$$J_2(\tau, a) = \dot{a}D(\tau, a) = -J_1(\tau, a).$$

Hence,

$$\frac{dJ(\tau, a)}{dt} = J_1(\tau, a) + J_2(\tau, a) = 0,$$

or

$$J(\tau, a) = \text{const}.$$

As an example, illustrating the application of the adiabatic invariant, let us consider the equation of oscillations of a system with a single degree of freedom:

$$\frac{d^2x}{dt^2} + c(\tau)x = 0. \tag{6.41}$$

We have

$$\omega(\tau) = \sqrt{c(\tau)}, \quad z(\tau, a, \psi) = a\cos\psi, \quad z'_\psi(\tau, a, \psi) = -a\sin\psi.$$

Substituting these values into (6.37), we obtain

$$J(\tau, a) = \frac{1}{2\pi}\int_0^{2\pi} \omega z'^2_\psi \, d\psi = \frac{1}{2\pi}\int_0^{2\pi} \sqrt{c(\tau)}a^2 \sin^2\psi \, d\psi = \frac{1}{2}\sqrt{c(\varepsilon t)}\, a^2.$$

Since $J(\tau, a)$ is a constant, we have:

$$a = \frac{\text{const}}{\sqrt[4]{c(\varepsilon\tau)}} \cdot$$

This formula shows that the amplitude of oscillation is inversely proportional to the fourth root of $c(\varepsilon\tau)$.

The results obtained above can be generalized for systems, close to the Hamiltonian ones.

Let us consider the equation

$$\frac{d}{dt}\left[m(\tau)\frac{dx}{dt}\right] + f(\tau, x) = \varepsilon F\left(\tau, x, \frac{dx}{dt}\right), \tag{6.42}$$

and suppose that, the solution of the unexcited equation

$$m(\tau)\frac{d^2x}{dt^2} + f(\tau, x) = 0, \quad \tau = \text{const},$$

is known:

$$x = z(\tau, a, \psi).$$

The equation for determination of $a$ and $\psi$ in the first approximation will be

$$\begin{aligned}\frac{da}{dt} &= \frac{-\varepsilon}{2\pi D(\tau,a)m(\tau)}\int_0^{2\pi}\left[F(\tau,z,z'_\psi)z'_\psi + m(\tau)\xi_1 - m'_\tau\omega(\tau,a)z'^2_\psi\right]d\psi,\\ \frac{d\psi}{dt} &= \omega(\tau,a) + \frac{\varepsilon}{2\pi m(\tau)D(\tau,a)}\times\\ &\times\int_0^{2\pi}\left[F(\tau,z,z'_\psi)z'_a + m(\tau)\xi_2 - m'_\tau\omega(\tau,a)z'_\psi z'_a\right]d\psi,\end{aligned} \tag{6.43}$$

where

$$\begin{aligned}D(\tau,a) &= \omega(\tau,a)z'_a z''_{\psi^2} - z'_\psi\left[\omega(\tau,a)z'_\psi\right]'_a,\\ \xi_1 &= \omega(\tau,a)\left[z'_\tau z''_{\psi^2} - z''_{\psi\tau}z'_\psi\right] - \omega'_\tau z'^2_\psi,\\ \xi_2 &= \omega(\tau,a)\left[z'_\tau z''_{a\psi} - z'_a z''_{\psi\tau}\right] + \omega'_a z'_\psi z'_\tau - \omega'_\tau z'_a z'_\psi.\end{aligned}$$

It will be shown that, for the equation (6.42), the action integral

$$J(\tau,a) = \frac{m(\tau)}{2\pi}\int_0^{2\pi}\omega(\tau,a)z'^2_\psi d\psi$$

is an adiabatic invariant. For this purpose, we calculate the derivative

$$\frac{dJ(\tau,a)}{dt} = J'_a(\tau,a) + J'_\tau(\tau,a)\frac{d\tau}{dt}. \tag{6.44}$$

Further, we can write

$$\dot{a}J'_a(\tau,a) = \frac{\dot{a}m(\tau)}{2\pi}\int_0^{2\pi}\left(\omega'_a z'^2_\psi + 2\omega(\tau,a)z'_\psi z''_{\psi a}\right)d\psi = -\dot{a}m(\tau)D(\tau,a),$$

$$\begin{aligned}\varepsilon J'_\tau(\tau,a) &= \varepsilon\frac{m'_\tau(\tau)}{2\pi}\int_0^{2\pi}\omega z'_\psi d\psi + \frac{\varepsilon m(\tau)}{2\pi}\int_0^{2\pi}\left[\omega'_\tau z'^2_\psi + 2\omega(\tau,a)z'_\psi z''_{\psi\tau}\right]d\psi =\\ &= -\frac{\varepsilon m(\tau)}{2\pi}\int_0^{2\pi}\xi_1 d\psi + \frac{\varepsilon m'_\tau(\tau)}{2\pi}\int_0^{2\pi}\omega(\tau,a)z'^2_\psi d\psi.\end{aligned}$$

Replacing here $\dot{a}$ by the first equation of (6.43), we find that the expression (6.44) becomes

$$\frac{dJ(\tau,a)}{dt} = \frac{\varepsilon}{2\pi}\int_0^{2\pi} F(\tau,z,z'_\psi)z'_\psi\,d\psi + \frac{\varepsilon m(\tau)}{2\pi}\int_0^{2\pi}\xi_1\,d\psi -$$
$$-\frac{\varepsilon m'_\tau(\tau)}{2\pi}\int_0^{2\pi}\omega(\tau,a)z'^2_\psi\,d\psi - \frac{\varepsilon m(\tau)}{2\pi}\int_0^{2\pi}\xi_1\,d\psi +$$
$$+\frac{\varepsilon m'_\tau(\tau)}{2\pi}\int_0^{2\pi}\omega(\tau,a)z'^2_\psi\,d\psi = \frac{\varepsilon}{2\pi}\int_0^{2\pi} F(\tau,z,z'_\psi)z'_\psi\,d\psi.$$

Thus, $J(\tau,a)$ remains constant not only when $F\left(\tau,x,\frac{dx}{dt}\right) \equiv 0$, but also when there exists orthogonality between $z'_\psi$ and $F(\tau,z,z'_\psi)$.

As an example, we examine the small oscillations of a mathematical pendulum, whose mass is slowly changing in time. The equation of motion of the pendulum is then

$$\frac{d}{dt}\left[m(\tau)\frac{dx}{dt}\right] + m(\tau)g\ell x = 0. \tag{6.45}$$

To find the amplitude of oscillation of the pendulum, it is not necessary to integrate the equation (6.45). We can use the adiabatic invariant

$$J(\tau,a) = \frac{m(\tau)}{2\pi}\int_0^{2\pi}\omega z'^2_\psi\,d\psi.$$

In this case, we have

$$\omega = \sqrt{g\ell}, \quad x = z(\tau,a,\psi) = a\cos\psi, \quad z'_\psi = -a\sin\psi$$

and the expression $J(\tau,a) = \text{const}$ gives

$$\frac{m(\tau)\sqrt{g\ell}}{2}a^2 = \text{const},$$

or

$$a = \frac{a_0}{\sqrt{m(\tau)}}\,.$$

In fact, the amplitude of oscillation of the pendulum does not change, even if there is a small perturbation $\varepsilon F(\tau,x)$, because in this case, the equation of motion will be

$$\frac{d}{dt}\left[m(\tau)\frac{dx}{dt}\right] + m(\tau)g\ell x = \varepsilon F(\tau,x),$$

and for which

$$\frac{dJ(\tau,a)}{dt} = \frac{\varepsilon}{2\pi}\int_0^{2\pi}F(\tau,z)z'_\psi\,d\psi = -\frac{\varepsilon a}{2\pi}\int_0^{2\pi}F(\tau,a\cos\psi)\sin\psi\,d\psi = 0.$$

## 7. Free Oscillations of Systems Governed by a Third Order Differential Equation

**Visco-elastic Hooke-Maxwell model of an oscillatory system** [37,b]

Consider a mass $m$ attached to a beam which is made of visco-elastic Hooke-Maxwell material (Fig. 6). The physical model of this system is represented in Fig. 7 and consists of a mass $m$, fastened on a main spring with stiffness $c$, and a weak damper continued by a weak supplementary spring with the coefficient of damping $\varepsilon h$ and stiffness $\varepsilon k$. We have the following differential equations of motion:

$$\begin{aligned} &m\ddot{x} + cx + \varepsilon h(\dot{x} - \dot{z}) = \varepsilon F(x, \dot{x}), \\ &\varepsilon h(\dot{z} - \dot{x}) + \varepsilon kz = 0, \end{aligned} \tag{7.1}$$

where $\varepsilon$ is a small positive parameter, $m$, $h$, $k$ are positive constants and $\varepsilon F(x, \dot{x})$ is the force acting on the mass. The overdots denote differentiation with respect to $t$.

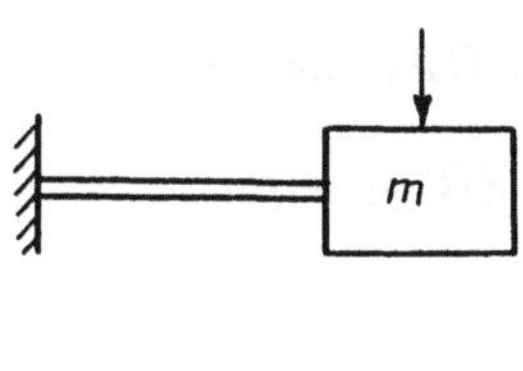


*Fig. 6.* Oscillating mass.

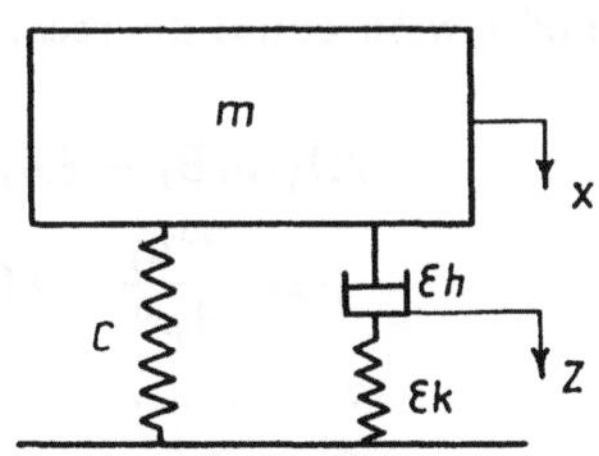


*Fig. 7.* Hooke-Maxwell model.

Adding equations (7.1) gives:

$$m\ddot{x} + cx + \varepsilon kz = \varepsilon F(x, \dot{x}).$$

By differentiating this equation, one has

$$m\dddot{x} + c\dot{x} + \varepsilon k\dot{z} = \varepsilon \dot{F}(x, \dot{x}).$$

Substituting $\dot{z}$ from equation (7.1) into the last equation, we obtain

$$m\dddot{x} + \frac{k}{h}m\ddot{x} + c\dot{x} + c\frac{k}{h}x = -\varepsilon\big[k\dot{x} - \dot{F}(x, \dot{x}) - \frac{k}{h}F(x, \dot{x})\big],$$

or

$$\dddot{x} + \xi\ddot{x} + \Omega^2\dot{x} + \xi\Omega^2 x = \varepsilon R(x, \dot{x}, \ddot{x}), \tag{7.2}$$

where

$$\xi = \frac{k}{h}, \qquad \Omega^2 = \frac{c}{m},$$
$$R(x, \dot{x}, \ddot{x}) = \frac{1}{m}\left[\xi F(x, \dot{x}) + \dot{F}(x, \dot{x}) - k\dot{x}\right].$$

**Construction of the solution**

We shall find a solution of equation (7.2) in the form

$$x = a\cos\varphi + \varepsilon u_1(a, \varphi) + \varepsilon^2 u_2(a, \varphi) + \dots, \tag{7.3}$$

where $\varphi = \Omega t + \psi$. Here $a$ and $\psi$ are functions, satisfying the following differential equations

$$\begin{aligned} \frac{da}{dt} &= \varepsilon A_1(a) + \varepsilon^2 A_2(a) + \dots, \\ \frac{d\psi}{dt} &= \varepsilon B_1(a) + \varepsilon^2 B_2(a) + \dots \end{aligned} \tag{7.4}$$

Substituting the expressions (7.3) and (7.4) into (7.2) and comparing the coefficients of $\varepsilon$ with equal degrees, we have

$$\begin{aligned} &2\Omega(\Omega a B_1 - \xi A_1)\sin\varphi - 2\Omega(\xi a B_1 + \Omega A_1)\cos\varphi + \\ &+ \Omega^3\frac{\partial^3 u_1}{\partial\varphi^3} + \xi\Omega^2\frac{\partial^2 u_1}{\partial\varphi^2} + \Omega^3\frac{\partial u_1}{\partial\varphi} + \xi\Omega^2 u_1 = R_0, \end{aligned} \tag{7.5}$$

..........................

$$R_0 = R(a\cos\varphi, -\Omega a\sin\varphi, -\Omega^2 a\cos\varphi). \tag{7.6}$$

To find the unknown functions $A_1$, $B_1$ and $u_1$ from equation (7.5), we first expand $R_0$ in the Fourier series:

$$R_0 = \sum_{n=0}^{\infty}\left[g_n(a)\cos n\varphi + h_n(a)\sin n\varphi\right], \tag{7.7}$$

$$g_0(a) = \frac{1}{2\pi}\int_0^{2\pi} R_0 d\varphi = \langle R_0\rangle,$$

$$g_n(a) = \frac{1}{\pi}\int_0^{2\pi} R_0\cos n\varphi d\varphi = 2\langle R_0\cos n\varphi\rangle,$$

$$h_n(a) = \frac{1}{\pi}\int_0^{2\pi} R_0\sin n\varphi d\varphi = 2\langle R_0\sin n\varphi\rangle,$$

in which $n = 1, 2, \ldots$ and $\langle \quad \rangle$ is a time averaging operator.

Function $u_1$ can be represented by

$$u_1 = \sum_m \left[u_{1m}(a)\cos m\varphi + v_{1m}(a)\sin m\varphi\right]. \tag{7.8}$$

By substituting equations (7.7) and (7.8) into equation (7.5), we obtain

$$\begin{aligned}&\Omega^2 \sum_m (1-m^2)\big[(\xi u_{1m} + m\Omega v_{1m})\cos m\varphi + \\ &+ (\xi v_{1m} - m\Omega u_{1m})\sin m\varphi\big] - 2\Omega(\Omega A_1 + \xi a B_1)\cos\varphi + \\ &+ 2\Omega(a\Omega B_1 - \xi A_1)\sin\varphi = \sum_{n=0}^{\infty}\left[g_n(a)\cos n\varphi + h_n(a)\sin n\varphi\right].\end{aligned} \tag{7.9}$$

Comparing the coefficients of the fundamental harmonics in equation (7.9) gives

$$\begin{aligned}2\Omega(\Omega A_1 + a\xi B_1) &= -g_1(a),\\ 2\Omega(-\xi A_1 + a\Omega B_1) &= h_1(a).\end{aligned}$$

From here, one obtains

$$\begin{aligned}A_1(a) &= -\frac{\xi h_1(a) + \Omega g_1(a)}{2\Omega(\xi^2+\Omega^2)} = -\frac{\xi\langle R_0\sin\varphi\rangle + \Omega\langle R_0\cos\varphi\rangle}{\Omega(\xi^2+\Omega^2)},\\ B_1(a) &= \frac{\Omega h_1(a) - \xi g_1(a)}{2\Omega a(\xi^2+\Omega^2)} = \frac{\Omega\langle R_0\sin\varphi\rangle - \xi\langle R_0\cos\varphi\rangle}{\Omega a(\xi^2+\Omega^2)}.\end{aligned} \tag{7.10}$$

The comparison of the other harmonics gives

$$\begin{aligned}&\Omega^2(1-m^2)(\xi u_{1m} + m\Omega v_{1m}) = g_m(a),\\ &\Omega^2(1-m^2)(-m\Omega u_{1m} + \xi v_{1m}) = h_m(a), \quad m \neq 1.\end{aligned}$$

From here, it follows

$$\begin{aligned}u_{1m} &= \frac{\xi g_m(a) - m\Omega h_m(a)}{\Omega^2(1-m^2)(\xi^2+m^2\Omega^2)},\\ v_{1m} &= \frac{\xi h_m(a) + m\Omega g_m(a)}{\Omega^2(1-m^2)(\xi^2+m^2\Omega^2)}.\end{aligned} \tag{7.11}$$

Thus, in the first approximation, we have

$$x = a\cos(\Omega t + \psi), \tag{7.12}$$

here $a$ and $\psi$ are the solutions of equations

$$\begin{aligned} \frac{da}{dt} &= -\varepsilon \frac{\xi \langle R_0 \sin \varphi \rangle + \Omega \langle R_0 \cos \varphi \rangle}{\Omega (\xi^2 + \Omega^2)}, \\ \frac{d\psi}{dt} &= \varepsilon \frac{\Omega \langle R_0 \sin \varphi \rangle - \xi \langle R_0 \cos \varphi \rangle}{\Omega a (\xi^2 + \Omega^2)}. \end{aligned} \tag{7.13}$$

The refinement of the first approximation is

$$x = a \cos(\Omega t + \psi) + \sum_{\substack{m=0 \\ m \neq 1}}^{\infty} \big[ (\xi g_m - m\Omega h_m) \cos m(\Omega t + \psi) + \\ + (\xi h_m + m\Omega g_m) \sin m(\Omega t + \psi) \big] \frac{1}{\Omega^2 (1 - m^2)(\xi^2 + m^2 \Omega^2)}. \tag{7.14}$$

The calculations of higher approximations do not present difficulty, but are rather long, and are not reproduced here.

We can write the first equation of the system (7.13) as

$$\frac{da}{dt} = \varepsilon \Phi(a). \tag{7.15}$$

Obviously, the stability condition for the stationary solution $a = a_0$, $\Phi(a_0) = 0$ of equation (7.15) is of the form:

$$\Phi'(a_0) < 0. \tag{7.16}$$

**Example 9**

Let us consider the case $R(x, \dot{x}, \ddot{x}) = -\beta x^3$. Now, equation (7.2) has the form

$$\dddot{x} + \xi \ddot{x} + \Omega^2 \dot{x} + \xi \Omega^2 x + \varepsilon \beta x^3 = 0. \tag{7.17}$$

In this case, we have

$$R_0 = -\beta a^3 \cos^3 \varphi = -\frac{\beta}{4} a^3 (3 \cos \varphi + \cos 3\varphi)$$

and, therefore,

$$\begin{aligned} g_1(a) &= -\frac{3}{4} \beta a^3, \qquad h_1(a) = 0, \\ A_1(a) &= \frac{3\beta a^3}{8(\xi^2 + \Omega^2)}, \quad B_1(a) = \frac{3\xi \beta a^2}{8\Omega (\xi^2 + \Omega^2)}. \end{aligned}$$

The equations (7.13) are of the form

$$\frac{da}{dt} = \frac{1}{2}\beta^* a^3, \quad \frac{d\psi}{dt} = \frac{\xi}{2\Omega}\beta^* a^2, \quad \beta^* = \frac{3\varepsilon\beta}{4(\xi^2 + \Omega^2)}.$$

By integrating these equations with initial values $(t_0, a_0)$, we obtain

$$a^2 = \frac{1}{\alpha - \beta^*(t - t_0)}, \quad \alpha = \frac{1}{a_0^2}, \quad \psi = \psi_0 - \frac{\xi}{2\Omega}\ell_n\left|1 - \frac{\beta^*}{\alpha}(t - t_0)\right|.$$

From here, it follows that (see Fig. 8):

1) If $\beta > 0$, then the amplitude of oscillation $a$ increases for $t > t_0$ from $a = a_0$ to $a \to \infty$.

2) If $\beta < 0$, then the amplitude of oscillation decreases from $a = a_0$ to $a = 0$.

Thus, the nonlinear term has a great influence on the form of the response curve of free oscillation. To check the validity of the theoretical results, an analog-computer analysis was carried out. Examples of wave forms obtained by an analog-computer for equation (7.1) with the parameters $\xi = 10$, $\Omega = 1$ are shown in Fig. 9. In this figure, the first wave (a) shows the case $\beta = 0$, the second (b) $\beta = -10$, and the third (c) $\beta = 10$. The figure contains all types of oscillations that are predicted by the theory.

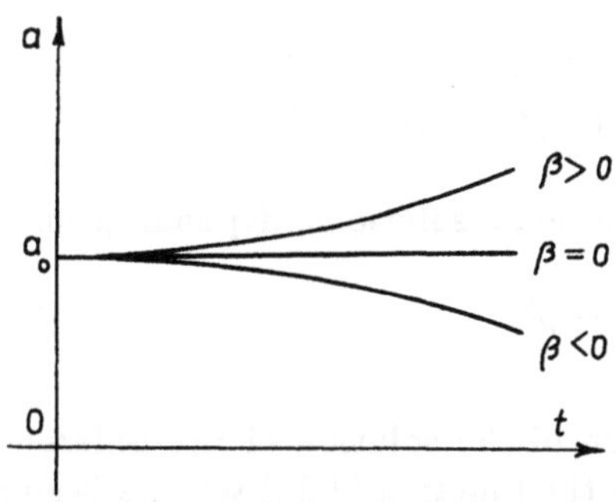


*Fig. 8.* Amplitude change in time.

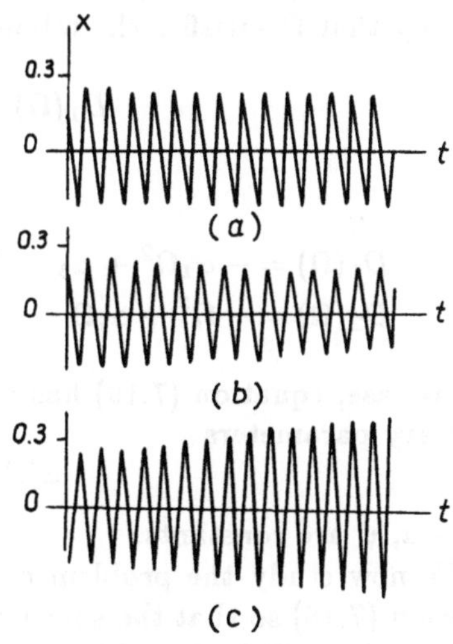


*Fig. 9.* Wave forms by an analog-computer.

**Quasi-linear differential equation of third order with delay**

Let us consider the equation of type [51]:

$$\frac{d^3x(t)}{dt^3} + \alpha_1\frac{d^2x(t)}{dt^2} + \alpha_2\frac{dx(t)}{dt} + \alpha_3 x(t) + \beta_1\frac{d^2x(t-\Delta)}{dt^2} + \beta_2\frac{dx(t-\Delta)}{dt} +$$
$$+ \beta_3 x(t-\Delta) = \varepsilon F\left(x(t), \frac{dx(t)}{dt}, \frac{d^2x(t)}{dt^2}, x(t-\Delta), \frac{dx(t-\Delta)}{dt}\right), \tag{7.18}$$

where $\alpha_i$, $\beta_i$ and $\Delta$ are constants and $F$ is an analytical function of its arguments.

Together with (7.18), we consider the corresponding degenerative linear equation:

$$\frac{d^3x(t)}{dt^3} + \alpha_1 \frac{d^2x(t)}{dt^2} + \alpha_2 \frac{dx(t)}{dt} + \alpha_3 x(t) + \\ + \beta_1 \frac{d^2x(t-\Delta)}{dt^2} + \beta_2 \frac{dx(t-\Delta)}{dt} + \beta_3 x(t-\Delta) = 0, \tag{7.19}$$

which is obtained from equation (7.18) by setting $\varepsilon = 0$. It is supposed that the characteristic equation for equation (7.19):

$$D(\lambda) = 0, \tag{7.20}$$

where

$$D(\lambda) = \lambda^3 + \alpha_1\lambda^2 + \alpha_2\lambda + \alpha_3 + \beta_1\lambda^2 e^{-\lambda\Delta} + \beta_2\lambda e^{-\lambda\Delta} + \beta_3 e^{-\lambda\Delta},$$

has a pair of pure imaginary simple roots $\lambda = \pm i\Omega$, and the remaining roots of equation (7.20) have negative real parts with sufficiently large values. It is easy to verify that $\Omega$ satisfies the following equations

$$D_1(\Omega) = 0, \qquad D_2(\Omega) = 0, \tag{7.21}$$

where

$$D_1(\Omega) = -\alpha_1\Omega^2 + \alpha_3 - \beta_1\Omega^2 \cos\Delta\Omega + \beta_2\Omega \sin\Delta\Omega + \beta_3 \cos\Delta\Omega,$$
$$D_2(\Omega) = -\Omega^3 + \alpha_2\Omega + \beta_1\Omega^2 \sin\Delta\Omega + \beta_2\Omega \cos\Delta\Omega - \beta_3 \sin\Delta\Omega.$$

In this case, equation (7.19) has a family of periodic solutions, depending on two arbitrary parameters

$$x(t) = a\cos(\Omega t + \psi), \tag{7.22}$$

where $a$, $\psi$ are constants.

We now study the problem of construction of the solution of the differential equation (7.18) so that the solution is close to the function (7.22) when $\varepsilon$ is small. The solution will be found in the form

$$x(t) = a\cos\varphi + \varepsilon u_1(a,\varphi) + \varepsilon^2 u_2(a,\varphi) + \ldots, \tag{7.23}$$

where $\varphi = \Omega t + \psi$, and functions $u_i(a,\varphi)$ are periodic with period $2\pi$ relative to $\varphi$. The quantities $a$ and $\psi$ are to be defined by differential equations

$$\frac{da}{dt} = \varepsilon A_1(a) + \varepsilon^2 A_2(a) + \ldots, \\ \frac{d\psi}{dt} = \varepsilon B_1(a) + \varepsilon^2 B_2(a) + \ldots \tag{7.24}$$

The problem leads to the determination of unknown functions

$$u_i(a,\varphi), \quad A_i(a), \quad B_i(a) \tag{7.25}$$

and to the integration of equations (7.24). For single-valued determination of functions (7.25), we require that there are no fundamental harmonics $\sin\varphi$, $\cos\varphi$ in the expressions for $u_i(a,\varphi)$. In other words, $a$ will be the full amplitude of the fundamental harmonic of the solution (7.23).

Substituting the function $x(t)$ (7.23) and its derivatives into equation (7.18), and comparing the like powers of $\varepsilon$ on both sides, one obtains a recursive system:

$$\begin{aligned}&\left[A_1\frac{dD_1(\Omega)}{d\Omega} - aB_1\frac{dD_2(\Omega)}{d\Omega}\right]\sin\varphi + \left[A_1\frac{dD_2(\Omega)}{d\Omega}+\right.\\ &\left.+ aB_1\frac{dD_1(\Omega)}{d\Omega}\right]\cos\varphi + \Omega^3\frac{\partial^3 u_1(a,\varphi)}{\partial\varphi^3} + \alpha_1\Omega^2\frac{\partial^2 u_1(a,\varphi)}{\partial\varphi^2}+\\ &+\alpha_2\Omega\frac{\partial u_1(a,\varphi)}{\partial\varphi} + \alpha_3 u_1(a,\varphi) + \beta_1\Omega^2\frac{\partial^2 u_1(a,\varphi-\Delta\Omega)}{\partial\varphi^2}+\\ &+\beta_2\Omega\frac{\partial u_1(a,\varphi-\Delta\Omega)}{\partial\varphi} + \beta_3 u_1(a,\varphi-\Delta\Omega) = F_0(a,\varphi),\end{aligned} \tag{7.26}$$

. . . . . . . . . . . . . . . . . . . . . . . . . .

where

$$\begin{aligned}F_0(a,\varphi) = F\big(&a\cos\varphi, -a\Omega\sin\varphi, -a\Omega^2\cos\varphi, a\cos(\varphi-\Delta\Omega),\\ &-a\Omega\sin(\varphi-\Delta\Omega), -a\Omega^2\cos(\varphi-\Delta\Omega), 0\big).\end{aligned}$$

Inasmuch as $F_0(a,\varphi)$ is a periodic function of $\varphi$ with period $2\pi$, one can expand it in the Fourier series

$$F_0(a,\varphi) = F_{00}(a) + \sum_{n=1}^{\infty}\big(F_{0nc}(a)\cos n\varphi + F_{0ns}(a)\sin n\varphi\big). \tag{7.27}$$

The unknown function $u_1(a,\varphi)$ will be found in the form of a series:

$$u_1(a,\varphi) = u_{10}(a) + \sum_{n=2}^{\infty}\big(u_{1nc}(a)\cos n\varphi + u_{1ns}(a)\sin n\varphi\big). \tag{7.28}$$

Substituting the expressions (7.27) and (7.28) into equation (7.26), we have

$$\begin{aligned}&(\alpha_3+\beta_3)u_{10} + \left[A_1\cdot\frac{dD_1(\Omega)}{d\Omega} - aB_1\cdot\frac{dD_2(\Omega)}{d\Omega}\right]\sin\varphi + \left[A_1\frac{dD_2(\Omega)}{d\Omega}+\right.\\ &\left.+ aB_1\cdot\frac{dD_1(\Omega)}{d\Omega}\right]\cos\varphi + \sum_{n=2}^{\infty}\Big[\big(D_2(n\Omega)u_{1ns} + D_1(n\Omega)u_{1nc}\big)\cos n\varphi+\\ &+\big(D_1(n\Omega)u_{1ns} - D_2(n\Omega)u_{1nc}\big)\sin n\varphi\Big] =\\ &= F_{00}(a) + \sum_{n=1}^{\infty}\big(F_{0nc}(a)\cos n\varphi + F_{0ns}(a)\sin n\varphi\big).\end{aligned} \tag{7.29}$$

Comparing the coefficients of equal harmonics in equation (7.29), we obtain after simple calculations:

$$A_1 = \frac{F_{01s}D_1'(\Omega) + F_{01c}D_2'(\Omega)}{D_1'^2(\Omega) + D_2'^2(\Omega)},$$
$$aB_1 = \frac{F_{01c}D_1'(\Omega) - F_{01s}D_2'(\Omega)}{D_1'^2(\Omega) + D_2'^2(\Omega)}, \tag{7.30}$$

$$u_{10} = \frac{F_{00}(a)}{\alpha_3 + \beta_3}, \quad u_{1ns} = \frac{D_1(n\Omega)F_{0ns} + D_2(n\Omega)F_{0nc}}{D_1^2(n\Omega) + D_2^2(n\Omega)},$$
$$u_{1nc} = \frac{D_1(n\Omega)F_{0nc} - D_2(n\Omega)F_{0ns}}{D_1^2(n\Omega) + D_2^2(n\Omega)}, \quad n \geq 2,$$

where a prime denotes the derivative with respect to $\Omega$. Hence,

$$u_1(a,\varphi) = \frac{F_{00}(a)}{\alpha_3 + \beta_3} + \sum_{n=2}^{\infty} \left\{ \frac{D_1(n\Omega)F_{0nc} - D_2(n\Omega)F_{0ns}}{D_1^2(n\Omega) + D_2^2(n\Omega)} \cos n\varphi + \right.$$
$$\left. + \frac{D_1(n\Omega)F_{0ns} + D_2(n\Omega)F_{0nc}}{D_1^2(n\Omega) + D_2^2(n\Omega)} \sin n\varphi \right\}. \tag{7.31}$$

In view of the assumptions for the roots of the characteristic equation, the denominators of expressions (7.30) are non-zero:

$$D_1'^2(\Omega) + D_2'^2(\Omega) \neq 0, \; D_1^2(n\Omega) + D_2^2(n\Omega) \neq 0, \; n \geq 2.$$

Finally, the first approximation is a harmonic oscillation

$$x = a\cos\varphi, \tag{7.32}$$

and the refinement of the first approximation will be:

$$x = a\cos\varphi + \varepsilon u_1(a,\varphi), \tag{7.33}$$

where $a$ and $\varphi$ satisfy the equations

$$\frac{da}{dt} = \varepsilon A_1(a),$$
$$\frac{d\varphi}{dt} = \Omega + \varepsilon B_1(a). \tag{7.34}$$

Functions $u_1(a,\varphi)$, $A_1(a)$, $B_1(a)$ are of the form given by (7.30) and (7.31).

**Example 10**

Consider the equation

$$\frac{d^3x(t)}{dt^3}+\xi\frac{d^2x(t)}{dt^2}+\Omega^2\frac{dx(t)}{dt}+\xi\Omega^2x(t-\varepsilon\Delta_1)=\varepsilon\mu(1-x^2(t))\frac{dx(t-\Delta_2)}{dt}, \quad (7.35)$$

where $\xi$, $\Omega$, $\Delta_1$, $\Delta_2$, $\mu$ are positive constants.

The characteristic equation (7.20) now has a pair of pure imaginary roots $\lambda = \pm i\Omega$. The expressions (7.33) and (7.34) are then

$$\begin{aligned} x(t) = {} & a\cos\varphi + \frac{\varepsilon\mu a^3}{32\Omega(9\Omega^2+\xi^2)}\Big[(\xi\sin\Delta_2\Omega+ \\ & + 3\Omega\cos\Delta_2\Omega)\cos\varphi + (3\Omega\sin\Delta_2\Omega - \xi\cos\Delta_2\Omega)\sin 3\varphi\Big], \end{aligned}$$

$$\begin{aligned} \frac{da}{dt} &= \frac{\varepsilon a}{2(\Omega^2+\xi^2)}\Big[\xi^2\Omega^2\Delta_1- \\ & - \mu\xi\Big(\frac{1}{4}a^2-1\Big)\cos\Delta_2\Omega + \mu\Omega\Big(\frac{3}{4}a^2-1\Big)\sin\Delta_2\Omega\Big], \\ \frac{d\varphi}{dt} &= \Omega + \frac{\varepsilon}{2(\Omega^2+\xi^2)}\Big[-\xi\Omega^3\Delta_1+ \\ & + \mu\Omega\Big(\frac{1}{4}a^2-1\Big)\cos\Delta_2\Omega + \xi\mu\Big(\frac{3}{4}a^2-1\Big)\sin\Delta_2\Omega\Big]. \end{aligned}$$

It is clear from these formulae that the small delay $\varepsilon\Delta_1$, intensifies the instability of the origin ($a = 0$) and has significant influence on the frequency of the oscillations.

## 8. Free Oscillations of Systems Governed by $N$ - Order Differential Equation

Let us consider the differential equation of the form [37 b, 52]

$$x^{(N)} + \alpha_1 x^{(N-1)} + \cdots + \alpha_{N-1}\dot{x} + \alpha_N x = \varepsilon F(x, \dot{x}, \ldots, x^{(N)}, \varepsilon), \quad (8.1)$$

where $\varepsilon$ is a small parameter, $\alpha_i$ are real constants, $F(x, \ldots, \varepsilon)$ is the known function, which is sufficiently differentiable for all finite values of its arguments. When $\varepsilon = 0$, the equation (8.1) reduces to

$$\begin{aligned} & x^{(N)} + \alpha_1 x^{(N-1)} + \cdots + \alpha_{N-1}\dot{x} + \alpha_N x = 0, \\ & x^{(k)} = \frac{d^k x}{dt^k}. \end{aligned} \quad (8.2)$$

The behaviour of the solution of this degenerate equation essentially depends on the roots of the characteristic equation

$$\lambda^N + \alpha_1\lambda^{N-1} + \cdots + \alpha_{N-1}\lambda + \alpha_N = 0. \tag{8.3}$$

It is supposed that the characteristic equation has a pair of simple imaginary roots $\lambda = \pm i\Omega$, and its other roots have negative real parts with sufficiently large values. In this case, the degenerate equation (8.2) has the periodic solution

$$x = a\cos(\Omega t + \psi), \tag{8.4}$$

where $a$ and $\psi$ are arbitrary real constants.

By virtue of the continuous dependence of the solution on the parameter $\varepsilon$, the equation (8.1) with sufficiently small $\varepsilon$ has a solution close to (8.1). Consequently, we shall find the solution of the equation (8.1) in the form

$$x = a\cos\varphi + \varepsilon u_1(a,\varphi) + \varepsilon^2 u_2(a,\varphi) + \dots, \tag{8.5}$$

where $u_i(a,\varphi)$ do not contain $\cos\varphi$, $\sin\varphi$. These functions are bounded for all finite values $a$ and real values $\varphi$, and are periodic functions of $\varphi$ with the period $2\pi$, $\varphi = \Omega t + \psi$. The quantities $a$ and $\psi$, being slowly varying functions of $t$, are determined by the differential equations

$$\begin{aligned} \frac{da}{dt} &= \varepsilon A_1(a) + \varepsilon^2 A_2(a) + \dots, \\ \frac{d\psi}{dt} &= \varepsilon B_1(a) + \varepsilon^2 B_2(a) + \dots \end{aligned} \tag{8.6}$$

Now the problem reduces to the determination of unknown functions $u_i(a,\varphi)$, $A_i(a)$, $B_i(a)$, and to the integration of equations (8.6).

Substituting the expressions (8.5), (8.6) into the equation (8.1) and comparing the coefficients of $\varepsilon$ with regard for

$$\begin{aligned} x^{(k)} = {} & a\Omega^k\cos\left(\varphi + k\frac{\pi}{2}\right) + \varepsilon\Big[\Omega^k\frac{\partial^k u_1}{\partial\varphi^k} + k\Omega^{k-1}A_1\sin\left(\varphi + k\frac{\pi}{2}\right) + \\ & + k\Omega^{k-1}aB_1\cos\left(\varphi + k\frac{\pi}{2}\right)\Big] + \varepsilon^2\Big\{\Omega^k\frac{\partial^k u_2}{\partial\varphi^k} + k\Omega^{k-1}A_2\sin\left(\varphi + k\frac{\pi}{2}\right) + \\ & + k\Omega^{k-1}aB_2\cos\left(\varphi + k\frac{\pi}{2}\right) + k\Omega^{k-1}A_1\frac{\partial^k u_1}{\partial a\,\partial\varphi^{k-1}} + k\Omega^{k-1}B_1\frac{\partial^k u_1}{\partial\varphi^k} + \\ & + \frac{1}{2}k(k-1)\Omega^{k-2}\left(aA_1\frac{dB_1}{da} + 2A_1B_1\right)\sin\left(\varphi + k\frac{\pi}{2}\right) - \\ & - \frac{1}{2}k(k-1)\Omega^{k-2}\left(A_1\frac{dA_1}{da} - aB_1^2\right)\cos\left(\varphi + k\frac{\pi}{2}\right)\Big\} + \varepsilon^3\dots, \end{aligned}$$

we have

$$\sum_{k=0}^{N} \alpha_{N-k}\Omega^k \frac{\partial^k u_1}{\partial \varphi^k} + A_1 \sum_{k=1}^{N} k\alpha_{N-k}\Omega^{k-1} \sin\left(\varphi + k\frac{\pi}{2}\right) +$$
$$+ aB_1 \sum_{k=1}^{N} k\alpha_{N-k}\Omega^{k-1} \cos\left(\varphi + k\frac{\pi}{2}\right) = F_0, \tag{8.7}$$

where $\alpha_0 \equiv 1$,

$$F_0 = F\left[a\cos\varphi, -a\Omega\sin\varphi, \ldots, a\Omega^N \cos\left(\varphi + N\frac{\pi}{2}\right), 0\right].$$

To determine $A_1(a)$, $B_1(a)$, $u_1(a,\varphi)$ from the equation (8.7), we require an additional condition, that $u_1(a,\varphi)$ does not contain the fundamental harmonics:

$$\begin{aligned} F_0 &= \sum_m \left(F_{mc}\cos m\varphi + F_{ms}\sin m\varphi\right), \\ u_1(a,\varphi) &= \sum_{m\neq 1} \left(u_{mc}\cos m\varphi + u_{ms}\sin m\varphi\right). \end{aligned} \tag{8.8}$$

Substituting these expressions into equation (8.7) and comparing the coefficients of $\sin\varphi$, $\cos\varphi$, we get

$$\begin{aligned} S_1 A_1 + aS_2 B_1 &= F_{1c}, \\ S_2 A_1 - aS_1 B_1 &= F_{1s}, \end{aligned} \tag{8.9}$$

where

$$S_1 = \sum_{k=1}^{N} k\alpha_{N-k}\Omega^{k-1} \sin k\frac{\pi}{2},$$

and

$$S_2 = \sum_{k=2}^{N} k\alpha_{N-k}\Omega^{k-1} \cos k\frac{\pi}{2}. \tag{8.10}$$

The equations (8.9) give

$$A_1 = \frac{S_1F_{1c} + S_2F_{1s}}{S_1^2 + S_2^2}, \quad B_1 = \frac{S_2F_{1c} - S_1F_{1s}}{a(S_1^2 + S_2^2)}. \tag{8.11}$$

Comparing the other harmonics in the equation (8.7), we have

$$\begin{aligned} L_1 u_{mc} + L_2 u_{ms} &= F_{mc}, \\ -L_2 u_{mc} + L_1 u_{ms} &= F_{ms}, \end{aligned} \tag{8.12}$$

where

$$L_1 = \sum_{k=0}^{N} \alpha_{N-k}\Omega^k m^k \cos\frac{k\pi}{2}, \quad L_2 = \sum_{k=1}^{N} \alpha_{N-k}\Omega^k m^k \sin\frac{k\pi}{2}.$$

Because $i\Omega$ is the root of the characteristic equation (8.3), then $L_1 = L_2 = 0$ when $m = 1$. From equations (8.12), we obtain

$$u_{ms} = \frac{L_1 F_{ms} + L_2 F_{mc}}{L_1^2 + L_2^2}, \quad u_{mc} = \frac{L_1 F_{mc} - L_2 F_{ms}}{L_1^2 + L_2^2}, \quad m \neq 1. \tag{8.13}$$

Thus, in the first approximation, we have the solution of the equation (8.1) in the form

$$x = a\cos(\Omega t + \psi), \tag{8.14}$$

where $a$ and $\psi$ are determined from the equations

$$\frac{da}{dt} = \varepsilon\frac{S_1 F_{1c} + S_2 F_{1s}}{S_1^2 + S_2^2}, \quad \frac{d\psi}{dt} = \varepsilon\frac{S_2 F_{1c} - S_1 F_{1s}}{a(S_1^2 + S_2^2)}. \tag{8.15}$$

The refinement of the first approximation is

$$x = a\cos(\Omega t + \psi) + \varepsilon \sum_{m\neq 1} \big[(L_1 F_{mc} - L_2 F_{ms})\cos m\varphi + + (L_1 F_{ms} + L_2 F_{mc})\sin m\varphi\big] \times \frac{1}{L_1^2 + L_2^2}, \tag{8.16}$$

$\varphi = \Omega t + \psi$. The functions $a$, $\psi$ are the solution of the equations (8.15).

In the same way, if we compare the terms $\varepsilon^2$ in the equation (8.1) and then we compare the coefficients of $\sin\varphi$, $\cos\varphi$, we obtain the following equations for $A_2(a)$ and $B_2(a)$:

$$\begin{aligned} S_1 A_2 + a S_2 B_2 &= G_{1c}, \\ S_2 A_2 - a S_1 B_2 &= G_{1s}. \end{aligned} \tag{8.17}$$

Here $G_{1c}$ and $G_{1s}$ are the Fourier coefficients of the expansion

$$G = \sum_m (G_{mc}\cos m\varphi + G_{ms}\sin m\varphi), \tag{8.18}$$

$$\begin{aligned} G = {} & \Big(\frac{\partial F}{\partial x}u_1 + \frac{\partial F}{\partial \dot{x}}\widetilde{\dot{u}_1} + \cdots + \frac{\partial F}{\partial x^{(N)}}\tilde{u}_1^{(N)} + \frac{\partial F}{\partial \varepsilon}\Big)_{\varepsilon=0} - \\ & - \sum_{k=1}^{N} k\alpha_{N-k}\Omega^{k-1}\Big(A_1\frac{\partial^k u_1}{\partial a\,\partial\varphi^{k-1}} + B_1\frac{\partial^k u_1}{\partial\varphi^k}\Big) - \\ & - \frac{1}{2}\sum_{k=2}^{N} k(k-1)\alpha_{N-k}\Omega^{k-2}\Big[\Big(A_1\frac{dA_1}{da} - aB_1^2\Big)\times \\ & \times \cos\Big(\varphi + k\frac{\pi}{2}\Big) - \Big(aA_1\frac{dB_1}{da} + 2A_1B_1\Big)\sin\Big(\varphi + k\frac{\pi}{2}\Big)\Big]. \end{aligned}$$

$$\tilde{u}_1^{(k)} = \Omega^k \frac{\partial^k u_1}{\partial \varphi^k} + k\Omega^{k-1} A_1 \sin\left(\varphi + k\frac{\pi}{2}\right) + k\Omega^{k-1} a B_1 \cos\left(\varphi + k\frac{\pi}{2}\right).$$

From the equations (8.17), it follows that

$$A_2 = \frac{S_1 G_{1c} + S_2 G_{1s}}{S_1^2 + S_2^2}, \quad B_2 = \frac{S_2 G_{1c} - S_1 G_{1s}}{a(S_1^2 + S_2^2)}. \tag{8.19}$$

Therefore, in the second approximation, we have the solution of the equation (8.1) in the form (8.16), with $a$ and $\psi$ determined by the system of equations

$$\begin{aligned} \frac{da}{dt} &= \frac{\varepsilon}{S_1^2 + S_2^2}\left[S_1 F_{1c} + S_2 F_{1s} + \varepsilon(S_1 G_{1c} + S_2 G_{1s})\right], \\ \frac{d\psi}{dt} &= \frac{\varepsilon}{a(S_1^2 + S_2^2)}\left[S_2 F_{1c} - S_1 F_{1s} + \varepsilon(S_2 G_{1c} - S_1 G_{1s})\right]. \end{aligned} \tag{8.20}$$

*Solution in the case of some pairs of imaginary characteristic roots.* Now, we reconsider the equation (8.1), but with the assumption that the characteristic equation (8.3) has some pairs of simple imaginary roots

$$\lambda = \pm i\Omega_s, \quad s = 1, 2, \ldots, \ell.$$

The other roots of the equation (8.3) have negative real parts with sufficiently large values. It is supposed that there is no relation of type

$$\sum_{s=1}^{\ell} q_s \Omega_s = 0, \tag{8.21}$$

where $q_s$ are integers. The opposite case will be considered in paragraph 8 of Chapter 3.

It is easily seen that the degenerate equation (8.2) has a family of quasi-periodic solutions:

$$x = \sum_{s=1}^{\ell} a_s \cos(\Omega_s t + \psi_s), \tag{8.22}$$

where $a_s$, $\psi_s$ are arbitrary real constants.

For sufficiently small values of $\varepsilon$, we shall find the solution of the equation (8.1) in the form of an asymptotic expansion

$$x = \sum_{s=1}^{\ell} a_s \cos(\Omega_s t + \psi_s) + \varepsilon u_1(a, \varphi) + \varepsilon^2 u_2(a, \varphi) + \ldots, \tag{8.23}$$

where $\varphi_s = \Omega_s t + \psi_s$, $a = (a_1, a_2, \ldots, a_\ell)$, $\varphi = (\varphi_1, \varphi_2, \ldots, \varphi_\ell)$ and $u_i(a, \varphi)$ do not contain $\sin\varphi_s$, $\cos\varphi_s$. These functions are bounded for all finite values of $a$, and are periodic with period $2\pi$ relative to each variable $\varphi_s$. The quantities $a_s$, $\psi_s$, being slowly varying functions of time, are determined from the equations

$$\begin{aligned}
\frac{da_s}{dt} &= \varepsilon A_{1s}(a) + \varepsilon^2 A_{2s}(a) + \ldots, \\
\frac{d\psi_s}{dt} &= \varepsilon B_{1s}(a) + \varepsilon^2 B_{2s}(a) + \ldots \\
s &= 1, 2, \ldots, \ell
\end{aligned} \tag{8.24}$$

To find the unknown functions $A_{js}(a)$, $B_{js}(a)$, $u_j(a,\varphi)$, we substitute the expressions (8.23), (8.24) into the equation (8.1) with regard for

$$\begin{aligned}
x^{(k)} &= \sum_{s=1}^{\ell} a_s \Omega_s^k \cos\left(\varphi_s + k\frac{\pi}{2}\right) + \varepsilon\Big[\Big(\sum_{s=1}^{\ell} \Omega_s \frac{\partial}{\partial \varphi_s}\Big)^k u_1 + \\
&+ k\sum_{s=1}^{\ell} A_{1s}\Omega_s^{k-1} \sin\left(\varphi_s + k\frac{\pi}{2}\right) + k\sum_{s=1}^{\ell} a_s B_{1s}\Omega_s^{k-1} \cos\left(\varphi_s + k\frac{\pi}{2}\right)\Big] + \\
&+ \varepsilon^2\Big\{\Big(\sum_{s=1}^{\ell} \Omega_s \frac{\partial}{\partial \varphi_s}\Big)^k u_2 + k\sum_{s=1}^{\ell} A_{2s}\Omega_s^{k-1} \sin\left(\varphi_s + k\frac{\pi}{2}\right) + \\
&+ k\sum_{s=1}^{\ell} a_s B_{2s}\Omega_s^{k-1} \cos\left(\varphi_s + k\frac{\pi}{2}\right) + k\sum_{s=1}^{\ell} B_{1s}\frac{\partial}{\partial \varphi_s}\Big(\sum_{j=1}^{\ell} \Omega_j \frac{\partial}{\partial \varphi_j}\Big)^{k-1} u_1 + \\
&+ k\sum_{s=1}^{\ell} A_{1s}\frac{\partial}{\partial a_s}\Big(\sum_{j=1}^{\ell} \Omega_j \frac{\partial}{\partial \varphi_j}\Big)^{k-1} u_1 - \frac{k}{2}(k-1)\Big[\sum_{s=1}^{\ell}\sum_{j=1}^{\ell} A_{1j}\frac{\partial A_{1s}}{\partial a_j} - \\
&- \sum_{s=1}^{\ell} a_s B_{1s}^2\Big]\Omega_s^{k-2} \cos\left(\varphi_s + k\frac{\pi}{2}\right) + \frac{k}{2}(k-1)\Big[\sum_{s=1}^{\ell}\sum_{j=1}^{\ell} a_s A_{1j}\frac{\partial B_{1s}}{\partial a_j} + \\
&+ 2\sum_{s=1}^{\ell} A_{1s}B_{1s}\Big]\Omega_s^{k-2} \sin\left(\varphi_s + k\frac{\pi}{2}\right)\Big\} + \varepsilon^3 \ldots \quad (k = 0, 1, 2, \ldots, N)
\end{aligned} \tag{8.25}$$

Comparing the coefficients of $\varepsilon$ in the equation (8.1), we have

$$\begin{aligned}
&\sum_{k=0}^{N} \alpha_{N-k}\Big(\sum_{s=1}^{\ell} \Omega_s \frac{\partial}{\partial \varphi_s}\Big)^k u_1 + \sum_{s=1}^{\ell}(L_{1s}A_{1s} + L_{2s}a_s B_{1s}) \cos\varphi_s + \\
&+ \sum_{s=1}^{\ell}(L_{2s}A_{1s} - L_{1s}a_s B_{1s}) \sin\varphi_s = F_0,
\end{aligned} \tag{8.26}$$

where

$$
\begin{aligned}
F_0 &= F(x_0, \dot{x}_0, \dots, x_0^{(N)}, 0), \\
x_0 &= \sum_{s=1}^{\ell} a_s \cos\varphi_s, \dots, \\
x_0^{(N)} &= \sum_{s=1}^{\ell} a_s \Omega_s^N \cos\left(\varphi_s + N\frac{\pi}{2}\right), \\
L_{1s} &= \sum_{k=1}^{N} k\alpha_{N-k}\Omega_s^{k-1} \sin k\frac{\pi}{2}, \\
L_{2s} &= \sum_{k=2}^{N} k\alpha_{N-k}\Omega_s^{k-1} \cos k\frac{\pi}{2}.
\end{aligned}
\tag{8.27}
$$

Now we expand the functions $F_0$ and $u_1$ in the Fourier series

$$
\begin{aligned}
F_0 &= \sum_q \frac{1}{(2\pi)^\ell} e^{i(q_1\varphi_1+\dots+q_\ell\varphi_\ell)} \int_0^{2\pi}\dots\int_0^{2\pi} \overline{F}_0 e^{-i(q_1\overline{\varphi}_1+\dots+q_\ell\overline{\varphi}_\ell)}\, d\overline{\varphi}_1\dots d\overline{\varphi}_\ell, \\
\overline{F}_0 &= F_0\Big|_{\varphi=\overline{\varphi}},
\end{aligned}
\tag{8.28}
$$

$$
u_1 = \sum_q u_{1q} e^{i(q_1\varphi_1+\dots+q_\ell\varphi_\ell)}. \tag{8.29}
$$

By substituting these expressions into the equation (8.26), we obtain

$$
\begin{aligned}
&\sum_q u_{1q}\Big[\sum_{k=0}^{N} \alpha_{N-k} i^k (q_1\Omega_1 + \dots + q_\ell\Omega_\ell)^k\Big] e^{i(q_1\varphi_1+\dots+q_\ell\varphi_\ell)} + \\
&+ \sum_{s=1}^{\ell}(L_{1s}A_{1s} + L_{2s}a_sB_{1s})\cos\varphi_s + \sum_{s=1}^{\ell}(L_{2s}A_{1s} - L_{1s}a_sB_{1s})\sin\varphi_s = \\
&= \frac{1}{(2\pi)^\ell}\sum_q e^{i(q_1\varphi_1+\dots+q_\ell\varphi_\ell)} \int_0^{2\pi}\dots\int_0^{2\pi} \overline{F}_0 e^{-i(q_1\overline{\varphi}_1+\dots+q_\ell\overline{\varphi}_\ell)}\, d\overline{\varphi}_1\dots d\overline{\varphi}_\ell.
\end{aligned}
\tag{8.30}
$$

It is noted that because there is no relation of type (8.21), the fundamental harmonics $\sin\varphi_s$, $\cos\varphi_s$, in $u_1$ correspond to $q_s = \pm1$, $q_j = 0 (j \neq s)$ and the corresponding coefficients of $u_1$ are equal to zero:

$$
\sum_{k=0}^{N} \alpha_{N-k} i^k (q_1\Omega_1 + \dots + q_\ell\Omega_\ell)^k = \sum_{k=0}^{N} \alpha_{N-k}(\pm i\Omega_s)^k = 0.
$$

Therefore, comparing the coefficients of $\sin\varphi_s$, $\cos\varphi_s$ in (8.30), we have

$$\begin{aligned} L_{1s}A_{1s} + L_{2s}a_sB_{1s} &= \frac{2}{(2\pi)^\ell}\int_0^{2\pi}\cdots\int_0^{2\pi}\overline{F}_0\cos\overline{\varphi}_s\,d\overline{\varphi}_1\ldots d\overline{\varphi}_\ell, \\ L_{2s}A_{1s} - L_{1s}a_sB_{1s} &= \frac{2}{(2\pi)^\ell}\int_0^{2\pi}\cdots\int_0^{2\pi}\overline{F}_0\sin\overline{\varphi}_s\,d\overline{\varphi}_1\ldots d\overline{\varphi}_\ell. \end{aligned} \tag{8.31}$$

From here it follows

$$\begin{aligned} A_{1s} &= \frac{2\int_0^{2\pi}\cdots\int_0^{2\pi}\overline{F}_0(L_{1s}\cos\overline{\varphi}_s + L_{2s}\sin\overline{\varphi}_s)\,d\overline{\varphi}_1\ldots d\overline{\varphi}_\ell}{(2\pi)^\ell(L_{1s}^2 + L_{2s}^2)}, \\ B_{1s} &= \frac{2\int_0^{2\pi}\cdots\int_0^{2\pi}\overline{F}_0(L_{2s}\cos\overline{\varphi}_s - L_{1s}\sin\overline{\varphi}_s)\,d\overline{\varphi}_1\ldots \overline{\varphi}_\ell}{(2\pi)^\ell(L_{1s}^2 + L_{2s}^2)a_s}. \end{aligned} \tag{8.32}$$

Comparing the other harmonics in (8.30), we get

$$u_{1q} = \frac{\int_0^{2\pi}\cdots\int_0^{2\pi}\overline{F}_0\, e^{-i(q_1\overline{\varphi}_1+\cdots+q_\ell\overline{\varphi}_\ell)}\,d\overline{\varphi}_1\ldots d\overline{\varphi}_\ell}{(2\pi)^\ell\sum_{k=0}^{N}\alpha_{N-k}i^k(q_1\Omega_1+\cdots+q_\ell\Omega_\ell)^k}. \tag{8.33}$$

$$q_1^2+\cdots+(q_s\pm1)^2+\cdots+q_\ell^2\neq0$$

Thus, we have obtained the expressions for $A_{1s}$, $B_{1s}$, $u_{1q}$ and, therefore, the solution of equation (8.1) in the first approximation. By continuing this process, we shall find the higher approximations. For example, if we compare the coefficients of $\varepsilon^2$ in (8.1), we get :

$$\begin{aligned} &\sum_{s=1}^{\ell} A_{2s}\Big[\sum_{k=1}^{N} k\alpha_{N-k}\Omega_s^{k-1}\sin\big(\varphi_s+k\frac{\pi}{2}\big)\Big]+ \\ &+\sum_{s=1}^{\ell} a_sB_{2s}\Big[\sum_{k=1}^{N} k\alpha_{N-k}\Omega_s^{k-1}\cos\big(\varphi_s+k\frac{\pi}{2}\big)\Big]+ \\ &+\sum_{k=0}^{N}\alpha_{N-k}\Big(\sum_{s=1}^{\ell}\Omega_s\frac{\partial}{\partial\varphi_s}\Big)^k u_2 = G(a,\varphi), \end{aligned} \tag{8.34}$$

here

$$G(a,\varphi)=\Big(\frac{\partial F}{\partial x}u_1+\frac{\partial F}{\partial \dot{x}}\tilde{\dot{u}}_1+\cdots+\frac{\partial F}{\partial x^{(N)}}\tilde{u}_1^{(N)}+\frac{\partial F}{\partial \varepsilon}\Big)_{\varepsilon=0}-$$
$$-\sum_{k=1}^{N}k\alpha_{N-k}\Big\{\sum_{s=1}^{\ell}B_{1s}\frac{\partial}{\partial\varphi_s}\Big(\sum_{j=1}^{\ell}\Omega_j\frac{\partial}{\partial\varphi_j}\Big)^{k-1}u_1+$$
$$+\sum_{s=1}^{\ell}A_{1s}\frac{\partial}{\partial a_s}\Big(\sum_{j=1}^{\ell}\Omega_j\frac{\partial}{\partial\varphi_j}\Big)^{k-1}u_1-\frac{1}{2}(k-1)\Big[\sum_{s=1}^{\ell}\sum_{j=1}^{\ell}A_{1j}\frac{\partial A_{1s}}{\partial a_j}-$$
$$-\sum_{s=1}^{\ell}a_sB_{1s}^2\Big]\Omega_s^{k-2}\cos\big(\varphi_s+k\frac{\pi}{2}\big)+\frac{1}{2}(k-1)\Big[\sum_{s=1}^{\ell}\sum_{j=1}^{\ell}a_sA_{1j}\frac{\partial B_{1s}}{\partial a_j}+$$
$$+2\sum_{s=1}^{\ell}A_{1s}B_{1s}\Big]\Omega_s^{k-2}\sin\big(\varphi_s+k\frac{\pi}{2}\big)\Big\},$$

$$\tilde{u}_1^{(k)}=\Big(\sum_{s=1}^{\ell}\Omega_s\frac{\partial}{\partial\varphi_s}\Big)^k u_1+k\sum_{s=1}^{\ell}A_{1s}\Omega_s^{k-1}\sin\big(\varphi_s+k\frac{\pi}{2}\big)+$$
$$+k\sum_{s=1}^{\ell}a_sB_{1s}\Omega_s^{k-1}\cos\big(\varphi_s+k\frac{\pi}{2}\big). \tag{8.35}$$

By a similar procedure as in the first approximation, we obtain

$$A_{2s}=\frac{2\int_0^{2\pi}\cdots\int_0^{2\pi}G(a,\overline{\varphi})(L_{1s}\cos\overline{\varphi}_s+L_{2s}\sin\overline{\varphi}_s)d\overline{\varphi}_1\ldots d\overline{\varphi}_\ell}{(2\pi)^\ell(L_{1s}^2+L_{2s}^2)},$$

$$B_{2s}=\frac{2\int_0^{2\pi}\cdots\int_0^{2\pi}G(a,\overline{\varphi})(L_{2s}\cos\overline{\varphi}_s-L_{1s}\sin\overline{\varphi}_s)d\overline{\varphi}_1\ldots d\overline{\varphi}_\ell}{(2\pi)^\ell(L_{1s}^2+L_{2s}^2)a_s},$$

$$u_2=\sum_q\frac{e^{i(q_1\varphi_1+\cdots+q_\ell\varphi_\ell)}}{(2\pi)^\ell\sum_{k=0}^{N}\alpha_{N-k}i^k(q_1\Omega_1+\cdots+q_\ell\Omega_\ell)^k}\times$$
$$\times\int_0^{2\pi}\cdots\int_0^{2\pi}G(a,\overline{\varphi})e^{-i(q_1\overline{\varphi}_1+\cdots+q_\ell\overline{\varphi}_\ell)}d\overline{\varphi}_1\ldots d\overline{\varphi}_\ell. \tag{8.36}$$

# CHAPTER 2

## SELF-EXCITED OSCILLATIONS

A *self-excited* system is a nonconservative system which possesses the ability to accomplish an undamped periodic oscillation, and is characterized by the presence of the following components:

1. a constant energy source,
2. an oscillatory system,
3. a device for control of the introduction of energy into the oscillatory system,
4. a feedback from the oscillatory system to the control device (see Fig. 10).

So, the distinguishing feature of self-excited oscillatory systems, as opposed to free oscillatory systems, is that there is an input of energy.

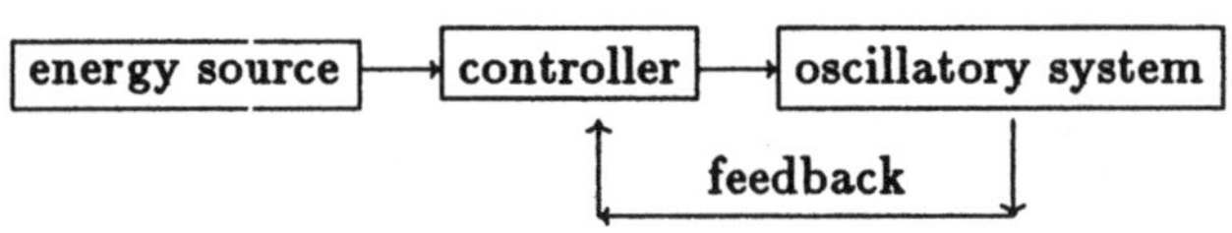


*Fig. 10.* Schema of a self-excited system.

Sometimes, the self-excited oscillatory character of some simple systems is recognized from the form of their equations of motion.

### 1. Lienard and Routh - Hurwitz Criteria. Stability of Equilibrium States

The existence of closed trajectories in nonlinear systems is of special interest, as such trajectories imply periodic motion. Closed trajectories can occur also in nonlinear nonconservative systems. Such closed trajectories are referred to as limit cycles.

There are many criteria for proving the existence of undamped periodic oscillations, based on the properties of the coefficients of differential equation of motion. Here, we give Lienard's Criterion, without proof.

Consider a differential equation with a dissipative middle term $f(x)\dfrac{dx}{dt}$ and a restoring force $g(x)$:

$$\frac{d^2x}{dt^2} + f(x)\frac{dx}{dt} + g(x) = 0, \tag{1.1}$$

which we shall call Lienard's equation. Lienard showed that the dynamic system described by the equation (1.1) possesses a stable limit cycle if the functions $f(x)$ and $g(x)$ satisfy the following conditions:

1) $f(x)$ is even, and $g(x)$ is odd,
2) $f(0) < 0$ ,
3) $xg(x) > 0$ for all $x \neq 0$, (1.2)
4) $F(x) = \int_0^x f(x)dx \to \pm\infty$ with $x$,
5) $F(x)$ has a single positive zero $x = a$ and is monotonously increasing for $x \geq a$.

The oddness of $g(x)$ and the condition $xg(x) > 0$ mean that the force $-g(x)$ always has the opposite sign to $x$. So, this force acts as a restoring force in a linear elastic system, and it is always directed towards the position of equilibrium.

The evenness of $f(x)$ together with the condition $f(0) < 0$ means that the coefficient of friction is negative for small values of $|x|$. Hence, for small deviations, the friction excites the system. For large values of $|x|$, the friction becomes positive and it damps the oscillations.

The last two conditions 4) and 5) mean that, in the neighbourhood of the zero position, the system (1.1) is not a dissipative one and that it is a self-excited system.

In the neighbourhood of the origin of coordinates, the graph of the function $f(x)$ has a parabolic form (Fig. 11). At the point $x = 0$, the function $F(x)$ has a decreasing part (Fig. 12).

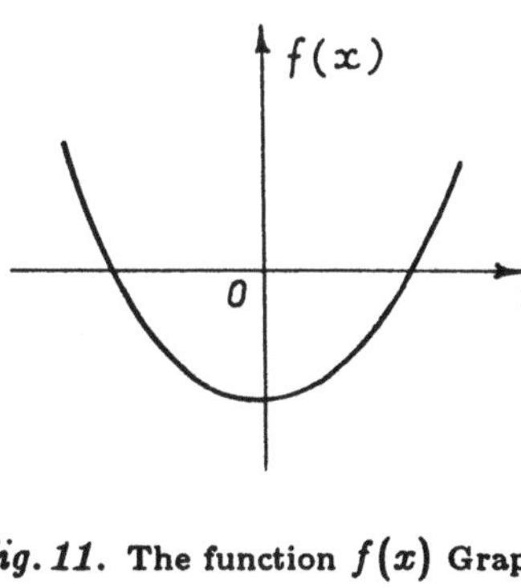


*Fig. 11.* The function $f(x)$ Graph.

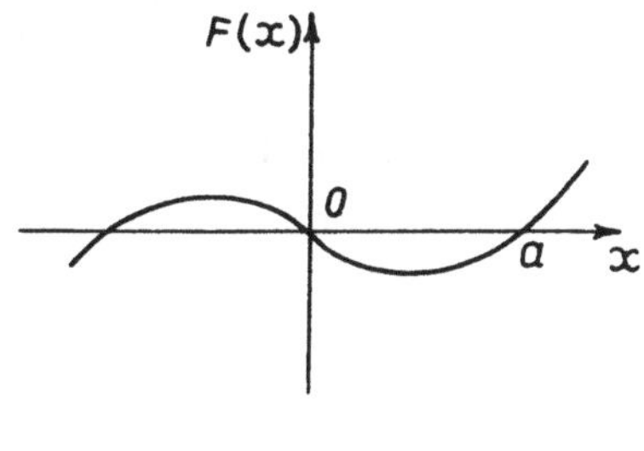


*Fig. 12.* Decreasing part of the function $F(x)$.

**Example 11**

Van der Pol's equation

$$\frac{d^2x}{dt^2} - \varepsilon(1 - x^2)\frac{dx}{dt} + x = 0, \quad \varepsilon > 0 \tag{1.3}$$

has a stable limit cycle, because the conditions (1.2) are satisfied:
1) $f(x) = -\varepsilon(1 - x^2)$ is even, and $g(x) = x$ is odd,
2) $f(0) = -\varepsilon < 0$,
3) $xg(x) = x^2 > 0$ for all $x \neq 0$,

4) $F(x) = -\varepsilon \int_0^x (1 - x^2)dx = -\varepsilon x + \frac{\varepsilon x^3}{3} \to \pm\infty$ with $x$,

5) $F(x) = 0$ for $x = +\sqrt{3}$, and $F(x)$ increases monotonously for $x > \sqrt{3}$.
Using the formulae (3.5) - (3.7) from Chapter 1, we have

$$f(x) = 1 - x^2, \quad F^*(x) = x - \frac{x^3}{3},$$

which gives

$$F^*(a\cos\psi) = a\Big(1 - \frac{a^2}{4}\Big)\cos\psi - \frac{a^3}{12}\cos 3\psi.$$

We have thus

$$F_1^*(a) = a\Big(1 - \frac{a^2}{4}\Big), \quad F_3^* = -\frac{a^3}{12},$$
$$F_n^*(a) = 0 \quad \text{for} \quad n \neq 1, \ n \neq 3.$$

For the first approximation, one has

$$x = a\cos\psi,$$
$$\frac{da}{dt} = \frac{\varepsilon a}{2}\Big(1 - \frac{a^2}{4}\Big); \quad \frac{d\psi}{dt} = 1, \tag{1.4}$$

or after multiplication by $2a$

$$da^2 = \varepsilon a^2\Big(1 - \frac{a^2}{4}\Big)dt,$$

or

$$\varepsilon dt = d\ln\frac{a^2}{4\Big(1 - \frac{a^2}{4}\Big)}.$$

Integrating this equation from $t = 0$ to $t$, we have:

$$\varepsilon t = \ln\left(\frac{a^2}{a_0^2} \cdot \frac{4 - a_0^2}{4 - a^2}\right).$$

From here, one obtains

$$a = \frac{a_0\exp\Big(\frac{\varepsilon}{2}t\Big)}{\sqrt{1 + \frac{a_0^2}{4}\big[\exp(\varepsilon t) - 1\big]}}.$$

For $t \to \infty$, $a \to a_0 = 2$, so that in the first approximation

$$x = 2\cos(t + \psi_0).$$

For the second approximation, one has

$$x = a\cos\psi - \frac{\varepsilon a^3}{32}\sin 3\psi,$$

where $a$ and $\psi$ are given by the differential equations

$$\frac{da}{dt} = \frac{\varepsilon a}{2}\Big(1 - \frac{a^2}{4}\Big),$$
$$\frac{d\psi}{dt} = 1 - \frac{\varepsilon^2}{8}\Big(1 - a^2 + \frac{7a^4}{32}\Big).$$

If we use the result of the first approximation $a = a_0 = 2$, we find the second approximation

$$x = 2\cos\psi - \frac{\varepsilon}{4}\sin 3\psi,$$
$$\psi = \omega t + \psi_0,$$
$$\omega = 1 - \frac{\varepsilon^2}{16}.$$

**Note:**
The equation

$$\ddot{z} + F(\dot{z}) + z = 0, \tag{1}$$
$$\dot{z} = \frac{dz}{dt}, \quad \ddot{z} = \frac{d^2z}{dt^2},$$

may be reduced to the form (1.1). In fact, by differentiating (1) and by setting $\dot{z} = x$, one obtains:

$$\ddot{x} + F'(x)\dot{x} + x = 0.$$

Thus, one has an equation of type (1.1) with $f(x) = F'(x)$ and $g(x) = x$.

**Example 12**
Rayleigh's equation

$$\ddot{z} - \varepsilon(1 - \dot{z}^2)\dot{z} + z = 0$$

has a limit cycle, because

$$F(\dot{z}) = -\varepsilon(1 - \dot{z}^2)\dot{z},$$

hence

$$F(x) = -\varepsilon(1 - x^2)x$$

and

$$f(x) = F'(x) = -\varepsilon(1 - 3x^2); \quad g(x) = x.$$

These functions satisfy all the conditions (1.2).

We shall consider more particularly an autonomous system

$$\frac{dx}{dt} = X(x, y), \qquad \frac{dy}{dt} = Y(x, y), \tag{1.5}$$

where $X(x, y)$, $Y(x, y)$ are generally nonlinear analytic functions of the variables $x$ and $y$. Our main goal is to obtain the phase-portrait in the $x, y$ plane or phase-plane. The equation of a phase-trajectory is given by $y = y(x)$. The important inferences about the properties of a given oscillation can be drawn solely from the pure geometrical configuration of the phase-trajectories.

A singular point $x_0$, $y_0$ or a critical point is a *point trajectory* i.e., a point such that $X(x_0, y_0) = 0$, $Y(x_0, y_0) = 0$. Equations (1.5) are assumed to possess only a finite number of singular points. We shall be concerned with one particular case, namely, that in which the equilibrium point coincides with the origin of the phase plane. There is no loss of generality in this, because the origin can always be translated by means of a coordinate transformation so that it coincides with an equilibrium point. Expanding Taylor's series for $P$ and $Q$ in the neighbourhood of the origin, we get

$$\begin{aligned} X(x, y) &= ax + by + X_2(x, y), \\ Y(x, y) &= cx + dy + Y_2(x, y), \end{aligned} \tag{1.6}$$

where $X_2$, $Y_2$ are functions, the terms contained in which are at least of the second order,

$$a = X'_x(0,0), \quad b = X'_y(0,0), \quad c = Y'_x(0,0), \quad d = Y'_y(0,0).$$

Neglecting nonlinear terms in (1.5), we have a linearized system

$$\frac{dx}{dt} = ax + by, \quad \frac{dy}{dt} = cx + dy. \tag{1.7}$$

The characteristic equation of (1.7 ) is

$$(a - \lambda)(d - \lambda) - bc = 0$$

or

$$\lambda^2 - (a + d)\lambda + ad - bc = 0. \tag{1.8}$$

We shall consider more particularly the case, when the characteristic roots $\lambda_1$, $\lambda_2$ are non-zero, i.e. when $ad - bc \neq 0$. The corresponding singular points are known as elementary points.

The behaviour of the system in the neighbourhood of the origin depends on the roots $\lambda_1$, $\lambda_2$ of the characteristic equation (1.8). To show this, let the solution of the equation (1.7) be written in the form

$$\begin{aligned} x &= C_1 e^{\lambda_1 t} + C_2 e^{\lambda_2 t}, \\ y &= C_1 \alpha_1 e^{\lambda_1 t} + C_2 \alpha_2 e^{\lambda_2 t}, \end{aligned} \tag{1.9}$$

where $\lambda_1$ and $\lambda_2$ are determined by

$$\begin{aligned}\lambda_1 &= \frac{1}{2}\left[a+d+\sqrt{(a-d)^2+4bc}\right],\\ \lambda_2 &= \frac{1}{2}\left[a+d-\sqrt{(a-d)^2+4bc}\right],\end{aligned} \tag{1.10}$$

and $\alpha_1$ and $\alpha_2$ are the roots of the equation:

$$b\alpha^2 + (d-a)\alpha - c = 0.$$

**First case.** $\lambda_1$ and $\lambda_2$ are real roots of the same sign. This occurs when

$$(a-d)^2 + 4bc > 0, \quad ad - bc > 0.$$

Suppose first $\lambda_1$ and $\lambda_2$ are both negative $(a+d<0)$. Then, $x \to 0$, and $y \to 0$ as $t \to +\infty$. Evidently, the phase trajectory tends to the origin as $t \to +\infty$. The form of the trajectories is shown in Fig. 13. The arrows indicate the direction of motion on the trajectories. The singular point O thus arising is called a stable node (or nodal point) .

If $\lambda_1$ and $\lambda_2$ are real and positive $a+d>0$, the preceding behaviour corresponds to $t \to -\infty$. We have an similar situation of Fig. 13, but all the arrows are reversed. The singular point is an unstable nodal point (Fig. 14).

When $\lambda_1 = \lambda_2$ or when $(a-d)^2 + 4bc = 0$, all the trajectories are straight lines through the origin. This case yields what is referred to as a degenerate nodal point.

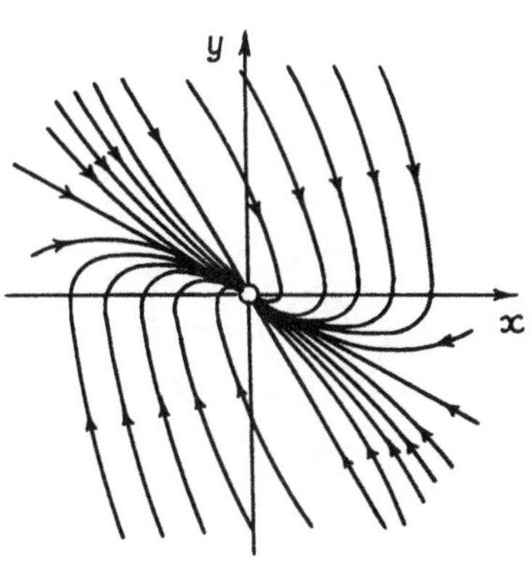


*Fig. 13.* Stable node.

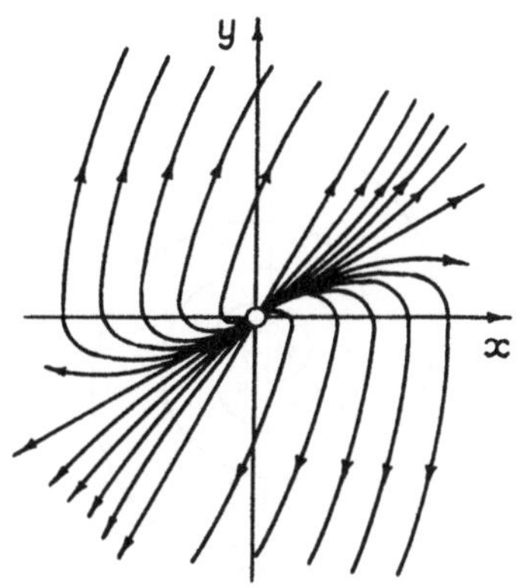


*Fig. 14.* Unstable node.

**Second case.** $\lambda_1$ and $\lambda_2$ are real roots of opposite sign. This occurs when

$$(a-d)^2 + 4bc > 0, \quad ad - bc < 0.$$

The phase trajectories have the general form shown in Fig. 15. The origin is then called a saddle point, and the equilibrium is unstable.

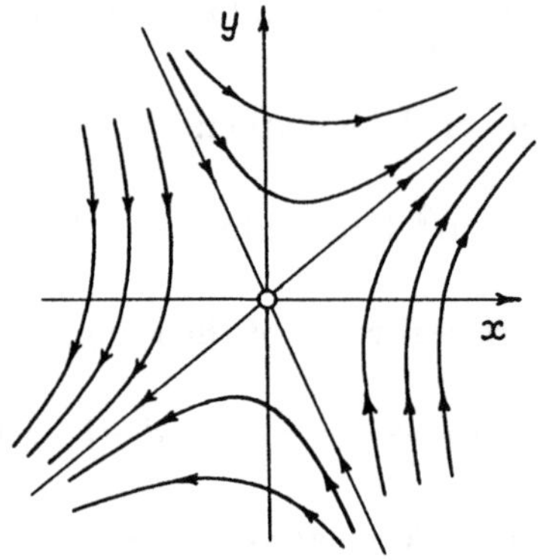


*Fig. 15.* Saddle point.

**Third case.** $\lambda_1$ and $\lambda_2$ are complex roots with a non-zero real part. This occurs when:

$$(a-d)^2 + 4bc < 0, \quad a+d \neq 0.$$

The aspect of the phase trajectories is illustrated in Fig. 16. The equilibrium point is known as a spiral point, focus, or focal point. When $a+d<0$, the singular point is called a stable focal point (Fig. 16). When $a+d>0$, the situation is the same with the arrows reversed, one has an unstable focal point (Fig. 17).

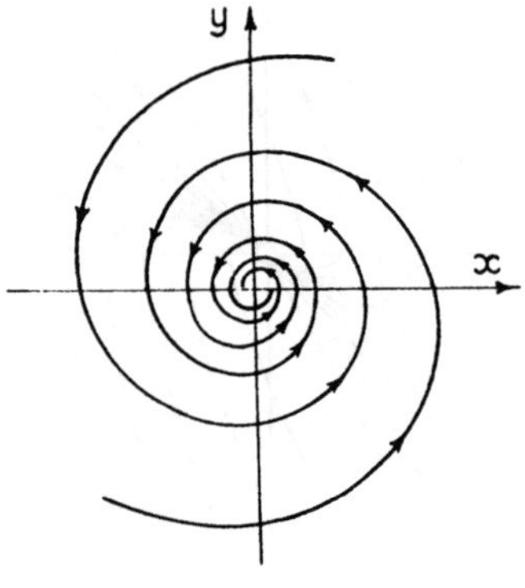


*Fig. 16.* Stable focal point.

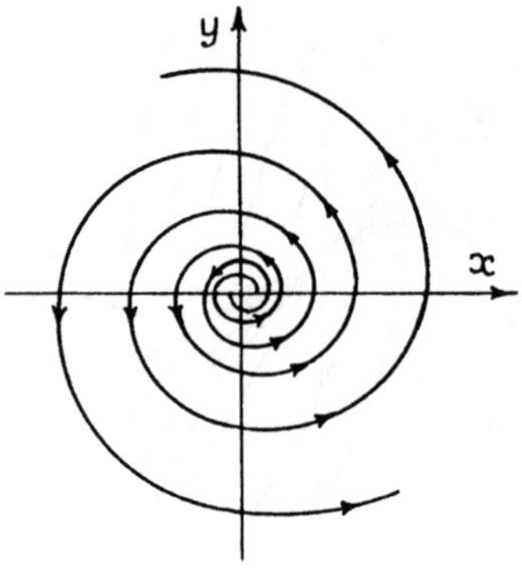


*Fig. 17.* Unstable focal point.

**Fourth case.** $\lambda_1$ and $\lambda_2$ are pure complex roots, i.e.

$$(a-d)^2 + 4bc < 0, \quad a+d = 0.$$

In the phase plane, the trajectories are circles with the singular point as the centre. This point is then known as a central point (Fig. 18). The motion is periodic, and hence stable. However, it is merely stable and not asymptotically stable.

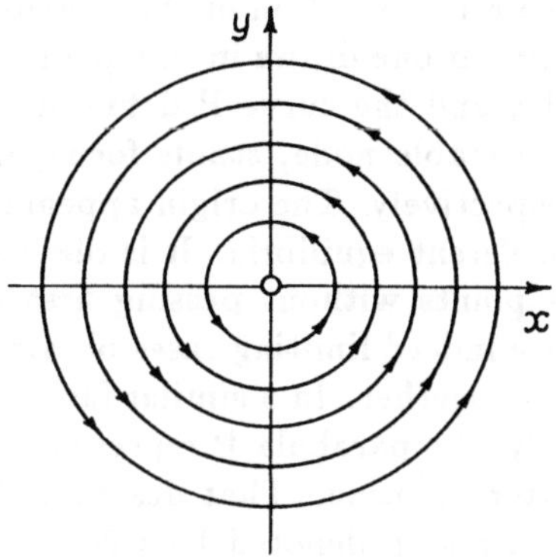


*Fig. 18.* Central point.

It appears that nodes and foci are either unstable or asymptotically stable, whereas saddle points are always unstable. On the other hand, centres are merely stable.

We recall that for asymptotic stability, either the characteristic roots are real and negative, or they are complex conjugates with negative real parts. For instability, at least one of the roots is real and positive, or complex with a positive real part.

It was proved by Liapunov that in the cases of asymptotic stability, and instability, the nature of the equilibrium of the original nonlinear system (1.5) is the same as that of the linearized system (1.7). In the case of mere stability (critical case), the original nonlinear system yields either a centre or a focus (stable or unstable). In the critical case, the linearized system (1.7) cannot be used to draw any conclusions about the behaviour of the original nonlinear system (1.5) in the neighbourhood of the equilibrium point, and higher-order terms contained in $X_2(x, y)$ and $Y_2(x, y)$ must be examined.

So, the type of equilibrium points obtained for a given system (1.7) can be determined more directly, by examining the coefficients $a, b, c, d$.

It will prove convenient to introduce the parameters

$$\begin{aligned} p &= a + d, \\ q &= ad - bc, \end{aligned} \tag{1.11}$$

and hence $p^2 - 4q = (a - d)^2 + 4bc$.

In summary, we have:

1. $p^2 - 4q > 0$ 1a) $q > 0$ : node (N)
   1b) $q < 0$ : saddle point (S)

2. $p^2 - 4q = 0$ : node
3. $p^2 - 4q < 0$ 3a) $p \neq 0$ : focus (F)
3b) $p = 0$ : centre (C)

Moreover, if $p < 0$, the node (or focus) is stable. If $p > 0$, the node (or focus) is unstable.

This suggests a simple representation of the distribution of equilibrium points in the $(p, q)$ plane (Fig. 19). If one draws in this plane a parabola P with equation $p^2 = 4q$, the axes $p$ and $q$ and the curve P define five regions: SN, SF, UN, UF and SP, corresponding to stable node, stable focus, unstable node, unstable focus and saddle points respectively. The origin appears thus as a trajectory point, around which permute different equilibria. It is observed that the foci cannot be transformed into saddle points without passing first through the regions of the nodes and that centres are indeed limiting cases obtained as the weakly stable and weakly unstable foci draw together. In a similar fashion, the parabola P separates nodes and foci. Physically, the parabola P represents the curve separating aperiodic motion from oscillatory motion. That designated by SF is characterized by damped oscillation; in the region denoted by UN, the motion is divergently aperiodic, whereas in the region marked by UF, the motion is divergently oscillatory. In the region denoted by C, consisting of the positive q axis alone, the motion is harmonic.

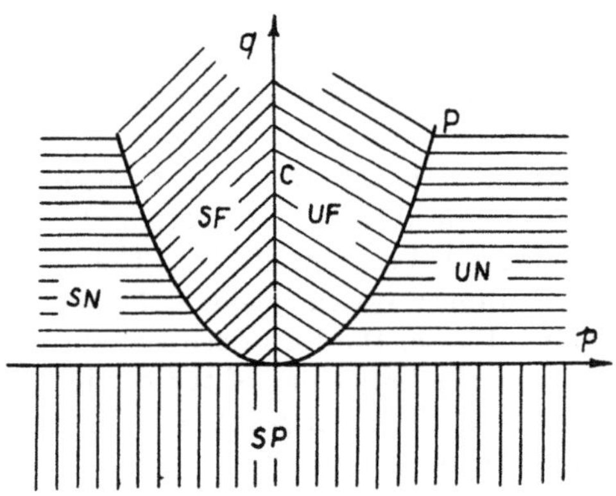


*Fig. 19.* Distribution of equilibrium points.

From Fig. 19, we conclude that the singular point is stable if $p \leq 0$ and $q > 0$; and unstable for any other combination of $p$ and $q$.

**Example 13**

If $x = x_1$, $\dot{x} = x_2$, Van der Pol's equation

$$\ddot{x} + \sigma(x^2 - 1)\dot{x} + x = 0$$

can be replaced by the two first-order differential equations:

$$\begin{aligned}\dot{x}_1 &= x_2,\\ \dot{x}_2 &= -x_1 + \sigma(1 - x_1^2)x_2.\end{aligned} \tag{1.12}$$

Clearly, the origin $x_1 = x_2 = 0$ is a singular point. To determine the nature of the singular point, we form the matrix of the coefficients of the linearized system

$$\begin{pmatrix} 0 & 1 \\ -1 & \sigma \end{pmatrix}.$$

Following formula (1.11), we have:

$$p = \sigma, \quad q = 1, \quad p^2 - 4q = \sigma^2 - 4.$$

From Fig. 19, it is seen that, when $\sigma$ is positive, the origin is an unstable node if $\sigma^2 - 4 > 0$ (or $\sigma > 2$), and is an unstable focus if $\sigma < 2$. In any event, the origin is an unstable equilibrium point, and any motion initiated in its neighbourhood will tend to leave that neighbourhood and reach the limit cycle.

To obtain the equation of the phase trajectories, we divide the second of Equations (1.12) by the first, with the result:

$$\frac{dx_2}{dx_1} = \sigma(1 - x_1^2) - \frac{x_1}{x_2}.$$

A closed-form solution of this equation is not possible. The phase trajectories can be obtained by the method of isoclines, or by an analog computer. The plots of Fig. 20 were obtained by an analog computer for the values $\sigma = 0.2$ and $\sigma = 1.0$ (see [27]). It is clear from Fig. 20 that the shape of the limit cycle depends on the parameter $\sigma$.

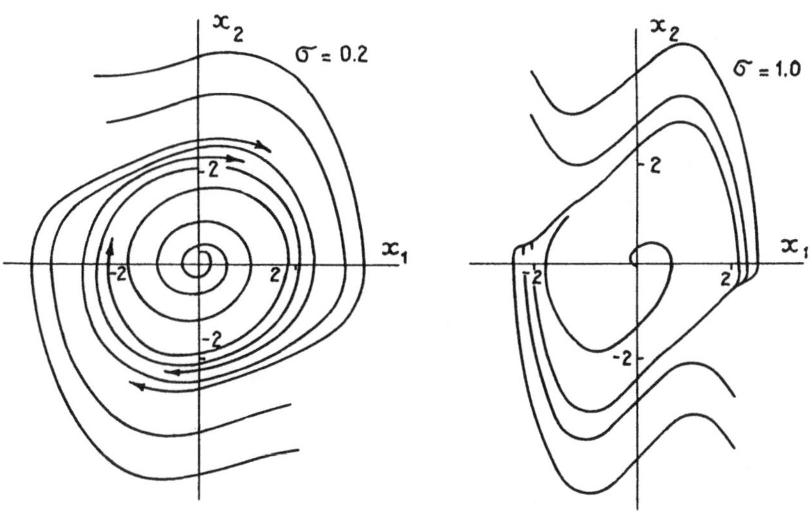


*Fig. 20.* Phase trajectories of a Van der Pol's oscillator.

**Routh-Hurwitz criterion**

From the above discussion, we see that the behaviour of a nonlinear system, in the neighbourhood of an equilibrium point, depends on the roots of the characteristic equation. It is necessary to know: whether all the real parts of these roots are negative, or at least one of the real parts is positive. In the first case, one has an asymptotically stable equilibrium point, and in the second case, one has an unstable equilibrium point.

For a second-order system, this presents no particular difficulty; but for higher-order systems, this becomes a problem of complexity. Moreover, in applied problems, the interest lies in the statement as to whether the system is asymptotically stable or not. To this end, it is sufficient to know whether all the characteristic roots possess negative real parts, and it is not necessary to actually evaluate the roots themselves. Necessary and sufficient conditions for the roots of a polynomial with real coefficients to possess negative real parts were derived by both Routh and Hurwitz.

Let us consider the characteristic equation in the form

$$a_0\lambda^n + a_1\lambda^{n-1} + \cdots + a_n = 0, \quad a_0 > 0. \tag{1.13}$$

The coefficients $a_i\,(i = 1, 2, \ldots, n)$ are all real. Next, we form the matrix:

$$Q = \begin{pmatrix} a_1 & a_3 & a_5 & \cdots & & 0 \\ a_0 & a_2 & a_4 & \cdots & & 0 \\ 0 & a_1 & a_3 & \cdots & & \cdot \\ \vdots & \vdots & \vdots & \ddots & \vdots & \vdots \\ 0 & \cdots & \cdots & \cdots & a_{n-2} & a_n \end{pmatrix},$$

where $a_i = 0$ for $i > n$ or for $i < 0$. Note that the $k$–th column of the matrix $Q$ is

$$\ldots a_{k+2}\ a_{k+1}\ a_k\ a_{k-1}\ a_{k-2} \ldots,$$

where the element $a_k$ is on the main diagonal. Next, we form the principal minor determinants associated with the matrix, and denote them by

$$\begin{gathered} \Delta_1 = a_1, \quad \Delta_2 = \begin{vmatrix} a_1 & a_3 \\ a_0 & a_2 \end{vmatrix}, \\ \Delta_3 = \begin{vmatrix} a_1 & a_3 & a_5 \\ a_0 & a_2 & a_4 \\ 0 & a_1 & a_3 \end{vmatrix}, \ldots, \Delta_n = a_n\Delta_{n-1}. \end{gathered} \tag{1.14}$$

Then, the Routh-Hurwitz criterion states that *the necessary and sufficient conditions for all the roots of the characteristic equation $\lambda_i$ to possess negative real parts is that all the determinants $\Delta_1, \Delta_2, \ldots, \Delta_n$ be positive* (For a proof, see [10]).

**Example 14**

For the case $n = 2$, we have:

$$a_0\lambda^2 + a_1\lambda + a_2 = 0, \quad a_0 > 0 \tag{1.15}$$

and $\Delta_1 = a_1$, $\Delta_2 = a_1a_2$. Hence, the necessary and sufficient conditions for $\mathrm{Re}\lambda_1 < 0$ and $\mathrm{Re}\lambda_2 < 0$ are $a_1 > 0$, $a_2 > 0$; i.e. all the coefficients of (1.15) must be positive, where Re denotes the real part.

**Example 15**

For the case $n = 3$, we have

$$a_0\lambda^3 + a_1\lambda^2 + a_2\lambda + a_3 = 0,$$

and $\Delta_1 = a_1$, $\Delta_2 = a_1a_2 - a_0a_3$, $\Delta_3 = a_3\Delta_2$. Hence, the necessary and sufficient conditions for $\mathrm{Re}\lambda_1 < 0$, $\mathrm{Re}\lambda_2 < 0$, $\mathrm{Re}\lambda_3 < 0$ are

$$\begin{aligned} &a_1 > 0, \quad a_2 > 0, \quad a_3 > 0, \\ &a_1a_2 > a_0a_3. \end{aligned} \tag{1.16}$$

In general, a necessary condition for all the roots of the characteristic equation to possess negative real parts is that all the coefficients $a_0, a_1, \ldots, a_n$ be positive.

**Note:**

The self-excited oscillatory process of a dynamical system can be studied simply by using the asymptotic methods in combination with an analog computer [34, 37,b].

**Example 16**

Consider the equation of the third order (7.2), Chapter 1, supposing that

$$R(x, \dot{x}, \ddot{x}) = (1 - x^2)\dot{x}. \tag{1.17}$$

Using the formulae (7.7) and (7.13) in Chapter 1, we see that

$$\begin{aligned} &\langle R_0 \sin\varphi\rangle = -\frac{a}{2}\Omega\Big(1 - \frac{a^2}{4}\Big), \\ &\langle R_0 \cos\varphi\rangle = 0, \\ &g_n = 0\ \forall n, \quad h_m = 0, \quad m \neq 1, 3, \\ &h_1 = -a\Omega\Big(1 - \frac{a^2}{4}\Big), \quad h_3 = \frac{\Omega}{4}a^3. \end{aligned}$$

Therefore, the equations of the first approximation are

$$
\begin{aligned}
x &= a\cos\varphi,\\
\frac{da}{dt} &= \frac{\varepsilon\xi}{2(\xi^2+\Omega^2)}a\left(1-\frac{a^2}{4}\right),\\
\frac{d\varphi}{dt} &= \Omega-\frac{\varepsilon\Omega}{2(\xi^2+\Omega^2)}\left(1-\frac{a^2}{4}\right).
\end{aligned}
\tag{1.18}
$$

Equation (1.18) has a stationary solution $a_0 = 2$. The stability condition (7.16), Chapter 1, for this solution is satisfied.

The refinement of the first approximation is of the form:

$$
x = 2\cos(\Omega t+\psi_0)\frac{\varepsilon}{4(\xi^2+9\Omega^2)}\left[3\cos 3(\Omega t+\psi_0)-\frac{\xi}{\Omega}\sin 3(\Omega t+\psi_0)\right].
$$

To verify the theoretical results, the original equation (7.2), Chapter 1 with $R$ from (1.17) has been modeled on the analog-computer for the case $\xi = \Omega = 1$, $\varepsilon = 0.1$. The oscillation diagrams are presented in Fig. 21: displacement-time and Fig. 22: phase trajectory. The experimental results agree well with the theoretical ones.

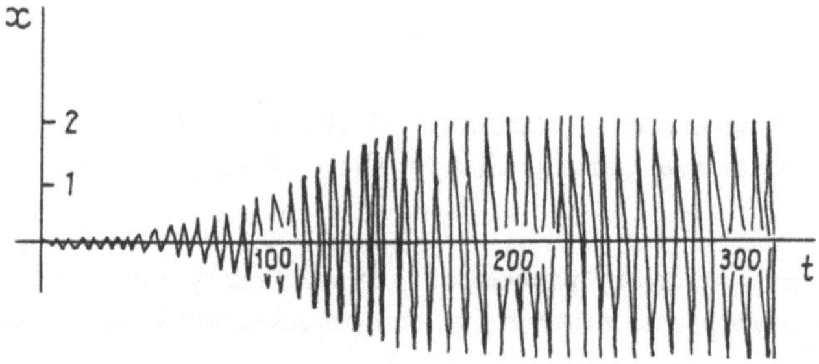


*Fig. 21.* Displacement-time diagram.

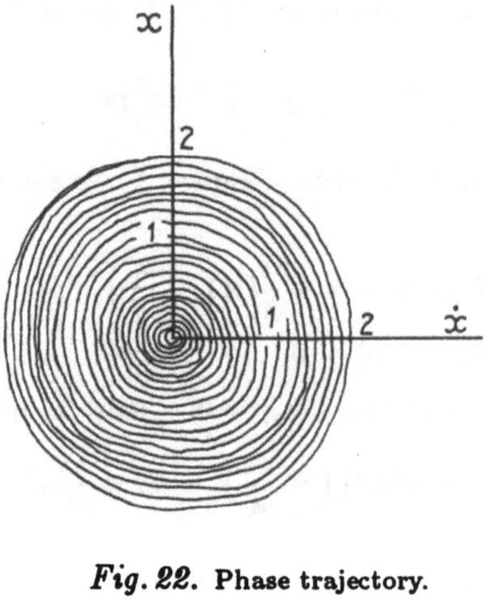


*Fig. 22.* Phase trajectory.

**Example 17**

It is supposed that the function $F(x, \dot{x})$ from (7.1), Chapter 1, is of the form

$$F(x, \dot{x}) = p\big(1 - \frac{r}{3}\dot{x}^2\big)\dot{x}, \tag{1.19}$$

here, $p$ and $r$ are positive constants. Now, the formulae (7.6), (7.7), Chapter 1, take the form

$$\begin{aligned} R_0 = \frac{1}{m}\big[ &- pa\Omega^2(1 - ra^2\Omega^2 \sin^2\varphi)\cos\varphi - \\ &- \xi pa\Omega\big(1 - \frac{r}{3}a^2\Omega^2\sin^2\varphi\big)\sin\varphi + ka\Omega\sin\varphi, \end{aligned}$$

hence,

$$\begin{aligned} \langle R_0 \sin\varphi\rangle &= \frac{a\Omega}{2m}\big[k - \xi p\big(1 - \frac{r}{4}\Omega^2 a^2\big)\big], \\ \langle R_0 \cos\varphi\rangle &= \frac{a\Omega^2}{2m}p\big(\frac{r}{4}\Omega^2 a^2 - 1\big), \end{aligned}$$

and the equation of the first approximation (7.12) and (7.13), Chapter 1, are

$$\begin{aligned} x &= a\cos(\Omega t + \psi), \\ \frac{da}{dt} &= \frac{\varepsilon a}{2m(\xi^2 + \Omega^2)}\big[p(\xi^2 + \Omega^2) - \xi k - \frac{1}{4}pr\Omega^2(\xi^2 + \Omega^2)a^2\big], \\ \frac{d\psi}{dt} &= \frac{\varepsilon k\Omega}{2m(\xi^2 + \Omega^2)}. \end{aligned}$$

From these, it follows that

1. If $\xi k > p(\xi^2 + \Omega^2)$, the equilibrium of the rigid body ($a = 0$) is stable and there is no self-excited oscillation. This may occur when negative friction is sufficiently small.
2. If $\xi k < p(\xi^2 + \Omega^2)$, the rigid body oscillates with amplitude

$$a_0 = 2\sqrt{\frac{(\xi^2 + \Omega^2) - \xi k}{pr\Omega^2(\xi^2 + \Omega^2)}}\,. \tag{1.20}$$

To verify the validity of the theoretical results, an experiment on an analog-computer was performed by modeling equations (7.1), Chapter 1, for the parameters $\varepsilon h/m = 0.1$, $\varepsilon p/m = 0.1$, $r = 3$, $\xi = \Omega = 1$ (Figs 23 and 24). In this case, the formula (1.20) gives $a_0 \approx 0.81$, while from the oscillation diagram (Fig. 24), we have $a_0 = \max|x| = 0.78$. Thus, the theoretical and experimental results are in good agreement.

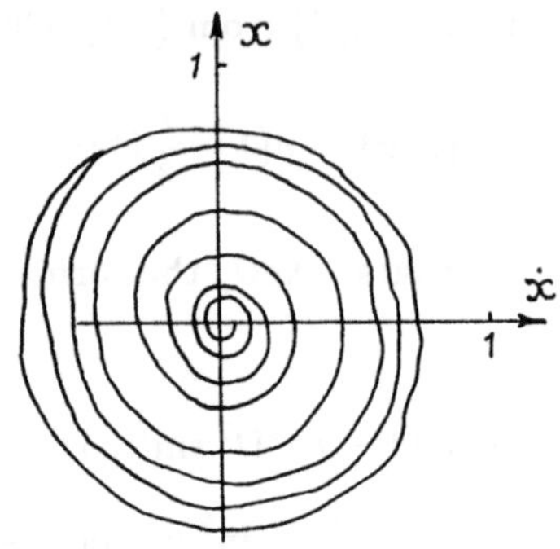


*Fig. 23.* Phase picture by analog computer.

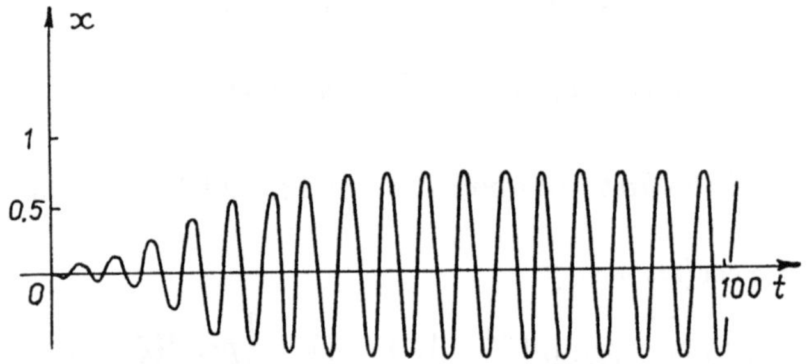


*Fig. 24.* Time dependence displacement.

## 2. Self-excited Oscillations of a Mechanical System

Self-excited oscillations occur in many technical systems, for instance, in machine tool flutter caused by cutting metals, and the oscillations produced by friction drive [18].

Let us consider the mechanical system, shown in Fig. 25.

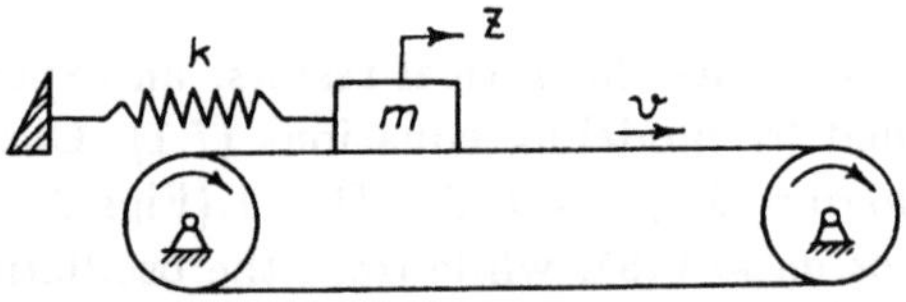


*Fig. 25.* Model of a self-excited oscillator.

A block of mass m rests on a rough conveyor belt, moving with a constant speed $v$, and is attached to a rigid support through a spring of stiffness $k$. There is friction between the belt surface and the block. The position of the block is denoted by $z$ which is chosen in such a way that, $z = 0$ corresponds to the unstretched state of the spring. The horizontal forces, acting on the block, are the spring force $kz$ and the friction force $F$, which is a function of the relative speed $v - \dot{z}$ between the belt and the block. Newton's equation of motion for the system in Fig. 25 is

$$m\ddot{z} + kz = F(v - \dot{z}). \tag{2.1}$$

It is convenient to introduce a new variable $x$, replacing $z$, by the relation

$$x = z - \frac{1}{k}F(v).$$

The differential equation for $x$ is then

$$m\ddot{x} + kx = F(v - \dot{x}) - F(v). \tag{2.2}$$

The frictional force is found experimentally and is assumed to be of the form shown in Fig. 26. The friction is at a maximum when the relative velocity is zero (static friction). As the relative velocity increase the friction diminishes and approaches a minimum. Then, the frictional force increases again with the advent of still greater velocities. The downward-sloping part of the frictional force can give rise to self-excited oscillations.

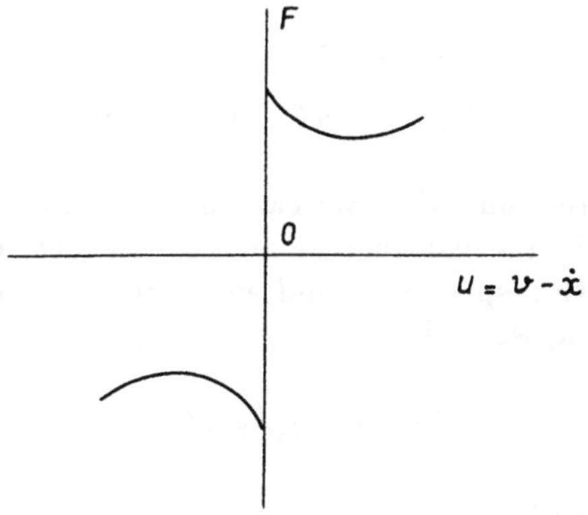


*Fig. 26.* The dependence of friction on relative velocity.

It is supposed that the frictional force may be expanded in a Taylor series in the range of $u = v - \dot{x}$, and that the velocity $v$ of the belt is always greater than that of the block, so that $u = v - \dot{x}$ is positive. Assuming also that $\dot{x}$ is small and that the series converges rapidly enough to justify the use of only the first three terms, we have

$$F(v-\dot{x})-F(v)=-\dot{x}F_v'+\frac{\dot{x}^2}{2}F_v''-\frac{\dot{x}^3}{6}F_v''', \tag{2.3}$$

where

$$F_v'=\frac{dF(v)}{dv}, \quad F_v''=\frac{d^2F(v)}{dv^2}, \quad F_v'''=\frac{d^3F(v)}{dv^3}.$$

Introducing the notations

$$\omega^2=\frac{k}{m}, \quad \tau=\omega t, \quad x'=\frac{dx}{dt}, \quad x''=\frac{d^2x}{dt^2},$$
$$h_1=\frac{-\omega}{k}F_v', \quad h_2=\frac{2\omega^2}{k}F_v'', \quad h_3=\frac{6\omega^3}{k}F_v''', \tag{2.4}$$

we may reduce equation (2.2) to

$$x''+x=h_1x'+h_2x'^2-h_3x'^3.$$

It is assumed that the coefficients $h_i$ are small. This smallness can be characterized, using a small positive dimensionless parameter $\varepsilon$, which will be set equal to unity in the final results. Then, we have the motion equation in the form

$$x''+x=\varepsilon\left(h_1x'+h_2x'^2-h_3x'^3\right). \tag{2.5}$$

Seeking a solution of the form

$$\begin{aligned} x&=a\cos\theta,\\ x'&=-a\sin\theta, \quad \theta=\tau+\psi, \end{aligned} \tag{2.6}$$

where $a$ and $\psi$ are functions of $\tau$, we can reduce equation (2.5) to a system of first order equations in the variables $a$ and $\psi$. In fact, differentiating the first expression of (2.6) with respect to $\tau$ and equating the second expression of (2.6) to the above expression, we find

$$a'\cos\theta-a\psi'\sin\theta=0. \tag{2.7}$$

Substituting (2.6) into (2.5) gives

$$-a'\sin\theta-a\psi'\cos\theta=\varepsilon f, \tag{2.8}$$

where

$$f=-h_1a\sin\theta+h_2a^2\sin^2\theta+h_3a^3\sin^3\theta.$$

From (2.7) and (2.8), one gets

$$\frac{da}{d\tau} = -\varepsilon f \sin\theta, \quad a\frac{d\psi}{d\tau} = -\varepsilon f \cos\theta. \tag{2.9}$$

Up to now, all the calculations have been mathematically exact. The approximation which is now to be introduced consists of averaging the right sides of equations (2.9) over $2\pi$ in $\theta$. It is based on the fact that, since $\varepsilon$ is a small parameter, $a$ and $\psi$ are slowly varying. They are essentially constant over an interval in which $\theta$ changes by $2\pi$. Carrying out the above averaging, we find the following results:

$$\frac{da}{d\tau} = -\frac{\varepsilon}{2\pi}\int_0^{2\pi} f \sin\theta d\theta = \frac{\varepsilon a}{2}\left(h_1 - \frac{3}{4}h_3 a^2\right), \tag{2.10}$$

$$a\frac{d\psi}{d\tau} = -\frac{\varepsilon}{2\pi}\int_0^{2\pi} f \cos\theta d\theta = 0. \tag{2.11}$$

To determine the steady-state response, we set $\frac{da}{dt} = 0$. There is a trivial solution $a = 0$. A non-trivial solution also exists in which $a$ is given by

$$\frac{3}{4}h_3 a^2 = h_1. \tag{2.12}$$

Denote the right side of (2.10) by $\varepsilon\Phi(a)$, where

$$\Phi(a) = \frac{a}{2}\left(h_1 - \frac{3}{4}h_3 a^2\right). \tag{2.13}$$

Inasmuch as

$$\left.\frac{d\Phi(a)}{da}\right|_{a=0} = \frac{h_1}{2},$$

is positive, the equilibrium point $a = 0$ is unstable. Otherwise,

$$\left.\frac{d\Phi(a)}{da}\right|_{\frac{3}{4}h_3 a^2 = h_1} = -h_1$$

which is negative, the corresponding stationary regime is stable.

## 3. Dynamic Absorber for Quenching Self-excited Oscillations of the Mechanical Systems having one Degree of Freedom

Self-excited oscillations, as occurring in many mechanical systems and structures, frequently lead to their critical operation (in terms of safety and working). So care in the design of such systems must be taken to suppress such self-excited oscillations.

The dynamic absorber is one of the effective devices to reduce undesirable forced oscillations in machines, which are subjected to a harmonic external excitation. Not long ago, Mansour [26], Clendening, Dubey [11], Tondl [50], Hagedorn [15], Nguyen Van Dao and Nguyen Van Dinh [38] and others proposed and studied the effect of the dynamic absorber for quenching self-excited oscillations of mechanical systems. It turned out that these oscillations may be quenched entirely by the aid of a dynamic absorber with a suitable value of the external friction force.

The theory of quenching for a linear system under the action of the external harmonic force is well known. Three methods have been used to quench the forced oscillation of this system:

1. Changing the frequency to move the natural frequency away from the forced one.
2. Introducing a damper to dissipate the oscillatory energy. However, the damping mechanism operates only when the oscillation exists. So, in this way (energy dissipation), the oscillation is not quenched completely.
3. Using the dynamic absorber - (method of dynamical balance) - an additional oscillatory system - consisting of a mass attached to an elastic element. Under ideal conditions, when there is no friction and when the natural frequency of the absorber is equal to that of the external force, the oscillation of the main mass disappears. The absorber continues to oscillate and creates an inertia force which upsets the balance with the excitational force. So, dynamical balance is the principle of action of the dynamical absorber.

For nonlinear systems, the first two methods (changing the frequency and energy dissipation) have been used to limit the growth of the amplitude of oscillation of the main mass. In the present paragraph, to clear up some questions about the principles of action of the dynamic absorber for a self-excited system, we limit ourselves by studying in depth a widespread class of nonlinear systems: the quasi-linear one, with the help of the asymptotic method which is mathematically well-founded. Here, it is necessary to identify two cases: the weak absorber when the mass and the spring stiffness of the absorber are small, and the strong absorber when those parameters are not small, but finite. In both cases, the damping mechanism connected with the absorber has decisive influence on the amplitude of self-excited oscillation. For the strong absorber, increasing the damping force leads to a decrease of the amplitude of self-excited oscillation. But for the weak absorber, the effect of quenching is achieved only with some intermediate values of the damping force.

Let us consider a self-excited mechanical system (Fig. 27), whose motion is described by the differential equation [38]:

$$m\ddot{x} + cx = \varepsilon R(x, \dot{x}). \tag{3.1}$$

The function $R(x, \dot{x})$, expressing the action of the self-excited force, is assumed to be of the form

$$R(x, \dot{x}) = h_1\dot{x} - h_3\dot{x}^3, \tag{3.2}$$

where $h_1$ and $h_3$ are positive constants. The other parameters are clear from the Figure 27; $\varepsilon$ is a small positive parameter, characterising the smallness of the terms standing behind it.

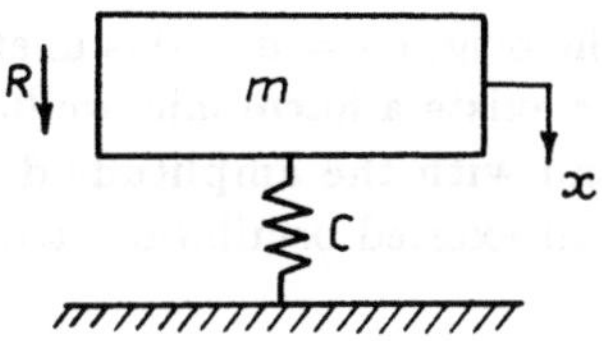


*Fig. 27.* A self-excited oscillating system.

Without loss of generality, one can assume that $m = 1$, $c = 1$. The solution of (3.1) will be found in the form

$$x = a \sin \varphi, \quad \dot{x} = a \cos \varphi, \quad \varphi = t + \psi, \tag{3.3}$$

where $a$ and $\psi$ are slowly varying functions of time. Substituting expressions (3.3) into equation (3.1) gives

$$\dot{a} \cos \varphi - a\dot{\psi} \sin \varphi = \varepsilon R(x, \dot{x}). \tag{a}$$

Taking into account

$$\dot{a} \sin \varphi + a\dot{\psi} \cos \varphi = 0, \tag{b}$$

we get from equations (a) and (b):

$$\begin{aligned} \dot{a} &= \varepsilon R(x, \dot{x}) \cos \varphi, \\ a\dot{\psi} &= -\varepsilon R(x, \dot{x}) \sin \varphi. \end{aligned} \tag{3.4}$$

Since $a$ and $\psi$ are slowly varying, the equation (3.4) may be averaged over one cycle as follows:

$$\begin{aligned} \dot{a} &= \frac{\varepsilon}{2\pi} \int_0^{2\pi} R(x, \dot{x}) \cos \varphi d\varphi, \\ a\dot{\psi} &= -\frac{\varepsilon}{2\pi} \int_0^{2\pi} R(x, \dot{x}) \sin \varphi d\varphi. \end{aligned} \tag{3.5}$$

Substituting (3.2) into (3.5) and calculating the integrals, one gets:

$$\begin{aligned} \dot{a} &= \frac{\varepsilon a}{2}(h_1 - A), \quad A = \frac{3}{4} h_3 a^2, \\ a\dot{\psi} &= 0. \end{aligned} \tag{3.6}$$

Equation (3.6) shows that $\psi$ is a constant. The steady state motion corresponds to $\dot{a} = 0$. Hence, equations (3.6) give

$$1) \qquad a = 0, \tag{3.7}$$

$$2) \qquad a \neq 0, \quad A = h_1. \tag{3.8}$$

It is easy to prove that the origin $a = \dot{a} = 0$ is unstable and the stationary solution (3.8) is stable. So, there exists a harmonic oscillation (self-excited oscillation) with the frequency $\dot{\varphi} = 1$ and with the amplitude determined by $A = h_1$.

In order to quench this self-excited oscillation, a dynamic absorber will be used (Fig. 28).

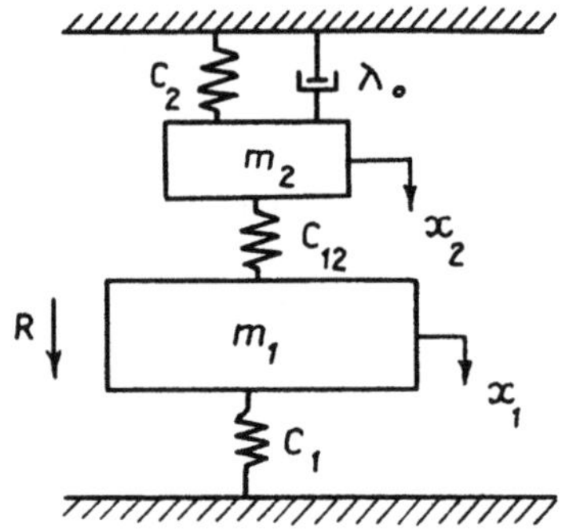


*Fig. 28.* Oscillating system with dynamic absorber.

**Weak dynamic absorber**

The absorber is called weak if its mass $m_2$ and spring stiffness $c_2$ are small in comparison with the main mass $m_1$ and spring $c_1$. The following equations are assumed for the system in Fig. 28:

$$\begin{aligned} \ddot{x}_1 + \omega_1^2 x_1 &= \frac{\varepsilon}{m_1}\left[h_1\dot{x}_1 - h_3\dot{x}_1^3 - c_{12}(x_1 - x_2)\right], \\ \ddot{x}_2 + \omega_2^2 x_2 + \lambda\dot{x}_2 &= \frac{c_{12}}{m_2}x_1, \end{aligned} \tag{3.9}$$

where

$$\omega_1^2 = \frac{c_1}{m_1}, \quad \omega_2^2 = \frac{1}{m_2}(c_2 + c_{12}),$$

$$\lambda_0 = m_2\lambda, \quad h_1 > 0, \quad h_3 > 0.$$

The first equation of (3.9) is quasi-linear, while the second is linear. In the first approximation, we shall find the solution of (3.9) in the form:

$$\begin{aligned} &x_1 = a\cos\theta, \ \dot{x}_1 = -a\omega_1\sin\theta, \ \theta = \omega_1 t + \psi, \\ &x_2 = a(M\cos\theta + N\sin\theta), \ \dot{x}_2 = a\omega_1(-M\sin\theta + N\cos\theta), \end{aligned} \tag{3.10}$$

where

$$M = \frac{c_{12}(\omega_2^2 - \omega_1^2)}{m_2\left[(\omega_2^2 - \omega_1^2)^2 + \omega_1^2\lambda^2\right]}, \quad N = \frac{c_{12}\omega_1\lambda}{m_2\left[(\omega_2^2 - \omega_1^2)^2 + \omega_1^2\lambda^2\right]}, \tag{3.11}$$

and $a$, $\psi$ are new unknown functions of time. Substituting (3.10) into (3.9) and solving for $\dot{a}$ and $\dot{\psi}$, we obtain:

$$\dot{a} = -\frac{\varepsilon}{\omega_1} F(x_1, \dot{x}_1, x_2) \sin\theta, \tag{3.12}$$

$$a\dot{\psi} = -\frac{\varepsilon}{\omega_1} F(x_1, \dot{x}_1, x_2) \cos\theta, \tag{3.13}$$

where

$$F(x_1, \dot{x}_1, x_2) = \frac{1}{m_1}\left[h_1\dot{x}_1 - h_3\dot{x}_1^3 - c_{12}(x_1 - x_2)\right]. \tag{3.14}$$

Averaging the right-hand sides of (3.12) on time, holding $a$ and $\psi$ constant, we have:

$$\begin{aligned} \dot{a} &= \frac{\varepsilon a}{2m_1}\left\{h_1 - \frac{c_{12}^2\lambda}{m_2\left[(\omega_2^2 - \omega_1^2)^2 + \omega_1^2\lambda^2\right]} - \frac{3}{4}h_3\omega_1^2 a^2\right\}, \\ \omega_1\dot{\psi} &= -\frac{\varepsilon c_{12}}{2m_1}\left\{\frac{c_{12}(\omega_2^2 - \omega_1^2)}{m_2\left[(\omega_2^2 - \omega_1^2)^2 + \omega_1^2\lambda^2\right]} - 1\right\}. \end{aligned} \tag{3.15}$$

It is easy to verify that the equilibrium $a = \dot{a} = 0$ is unstable and that there exists a stationary self-excited oscillation with amplitude $a_0$ determined by

$$\frac{3}{4}h_3\omega_1^2 a_0^2 = h_1 - \frac{c_{12}^2\lambda}{m_2\left[(\omega_2^2 - \omega_1^2)^2 + \omega_1^2\lambda^2\right]}, \tag{3.16}$$

if the right-hand side of (3.16) is positive. In Fig. 29 the dependence of amplitude

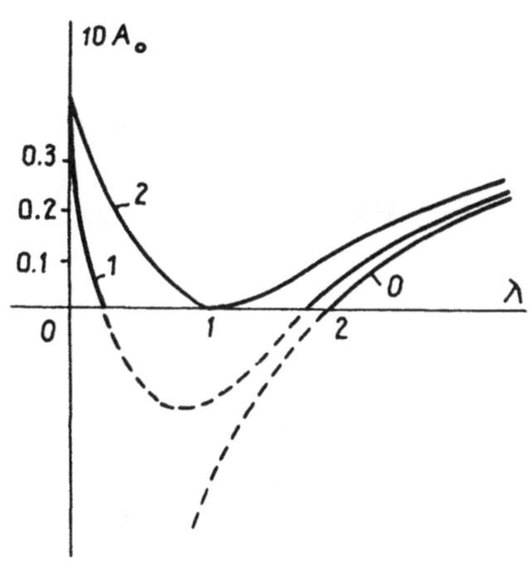


*Fig. 29.* The dependence of oscillation amplitude of main mass on the damping parameter for a weak absorber.

of oscillation on the damping parameter $\lambda$ is represented by the case $h_1 = 0.05$, $\omega_1 = 1$, $\dfrac{c_{12}^2}{m_2} = 0.1$, $A_0 = \dfrac{3}{4}h_3\omega_1^2 a_0^2$. The curve 1 corresponds to $\omega_2^2 = 1.5$ and the curve 2 to $\omega_2^2 = 2$. When $\omega_1$, $\omega_2$ are equal, the expression (3.16) takes the form:

$$\frac{3}{4}h_3\omega_1^2 a_0^2 = h_1 - \frac{c_{12}^2}{m_2\omega_1^2\lambda}\,, \tag{3.17}$$

and the resonance curve is shown in Fig. 29 by "0". In this case, the effect of quenching is highest.

**Strong dynamic absorber**

Let us consider a strong absorber for which the parameters $m_2$ and $c_2$ are not small, but finite. In this case, the equations of motion are

$$\begin{aligned} &m_1\ddot{x}_1 + c_1x_1 + c_{12}(x_1 - x_2) = \varepsilon(h_1\dot{x}_1 - h_3\dot{x}_1^3),\\ &m_2\ddot{x}_2 + c_2x_2 + c_{12}(x_2 - x_1) = -\varepsilon\lambda\dot{x}_2. \end{aligned} \tag{3.18}$$

They are not separable when $\varepsilon = 0$ as the equations (3.9) were. Using the transformation into the principal modes $\xi_1$, $\xi_2$ (see appendix 1):

$$\begin{aligned} x_1 &= \xi_1 + \xi_2, \quad x_2 = d_1\xi_1 + d_2\xi_2,\\ d_i &= \frac{c_{12}}{(c_{12} + c_2 - m_2\Omega_i^2)}\,, \quad i = 1,2,\\ \Omega_{1,2}^2 &= \frac{1}{2}\Big\{\frac{c_1 + c_{12}}{m_1} + \frac{c_2 + c_{12}}{m_2} \pm\\ &\pm\sqrt{\Big(\frac{c_1 + c_{12}}{m_1} - \frac{c_2 + c_{12}}{m_2}\Big)^2 + 4\frac{c_{12}}{m_1m_2}}\Big\}, \end{aligned} \tag{3.19}$$

we have

$$\begin{aligned} \ddot{\xi}_1 + \Omega_1^2\xi_1 &= \frac{\varepsilon}{M_1}[f_1 + d_1f_2],\\ \ddot{\xi}_2 + \Omega_2^2\xi_2 &= \frac{\varepsilon}{M_2}[f_1 + d_2f_2], \end{aligned} \tag{3.20}$$

where

$$\begin{aligned} &M_i = m_1 + d_i^2m_2,\\ &f_1 = h_1\dot{x}_1 - h_3\dot{x}_1^3, \quad f_2 = -\lambda\dot{x}_2. \end{aligned}$$

Introducing the new variables $a$, $b$, $\psi$, $\phi$ by means of the formulae

$$\begin{aligned} \xi_1 &= a\cos\theta, \quad \dot{\xi}_1 = -a\Omega_1\sin\theta, \quad \theta = \Omega_1t + \psi,\\ \xi_2 &= b\cos\varphi, \quad \dot{\xi}_2 = -b\Omega_2\sin\varphi, \quad \varphi = \Omega_2t + \phi, \end{aligned} \tag{3.21}$$

and imposing the conditions

$$\dot{a}\cos\theta - a\dot{\psi}\sin\theta = 0, \quad \dot{b}\cos\varphi - b\dot{\phi}\sin\varphi = 0,$$

we find that equations (3.20) become

$$\begin{aligned}
\Omega_1\dot{a} &= -\frac{\varepsilon}{M_1}(f_1 + d_1 f_2)\sin\theta,\\
\Omega_1 a\dot{\psi} &= -\frac{\varepsilon}{M_1}(f_1 + d_1 f_2)\cos\theta,\\
\Omega_2\dot{b} &= -\frac{\varepsilon}{M_2}(f_1 + d_2 f_2)\sin\varphi,\\
\Omega_2 b\dot{\phi} &= -\frac{\varepsilon}{M_2}(f_1 + d_2 f_2)\cos\varphi.
\end{aligned} \tag{3.22}$$

Since the new variables are slowly varying, the equations (3.22) may be averaged over one cycle as follows:

$$\Omega_1\dot{a} = -\frac{\varepsilon}{4\pi^2 M_1}\int_0^{2\pi}\int_0^{2\pi}(f_1 + d_1 f_2)\sin\theta\, d\theta d\varphi,$$

and so on. Calculating the integrals, holding $a$, $\psi$, $b$, $\phi$ constant, we obtain:

$$\begin{aligned}
\dot{a} &= -\frac{\varepsilon a}{2M_1}\left(-h_1 + d_1^2\lambda + \frac{3}{4}h_3\Omega_1^2 a^2 + \frac{3}{2}h_3\Omega_2^2 b^2\right),\\
\dot{b} &= -\frac{\varepsilon b}{2M_2}\left(-h_1 + d_2^2\lambda + \frac{3}{4}h_3\Omega_2^2 b^2 + \frac{3}{2}h_3\Omega_1^2 a^2\right),\\
\dot{\psi} &= \dot{\phi} = 0.
\end{aligned} \tag{3.23}$$

From the result of the last two equations of (3.23), it follows that $\psi$ and $\phi$ are constants. It is easy to verify the existence of the following steady state regimes of self-excited oscillation determined by the conditions $\dot{a} = \dot{b} = 0$:

1. Equilibrium $a = b = 0$, which is stable if $h_1 - d_1^2\lambda < 0$ and $h_1 - d_2^2\lambda < 0$.
2. Oscillation of the first principal mode $\xi_1$ with frequency $\Omega_1$ and amplitude $a_0$ determined by

$$b = 0,\ A = \frac{3}{4}h_3\Omega_1^2 a_0^2 = h_1 - d_1^2\lambda,\ h_1 - d_1^2\lambda > 0. \tag{3.24}$$

This regime is stable if

$$\frac{3}{2}h_3\Omega_1^2 a_0^2 > h_1 - d_2^2\lambda, \tag{3.25}$$

or

$$A > \frac{1}{2}B.$$

3. Oscillation of the second principal mode $\xi_2$ with frequency $\Omega_2$ and amplitude $b_0$ determined by

$$a = 0,\ B = \frac{3}{4}h_3\Omega_2^2 b_0^2 = h_1 - d_2^2\lambda,\ h_1 - d_2^2\lambda > 0. \tag{3.26}$$

This regime is stable if

$$\frac{3}{2}h_3\Omega_2^2 b_0^2 > h_1 - d_1^2\lambda, \tag{3.27}$$

or

$$B > \frac{1}{2}A$$

4. Oscillation of both principal modes $\xi_1$ and $\xi_2$ with two frequencies $\Omega_1$ and $\Omega_2$. This regime is always unstable.

From equations (3.24) and (3.26), it follows that the amplitude of self-excited oscillation decreases as the coefficient $\lambda$ of the damping force increases (Fig. 30).

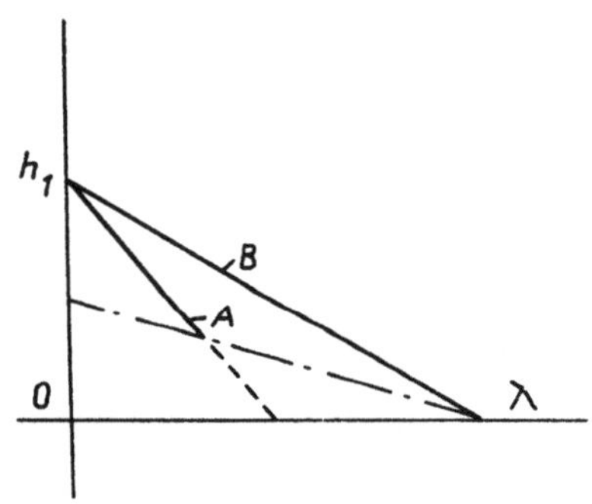


*Fig. 30.* The dependence of the oscillation amplitude of main mass on the damping parameter for a strong absorber.

**Some remarks**

1. We have seen above that the damping force ($\lambda$) plays an important role in quenching the self-excited oscillation by means of the dynamic absorber. The question posed now is the influence of the position of the damping mechanism. It was found that the effect of quenching is the same for other methods of attaching the damping mechanism, for example, for those shown in Fig. 31 and Fig. 32.

For weak absorbers, the quenching effect is high with suitable choice of the value of $\lambda$ (Fig. 29), and for strong absorbers, increasing the damping force ($\lambda$) leads to a decrease of the amplitude of self-excited oscillation of the main mass $m_1$ (Fig. 30).

2. Now, we discuss the role of the damping force ($\lambda$). Let us consider the case when the equilibrium $a = 0$ for the weak absorber and $a = b = 0$ for the strong

absorber, which is unstable without the damping force ($\lambda = 0$), becomes stable with a sufficiently large value of $\lambda$. Here, the role of the damping force is not in energy dissipation because both masses $m_1$ and $m_2$ are at rest. In this case, the damping force only stabilizes the equilibrium.

3. The action of the dynamic absorber for self-excited oscillation is not based on the dynamic balance principle, because when the self-excited oscillation is quenched completely, the absorber is at rest too. Moreover, the dynamic balance principle acts only for the system without damping force, while here damping is unavoidable. Thus, the action of the dynamic absorber for the self-excited system is different in principle from that for the forced system.

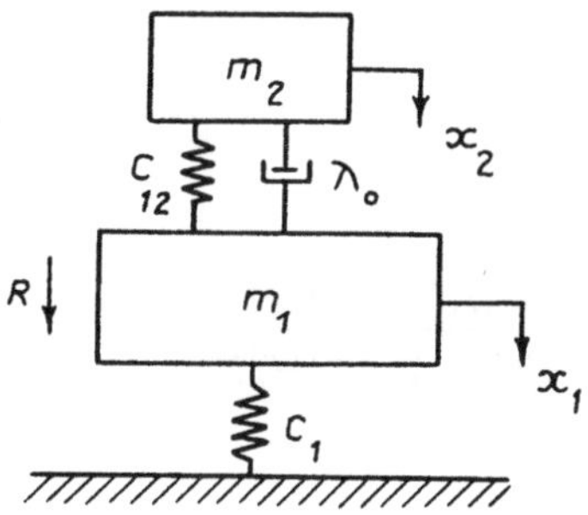


*Fig. 31.* Another method of attaching absorber.

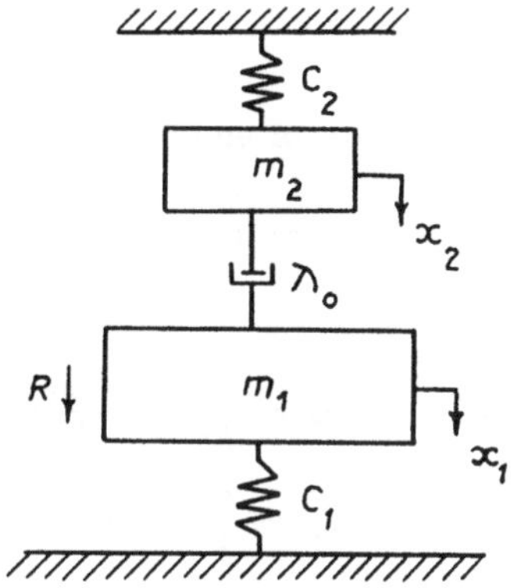


*Fig. 32.* Damper connected two masses.

**Dynamic absorber for drilling instrument**

For quenching the self-excited oscillation of the instrument (1) when drilling, a special mechanism which consists of an absorber tube (2) with vibrodamping layer (3) is used (Fig. 33).

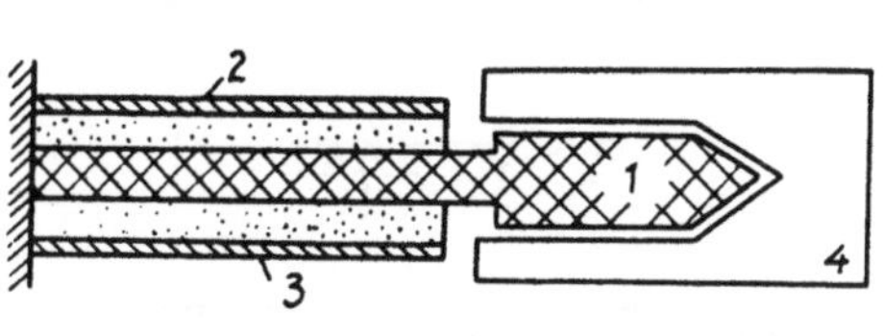


*Fig. 33.* Schema of a drilling instrument.

Let us consider a weak absorber with a small stiffness $c_2$ and small moment of inertia $I_2$. The motion of this system is governed by equations

$$I_1\ddot{\theta}_1 + c_1\theta_1 = \varepsilon\big[-h(\dot{\theta}_1 - \dot{\theta}_2) + \mathcal{M}(\dot{\theta}_1)\big],$$
$$I_2\ddot{\theta}_2 + c_2\theta_2 = -h(\dot{\theta}_2 - \dot{\theta}_1), \tag{3.28}$$

where $\mathcal{M}(\dot{\theta}_1) = \alpha\dot{\theta}_1 - \beta\dot{\theta}_1^3$, $\alpha > 0$, $\beta > 0$, is the moment of the friction force between instrument and detail (4). For a strong absorber, when $c_2$, $I_2$ are not small, but finite, the equations of motion take the form

$$I_1\ddot{\theta}_1 + c_1\theta_1 = \varepsilon\big[-h(\dot{\theta}_1 - \dot{\theta}_2) + \mathcal{M}(\dot{\theta}_1)\big],$$
$$I_2\ddot{\theta}_2 + c_2\theta_2 = -\varepsilon h(\dot{\theta}_2 - \dot{\theta}_1), \tag{3.29}$$

and the damping condition of oscillation of the instrument (1) is

$$h > \alpha. \tag{3.30}$$

The solution of equations (3.28) will be found in the form

$$\theta_1 = a\cos\phi,\ \theta_2 = a(M\cos\phi + N\sin\phi),\ \phi = \omega_1 t + \psi, \tag{3.31}$$

$$M = \frac{h_2^2\omega_1^2}{h_2^2\omega_1^2 + (\omega_2^2 - \omega_1^2)^2},\ N = \frac{h_2\omega_1(\omega_1^2 - \omega_2^2)}{h_2^2\omega_1^2 + (\omega_2^2 - \omega_1^2)^2},$$
$$\omega_i^2 = \frac{c_i}{I_i},\quad h_2 = \frac{h}{I_2}.$$

Substituting these expressions into equations (3.28); then solving them with respect to $\frac{da}{dt}, \frac{d\psi}{dt}$ and averaging the results over time, we obtain

$$\frac{da}{dt} = \frac{\varepsilon a}{2I_1}\Big(\alpha - \frac{h}{1+\sigma^2 h^2} - \frac{3}{4}\beta\omega_1^2 a^2\Big),$$
$$a\frac{d\psi}{dt} = -\frac{\varepsilon}{2I_1}h\omega_1 N a,\ \sigma^2 = \frac{\omega_1^2}{I_2^2(\omega_2^2 - \omega_1^2)^2}. \tag{3.32}$$

From these equations, we find the stationary amplitudes $a$ =const:

1) $a = 0$, stable if

$$\alpha < \frac{h}{1+\sigma^2 h^2}, \tag{3.33}$$

2)

$$a \neq 0,\quad \frac{3}{4}\beta\omega_1^2 a^2 = \alpha - \frac{h}{1+\sigma^2 h^2}. \tag{3.34}$$

Obviously, for optimal quenching of the oscillation of the instrument, the most advantageous value of $h$ will be $h = \frac{1}{\sigma}$.

## 4. Dynamic Absorber for Quenching Self-excited Oscillations of Systems having Two Degrees of Freedom

Let us consider the mechanical system presented in Fig. 34, which consists of two main masses $m_1$, $m_2$, and is subjected to a friction force

$$R = \ell_1 \dot{x}_1 - \ell_3 \dot{x}_1^3, \quad \ell_1 > 0,\ \ell_3 > 0. \tag{4.1}$$

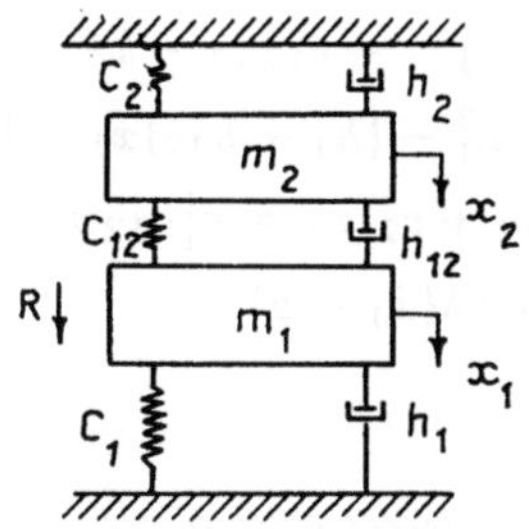


*Fig. 34.* Oscillating system with two degrees of freedom.

The differential equations for this system are [38,a]

$$\begin{aligned} m_1\ddot{x}_1 + (c_1 + c_{12})x_1 - c_{12}x_2 &= \\ = \varepsilon\big[\ell_1\dot{x}_1 - \ell_3\dot{x}_1^3 - (h_1 + h_{12})\dot{x}_1 &+ h_{12}\dot{x}_2\big], \\ m_2\ddot{x}_2 - c_{12}x_1 + (c_{12} + c_2)x_2 &= \varepsilon\big[h_{12}\dot{x}_1 - (h_{12} + h_2)\dot{x}_2\big]. \end{aligned} \tag{4.2}$$

Using the same technique as we used in the previous paragraph 3, we find the following stationary regimes:

1. The equilibrium $x_1 = \dot{x}_1 = x_2 = \dot{x}_2 = 0$, which is stable if

$$\ell_1 - H_i < 0, \tag{4.3}$$

$$H_i = h_1 + h_2 d_i^2 + h_{12}(1 - d_i)^2, \tag{4.4}$$

$$i = 1, 2.$$

2. The oscillation of the first principal mode with frequency $\omega_1$ and amplitude $a_1$, is determined by

$$A_1 = \frac{3}{4}\ell_3\omega_1^2 a_1^2 = \ell_1 - H_1, \tag{4.5}$$

which is stable if

$$A_1 > \frac{1}{2}(\ell_1 - H_2) = \frac{1}{2}A_2. \tag{4.6}$$

3. The oscillation of the second principal mode with frequency $\omega_2$ and amplitude $a_2$, is determined by

$$A_2 = \frac{3}{4}\ell_3\omega_2^2 a_2^2 = \ell_1 - H_2, \tag{4.7}$$

which is stable if

$$A_2 > \frac{1}{2}(\ell_1 - H_1) = \frac{1}{2}A_1. \tag{4.8}$$

4. The oscillation of both principal modes, which is always unstable.

**Strong absorber for the system with two degrees of freedom**

To quench the oscillation of masses $m_1$ and $m_2$, we use an absorber $(m, k, \lambda)$ (Fig. 35). It is assumed that, the absorber is attached to the first mass $m_1$. The equations of motion are now of the form:

$$\begin{aligned} m_1\ddot{x}_1 + (c_1 + c_{12} + k)x_1 - c_{12}x_2 - kz &= \\ = \varepsilon\big[\ell_1\dot{x}_1 - \ell_3\dot{x}_1^3 - (h_1 + h_{12})\dot{x}_1 + h_{12}\dot{x}_2 - \lambda(\dot{x}_1 - \dot{z})\big], \\ m_2\ddot{x}_2 - c_{12}x_1 + (c_{12} + c_2)x_2 = \varepsilon\big[h_{12}\dot{x}_1 - (h_{12} + h_2)\dot{x}_2\big], \\ m\ddot{z} - kx_1 + kz = \varepsilon\lambda(\dot{x}_1 - \dot{z}). \end{aligned} \tag{4.9}$$

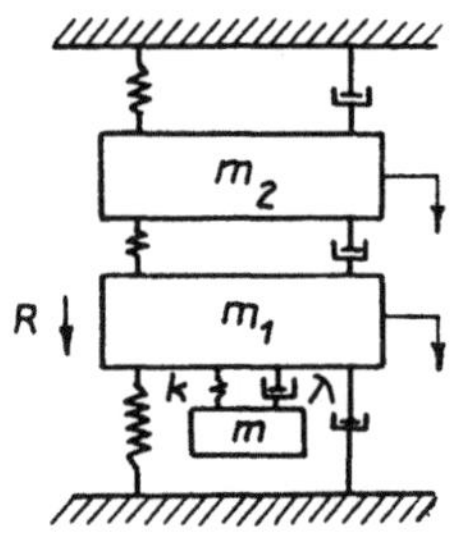


*Fig. 35.* A dynamic absorber is attached to oscillating system.

Denoting by $\omega_i$ the natural frequencies, and by $d_1^{(i)} = 1$, $d_2^{(i)}$, $d_3^{(i)}$ $(i = 1, 2, 3)$ the coefficients of distribution of the homogeneous system, and using the transformation into the principal coordinates $\xi_1$, $\xi_2$, $\xi_3$ (see appendix 1):

$$x_i = d_i^{(1)}\xi_1 + d_i^{(2)}\xi_2 + d_i^{(3)}\xi_3, \ (i = 1, 2, 3, \ x_3 = z), \tag{4.10}$$

we have

$$\ddot{\xi} + \omega_i^2\xi_i = \frac{\varepsilon}{M_i}\sum_{j=1}^{3} d_j^{(i)} f_j, \quad i = 1, 2, 3, \tag{4.11}$$

where

$$M_i = \sum_{j=1}^{3} m_j d_j^{(i)^2}, \quad i = 1, 2, 3.$$

By introducing amplitude-phase variables $a_i$, $\psi_i$

$$\begin{aligned} &\xi_i = a_i\cos\varphi_i, \ \dot{\xi}_i = -a_i\omega_i\sin\varphi_i, \ \varphi_i = \omega_i t + \psi_i, \\ &\dot{a}_i\cos\varphi_i - a_i\dot{\psi}_i\sin\varphi_i = 0, \quad i = 1, 2, 3, \end{aligned}$$

one gets:

$$\omega_i \dot{a}_i = \frac{-\varepsilon \sin \varphi_i}{M_i} \sum_{j=1}^{3} d_j^{(i)} f_j,$$
$$\omega_i a_i \dot{\psi}_i = \frac{-\varepsilon \cos \varphi_i}{M_i} \sum_{j=1}^{3} d_j^{(i)} f_j. \tag{4.12}$$

Since $a_i$ and $\psi_i$ are slowly varying functions of time, the change in their values during a period $(\varphi_i, \varphi_i + 2\pi)$ is very small. Hence, one may replace equations (4.12) by their averages over $(\varphi_i, \varphi_i + 2\pi)$, assuming $a_i$ and $b_i$ to be constant. If this is done, we have

$$\dot{a}_i = \frac{\varepsilon a_i}{2M_i}\left[\ell_1 - H_i^* - \frac{3}{4}\ell_3 \omega_i^2 a_i^2 - \frac{3}{2}\ell_3 \sum_{\substack{j=1 \\ j\neq i}}^{3} \omega_j^2 a_j^2\right],$$
$$a_i \dot{\psi}_i = 0. \tag{4.13}$$

From equations (4.13), we find the following stationary regimes:

1. The equilibrium $x_i = \dot{x}_i = 0$, which is stable if

$$\ell_1 - H_i^* < 0, \quad (i = 1, 2, 3) \tag{4.14}$$

where

$$H_i^* = h_1 + h_2 d_2^{(i)^2} + h_{12}(1 - d_2^{(i)})^2 + \lambda(1 - d_3^{(i)})^2. \tag{4.15}$$

2. Harmonic oscillation of the first principal mode $\xi_1$ with frequency $\omega_1$ and amplitude $a_1$, is determined by

$$A_1 = \frac{3}{4}\ell_3 \omega_1^2 a_1^2 = \ell_1 - H_1^*, \tag{4.16}$$

which is stable if

$$A_1 > \frac{1}{2}(\ell_1 - H_j^*) = \frac{1}{2}A_j, (j = 2, 3). \tag{4.17}$$

3. Harmonic oscillation of the second principal mode $\xi_2$ with frequency $\omega_2$ and amplitude $a_2$, is determined by

$$A_2 = \frac{3}{4}\ell_3 \omega_2^2 a_2^2 = \ell_1 - H_2^*, \tag{4.18}$$

which is stable if

$$A_2 > \frac{1}{2}A_j, \quad (j = 1, 3). \tag{4.19}$$

4. Harmonic oscillation of the third principal mode with frequency $\omega_3$ and amplitude $a_3$, is determined by

$$A_3 = \frac{3}{4}\ell_3 \omega_3^2 a_3^2 = \ell_1 - H_3^*, \tag{4.20}$$

which is stable if

$$A_3 > \frac{1}{2}A_j, \quad (j = 1, 2). \tag{4.21}$$

5. Simultaneous oscillation of two or three principal modes, which is always unstable.

Here, one can see some phenomena similar to the case of the system with one degree of freedom. The generalized damping force $H_i^*(i = 1, 2, 3)$ is linearly dependent on the damping coefficient $\lambda$. Increasing $\lambda$ leads to a decrease of the amplitude of oscillation. The oscillation disappears and the equilibrium becomes stable. It is to be noted that here, the damping force ($\lambda$) also does not dissipate energy (see remark 2, page 82).

**Weak absorber for a system with two degrees of freedom**

In contrast to strong absorber, here the mass $m$ and the stiffness $k$ are small. The differential equations of motion are:

$$\begin{aligned}
&m_1\ddot{x}_1 + (c_1 + c_{12})x_1 - c_{12}x_2 = \varepsilon f_1 = \\
&= \varepsilon\big[\ell_1\dot{x}_1 - \ell_3\dot{x}_1^3 - (h_1 + h_{12})\dot{x}_1 + h_{12}\dot{x}_2 - k(x_1 - z) - \lambda(\dot{x}_1 - \dot{z})\big], \\
&m_2\ddot{x}_2 - c_{12}x_1 + (c_{12} + c_2)x_2 = \varepsilon f_2 = \varepsilon\big[h_{12}\dot{x}_1 - (h_{12} + h_2)\dot{x}_2\big], \\
&\ddot{z} + \lambda_0\dot{z} + \omega^2 z = \omega^2 x_1 + \lambda_0\dot{x}_1,
\end{aligned} \tag{4.22}$$

where

$$\omega^2 = \frac{k}{m}, \quad \lambda_0 = \frac{\lambda}{m}.$$

Similarly in paragraph 3, transforming the subsystem ($x_1$, $x_2$) into the principal mode ($\xi_1$, $\xi_2$), by using the formulae (3.19) (see Appendix 1), we have

$$\begin{aligned}
&\ddot{\xi}_1 + \omega_1^2\xi_1 = \frac{\varepsilon}{M_1}(f_1 + d_1 f_2), \\
&\ddot{\xi}_2 + \omega_2^2\xi_2 = \frac{\varepsilon}{M_2}(f_1 + d_2 f_2), \\
&\ddot{z} + \lambda_0\dot{z} + \omega^2 z = \omega^2(\xi_1 + \xi_2) + \lambda_0(\dot{\xi}_1 + \dot{\xi}_2).
\end{aligned} \tag{4.23}$$

We now introduce the new variables $a_1$, $a_2$, $\varphi_1$, $\varphi_2$ by means of the formulae

$$\begin{aligned}
&\xi_1 = a_1 \sin\varphi_1, \quad \dot{\xi}_1 = a_1\omega_1 \cos\varphi_1, \\
&\xi_2 = a_2 \sin\varphi_2, \quad \dot{\xi}_2 = a_2\omega_2 \cos\varphi_2,
\end{aligned} \tag{4.24}$$

$$\begin{aligned}
&z = a_1(P_1 \sin\varphi_1 + Q_1 \cos\varphi_1) + a_2(P_2 \sin\varphi_2 + Q_2 \cos\varphi_2), \\
&\dot{z} = a_1\omega_1(P_1 \cos\varphi_1 - Q_1 \sin\varphi_1) + a_2\omega_2(P_2 \cos\varphi_2 - Q_2 \sin\varphi_2),
\end{aligned}$$

where

$$P_i = \frac{\omega^2\Delta_i + \lambda_0^2\omega_i^2}{\Delta_i^2 + \lambda_0^2\omega_i^2}, \quad Q_i = \frac{-\lambda_0\omega_i^3}{\Delta_i^2 + \lambda_0^2\omega_i^2}, \quad \Delta_i = \omega^2 - \omega_i^2. \tag{4.25}$$

Substituting (4.24) into (4.23), then solving for the variables $\frac{da_i}{dt}$, $\frac{d\varphi_i}{dt}$ and using the averaging techniques, we obtain the stationary regimes as those given by the formulae (4.3) - (4.8), but there exist additional damping forces

$$H_i^* = \frac{m\lambda_0\omega_i^4}{\Delta_i^2 + \lambda_0^2\omega_i^2}, \quad i = 1, 2. \tag{4.26}$$

As an example, taking $m_1 = m_2 = 1$, $c_2 = h_2 = 0$, $c_{12} = 2$, $h_1 = 0.13$, $h_{12} = 0.02$, we have $\Omega_1^2 = 1$, $\Omega_2^2 = 6$, $d_1 = 2$, $d_2 = -1/2$, $H_1 = 0.05$, $H_2 = 0.175$, $M_1 = 5$, $M_2 = 5/4$ and

$$A_1 = 0.05 - \frac{m\lambda_0}{\Delta_1^2 + \lambda_0^2}, \quad A_2 = 0.025 - 36\frac{m\lambda_0}{\Delta_2^2 + 6\lambda_0^2}. \tag{4.27}$$

Let $m = 0.05$, we draw the curves giving the dependence of $A_1$ and $A_2$ on the damping coefficient $\lambda_0$ in the case of exact resonance with low frequency $\omega^2 = \omega_1^2$, $\Delta_1 = 0$: $\omega^2 = 1$ (Fig. 36) and high frequency $\omega^2 = \omega_2^2$, $\Delta_2 = 0$: $\omega^2 = 6$ (Fig. 37). From these, it follows that: for a self-excited system with two degrees of freedom, the action of the weak absorber is effective only with a suitable choice of damping force ($\lambda$). With given values of the parameters, if $0.36 \le \lambda_0 \le 1$, both modes of oscillations with high and low frequency are quenched if the exact resonance with low frequency takes place (Fig. 36).

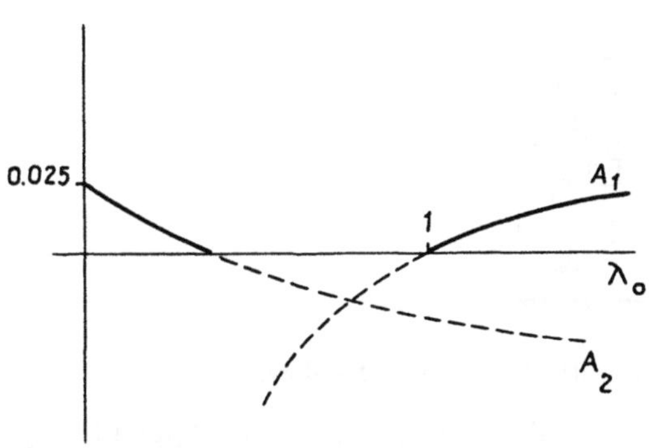


*Fig. 36.* Amplitude dependence on damping coefficient for low frequency.

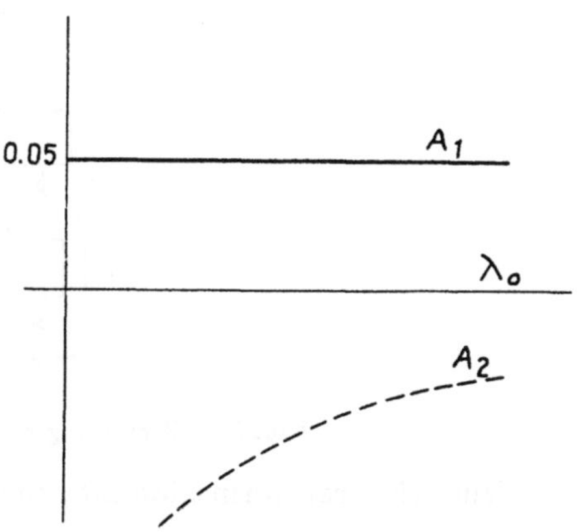


*Fig. 37.* Amplitude dependence on damping coefficient for high frequency.

## 5. Self-excited Oscillation of a System with N Degrees of Freedom

The system under consideration in this paragraph is presented in Fig. 38 and is described by the following differential equations

$$\begin{aligned}
&m_1\ddot{x}_1 + (c_1 + c_{12})x_1 - c_{12}x_2 = \varepsilon f_1 = \\
&= \varepsilon\big[\ell_1\dot{x}_1 - \ell_3\dot{x}_1^3 - (h_1 + h_{12})\dot{x}_1 + h_{12}\dot{x}_2\big], \\
&m_2\ddot{x}_2 - c_{12}x_1 + (c_{12} + c_2 + c_{23})x_2 - c_{23}x_3 = \varepsilon f_2 = \\
&= \varepsilon\big[h_{12}\dot{x}_1 - (h_{12} + h_2 + h_{23})\dot{x}_2 + h_{23}\dot{x}_3\big], \qquad (5.1)\\
&\dots\dots\dots\dots\dots\dots\dots\dots \\
&m_N\ddot{x}_N - c_{N-1,N}x_{N-1} + (c_{N-1,N} + c_N)x_N = \varepsilon f_N = \\
&= \varepsilon\big[h_{N-1,N}\dot{x}_{N-1} - (h_{N-1,N} + h_N)\dot{x}_N\big].
\end{aligned}$$

The characteristic determinant of the homogeneous part of this system is

$$D(\omega^2) = \begin{vmatrix} c_1 - c_{12} - m_1\omega^2 & -c_{12} & \dots \\ -c_{12} & c_{12} + c_2 + c_{23} - m_2\omega^2 & \dots \\ \vdots & \vdots & \ddots \\ 0 & 0 & \dots \end{vmatrix} . \qquad (5.2)$$

The roots of the equation $D(\omega^2) = 0$ are denoted by $\omega_1^2, \dots, \omega_N^2$.

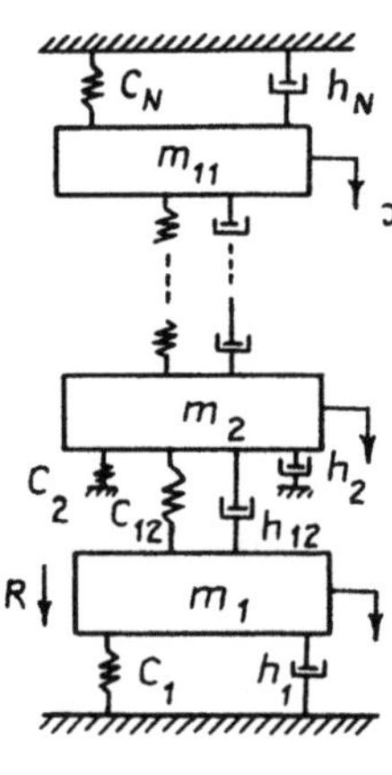


*Fig. 38.* Oscillating system with $N$ degrees of freedom.

Using the transformation into the principal mode $\xi_1, \dots, \xi_N$ (see Appendix 1)

$$x_i = \sum_{j=1}^{N} d_i^{(j)}\xi_j, \quad d_i^{(j)} = \frac{\sigma_i^{(j)}}{\sigma_1^{(j)}}, \qquad (5.3)$$

we have

$$\ddot{\xi}_i + \omega_i^2 \xi_i = \frac{\varepsilon}{M_i} \sum_{j=1}^{N} d_j^{(i)} f_j, \tag{5.4}$$

where

$$M_i = \sum_{j=1}^{N} m_j d_j^{(i)2}, \tag{5.5}$$

$d_i^{(j)}$ are the coefficients of distribution.

Introducing the amplitude and phase variables $(a_i, \psi_i)$:

$$\xi_i = a_i \sin \varphi_i, \quad \dot{\xi}_i = a_i \omega_i \cos \varphi_i, \quad \varphi_i = \omega_i t + \psi_i,$$
$$\dot{a}_i \sin \varphi_i + a_i \dot{\psi}_i \cos \varphi_i = 0 \quad (i = 1, 2, \dots, N),$$

we have

$$\begin{aligned} \omega_i \dot{a}_i &= \frac{\varepsilon \cos \varphi_i}{M_i} \sum_{j=1}^{N} d_j^{(i)} f_j, \\ \omega_i a_i \dot{\psi}_i &= \frac{-\varepsilon}{M_i} \sin \varphi_i \sum_{j=1}^{N} d_j^{(i)} f_j, \quad i = 1, 2, \dots, N \end{aligned} \tag{5.6}$$

Averaging the right-hand side of equations (5.6) over time, holding $a_i$ and $\psi_i$ constant, we obtain:

$$\begin{aligned} \dot{a}_i &= \frac{\varepsilon a_i}{2M_i} [(\ell_1 - H_i) - A_i - 2 \sum_{\substack{j=1 \\ j \neq i}}^{N} A_j], \\ a_i(\dot{\varphi}_i - \omega_i) &= 0, \quad i = 1, 2, \dots, n, \end{aligned} \tag{5.7}$$

where

$$\begin{aligned} A_i &= \frac{3}{4} \ell_3 \omega_i^2 a_i^2, \\ H_i &= \sum_{j=1}^{N} h_j d_j^{(i)2} + \sum_{\substack{k,j=1 \\ k<j}}^{N} h_{kj} (d_k^{(i)} - d_j^{(i)})^2. \end{aligned} \tag{5.8}$$

From (5.7), we find the following stationary regimes:

1. Equilibrium $a_i = 0 (i = 1, 2, \dots, N)$, which is stable if

$$\ell_1 - H_i < 0 \quad (i = 1, 2, \dots, N). \tag{5.9}$$

2. The harmonic oscillation of the single $i$ - principal mode with frequency $\omega_i$ and amplitude $a_i$, is determined by

$$A_i = \ell_1 - H_i \quad (i = 1, 2, \dots, N), \tag{5.10}$$

which is stable if

$$A_i > \frac{1}{2}(\ell_1 - H_j) = \frac{1}{2}A_j \quad (i, j = 1, 2, \ldots, N, i \neq j) \tag{5.11}$$

3. Simultaneous oscillation of some principal modes, which is always unstable.

Thus, in the system under consideration, there exist $N$ harmonic oscillations with the natural frequencies $\omega_1, \ldots, \omega_N$. Between these oscillations there is a stability condition: oscillations having amplitudes less than half of the amplitudes of the other oscillations are unstable.

By attaching a dynamic absorber, for example to the mass $m_1$, we have a system with $N+1$ degrees of freedom (Fig. 39). For a strong absorber, we obtain, by means of the same technique used in paragraph 4, the following results: besides the equilibrium and multifrequency unstable oscillations, there may be $N+1$ harmonic oscillations with natural frequencies. However, now the damping force acting on every mode is linearly dependent on the damping coefficient $\lambda$ of the absorber, and the equilibrium will be established when the damping force ($\lambda$) is sufficiently large. So, increasing the damping force is the principle of action of a strong absorber for the self-excited system with several degrees of freedom.

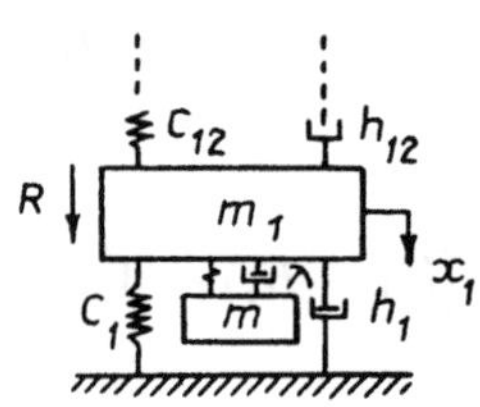


*Fig. 39.* An absorber is attached to oscillating system.

**Weak absorber for system with N degrees of freedom**

It is assumed that the mass $m$, the stiffness $k$ and the damping coefficient $\lambda$ of the absorber are small quantities of order $\varepsilon$. We have the following equations of motion [38]:

$$\begin{aligned} &m_1\ddot{x}_1 + (c_1 + c_{12})x_1 - c_{12}x_2 = \varepsilon F_1 = \varepsilon\big[\ell_1\dot{x}_1 - \ell_3\dot{x}_1^3 - \\ &- (h_1 + h_{12})\dot{x}_1 + h_{12}\dot{x}_2 - k(x_1 - z) - \lambda(\dot{x}_1 - \dot{z})\big], \\ &\cdots\cdots\cdots\cdots\cdots\cdots \\ &\ddot{z} + \lambda_0\dot{z} + \omega^2 z = \omega^2 x_1 + \lambda_0\dot{x}_1, \end{aligned} \tag{5.12}$$

where

$$\lambda_0 = \frac{\lambda}{m}, \quad \omega^2 = \frac{k}{m}.$$

Transforming the first N equations of (5.12) into the principal modes by means of the formulae (5.3), we have

$$\begin{aligned} &\ddot{\xi}_i + \omega_i^2 \xi_i = \frac{\varepsilon}{M_i} \sum_{j=1}^{N} d_j^{(i)} F_j, \\ &\ddot{z} + \lambda_0 \dot{z} + \omega^2 z = \omega^2 x_1 + \lambda_0 \dot{x}_1. \end{aligned} \tag{5.13}$$

Introduce the amplitude and phase variables $a_i$, $\varphi_i$ as follows

$$\begin{aligned} \xi_i &= a_i \sin\varphi_i, \quad \dot{\xi}_i = a_i \omega_i \cos\varphi_i, \\ z &= \sum_{i=1}^{N} a_i (P_i \sin\varphi_i + Q_i \cos\varphi_i), \\ \dot{z} &= \sum_{i=1}^{N} a_i \omega_i (P_i \cos\varphi_i - Q_i \sin\varphi_i), \end{aligned} \tag{5.14}$$

where

$$\begin{aligned} P_i &= \frac{\Delta_i \omega^2 + \lambda_0^2 \omega_i^2}{\Delta_i^2 + \lambda_0^2 \omega_i^2}, \quad \Delta_i = \omega^2 - \omega_i^2, \\ Q_i &= \frac{-\lambda_0 \omega_i^2}{\Delta_i^2 + \lambda_0^2 \omega_i^2}, \quad i = 1, 2, \ldots, N. \end{aligned} \tag{5.15}$$

Substituting the expressions (5.14) into (5.13) and then solving for the amplitude and phase variables, and averaging the results over the time, we get in the first approximation

$$\begin{aligned} \dot{a}_i &= \frac{\varepsilon a_i}{2M_i} [(\ell_1 - H_i - H_i^*) - A_i - 2 \sum_{\substack{j=1 \\ j \neq i}}^{N} A_j], \\ a_i(\dot{\varphi}_i - \omega_i) &= \frac{-\varepsilon}{2M_i} \Delta_i^*, \quad (i = 1, 2, \ldots, N), \end{aligned} \tag{5.16}$$

where

$$\begin{aligned} H_i^* &= \frac{m \lambda_0 \omega_i^4}{\Delta_i^2 + \lambda_0^2 \omega_i^2}, \\ \Delta_i^* &= m \Omega_i \frac{1 + \omega_i^2 \Delta_i^2}{\Delta_i^2 + \lambda_0^2 \omega_i^2}, \end{aligned} \tag{5.17}$$

A comparison of equations (5.16) and (5.7) shows that the dynamic absorber gives additional damping $H_i^*$ and frequency deviation $\Delta_i^*$, which are nonlinearly dependent on the damping coefficient $\lambda_0$. Thus, as in paragraph 4, the dynamic absorber works effectively with a suitable value of the damping force ($\lambda$).

## 6. Dynamic Absorber for a Beam undergoing Self-excited Oscillation

In this paragraph, a rectangular cantilever beam of length $\ell$, cross section $S$, Young's modulus $E$, moment of inertia $I$ and mass density $\rho$ is considered. The beam is subjected to a force

$$R\Big(\frac{\partial y}{\partial t}\Big) = \varepsilon\Big[h_1^0\frac{\partial y}{\partial t} - h_3^0\Big(\frac{\partial y}{\partial t}\Big)^3\Big],$$

distributed along its length as shown in Fig. 40. Another beam $(\ell, S_1, E_1, I_1, \rho_1)$ connected with the first through an elastic layer $(c, \lambda_0)$ is used as a dynamic absorber for quenching the self-excited oscillation of the first beam.

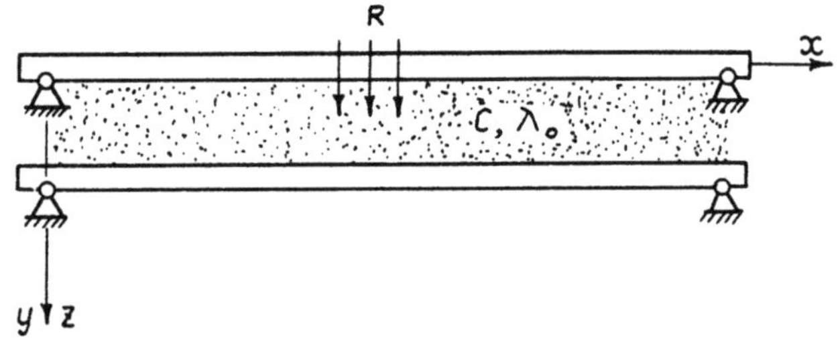


*Fig. 40.* Beam and beam absorber.

With the adoption of elementary beam theory, the equations of motion of the beams are [38,b]:

$$\begin{aligned}\rho S\frac{\partial^2 y}{\partial t^2} + EI\frac{\partial^4 y}{\partial x^4} &= \varepsilon\Big[h_1^0\frac{\partial y}{\partial t} - h_3^0\Big(\frac{\partial y}{\partial t}\Big)^3\Big] - c(y-z) - \lambda_0\Big(\frac{\partial y}{\partial t} - \frac{\partial z}{\partial t}\Big),\\ \rho_1 S_1\frac{\partial^2 z}{\partial t^2} + E_1 I_1\frac{\partial^4 z}{\partial x^4} &= -c(z-y) - \lambda_0\Big(\frac{\partial z}{\partial t} - \frac{\partial y}{\partial t}\Big),\end{aligned} \tag{6.1}$$

with boundary conditions:

$$\begin{aligned} y(t,x) = z(t,x) = 0, &\quad \text{at } x = 0,\ x = \ell,\\ \frac{\partial^2 y}{\partial x^2} = \frac{\partial^2 z}{\partial x^2} = 0, &\quad \text{at } x = 0,\ x = \ell,\end{aligned} \tag{6.2}$$

where $\varepsilon$ is a small positive parameter, characterising the smallness of the terms standing behind it.

Using the notations:

$$a^2 = \frac{EI}{\rho S},\ b^2 = \frac{E_1 I_1}{\rho_1 S_1},\ h_1 = \frac{h_1^0}{\rho S},\ h_3 = \frac{h_3^0}{\rho S},$$
$$\varsigma = \frac{\rho_1 S_1}{\rho S},\ \gamma = \frac{c}{\rho_1 S_1},\ \lambda = \frac{\lambda_0}{\rho S},$$

we have for a weak absorber ($c$, $\lambda_0$, $\rho_1 S_1$, $E_1 I_1$ are small)

$$\frac{\partial^2 y}{\partial t^2} + a^2 \frac{\partial^4 y}{\partial x^4} = \varepsilon\Big[h_1 \frac{\partial y}{\partial t} - h_3\Big(\frac{\partial y}{\partial t}\Big)^3 - \varsigma\Big(\frac{\partial^2 z}{\partial t^2} + b^2 \frac{\partial^4 z}{\partial x^4}\Big)\Big].$$
$$\frac{\partial^2 z}{\partial t^2} + b^2 \frac{\partial^4 z}{\partial x^4} + \gamma z + \lambda \frac{\partial z}{\partial t} = \gamma y + \lambda \frac{\partial y}{\partial t}. \tag{6.3}$$

When $\varepsilon = 0$, the system of equations (6.3) has a solution :

$$y = \sum_{n=1}^{\infty} \sin \frac{\pi}{\ell} nx (a_n \cos \omega_n t + b_n \sin \omega_n t),$$
$$z = \sum_{n=1}^{\infty} \sin \frac{\pi}{\ell} nx (\alpha_n \cos \omega_n t + \beta_n \sin \omega_n t), \tag{6.4}$$

where $a_n$ and $b_n$ are constants, and

$$\omega_n = a\Big(\frac{\pi}{\ell} n\Big)^2,$$

$$\alpha_n = \frac{\big[\gamma(b^2 - a^2)\big(\frac{\pi}{\ell}n\big)^4 + \gamma^2 + \lambda^2\omega_n^2\big]a_n}{\big[(b^2 - a^2)\big(\frac{\pi}{\ell}n\big)^4 + \gamma\big]^2 + a^2\big(\frac{\pi}{\ell}n\big)^4\lambda^2} +$$
$$+ \frac{(b^2 - a^2)\big(\frac{\pi}{\ell}n\big)^4\lambda\omega_n b_n}{\big[(b^2 - a^2)\big(\frac{\pi}{\ell}n\big)^4 + \gamma\big]^2 + a^2\big(\frac{\pi}{\ell}n\big)^4\lambda^2},$$
$$\beta_n = \frac{-(b^2 - a^2)\big(\frac{\pi}{\ell}n\big)^4\lambda\omega_n a_n}{\big[(b^2 - a^2)\big(\frac{\pi}{\ell}n\big)^4 + \gamma\big]^2 + a^2\big(\frac{\pi}{\ell}n\big)^4\lambda^2} +$$
$$+ \frac{\big[\gamma(b^2 - a^2)\big(\frac{\pi}{\ell}n\big)^4 + \gamma^2 + \lambda^2\omega_n^2\big]b_n}{\big[(b^2 - a^2)\big(\frac{\pi}{\ell}n\big)^4 + \gamma\big]^2 + a^2\big(\frac{\pi}{\ell}n\big)^4\lambda^2}. \tag{6.5}$$

We use now (6.4) as the formulae of transformation into the new variables $a_n$, $b_n$ with the additional relation

$$\dot{a}_n \cos \omega_n t + \dot{b}_n \sin \omega_n t = 0. \tag{6.6}$$

Substituting expressions (6.4) into (6.3) and comparing the coefficients of harmonics $\sin \frac{\pi}{\ell} nx$, we obtain equations for $\dot{a}_n$ and $\dot{b}_n$.Then combining them with (6.6), we get equations for $\dot{a}_n$, $\dot{b}_n$ in the standard form. For $\dot{a}_1$ and $\dot{b}_1$, we have

$$\omega_1 \dot{a}_1 = \varepsilon\Big\{\omega_1 h_1 a_1 - \frac{9}{16}\omega_1^3 h_3 a_1(a_1^2 + b_1^2) +$$
$$+ \varsigma\beta_1\Big[b^2\Big(\frac{\pi}{\ell}\Big)^4 - \omega_1^2\Big]\Big\} \sin^2 \omega_1 t + \ldots,$$
$$\omega_1 \dot{b}_1 = \varepsilon\Big\{\omega_1 h_1 b_1 - \frac{9}{16}\omega_1^3 h_3 b_1(a_1^2 + b_1^2) -$$
$$- \varsigma\alpha_1\Big[b^2\Big(\frac{\pi}{\ell}\Big)^4 - \omega_1^2\Big]\Big\} \cos^2 \omega_1 t + \ldots \tag{6.7}$$

Averaging the right-hand sides of (6.7) over time, we get:

$$\omega_1\dot{a}_1 = \frac{\varepsilon}{2}\Big\{\big[\omega_1 h_1 - \varsigma d_2\big(b^2\big(\frac{\pi}{\ell}\big)^4 - \omega_1^2\big)\big]a_1 + \\ + \varsigma d_1 b_1\big[b^2\big(\frac{\pi}{\ell}\big)^4 - \omega_1^2\big] - \frac{9}{16}\omega_1^3 h_3 a_1(a_1^2 + b_1^2)\Big\}, \tag{6.8}$$

$$\omega_1\dot{b}_1 = \frac{\varepsilon}{2}\Big\{ - \varsigma d_1\big[b^2\big(\frac{\pi}{\ell}\big)^4 - \omega_1^2\big]a_1 + \\ + b_1\big[\omega_1 h_1 - \varsigma d_2\big(b^2\big(\frac{\pi}{\ell}\big)^4 - \omega_1^2\big)\big] - \frac{9}{16}\omega_1^3 h_3 b_1(a_1^2 + b_1^2)\Big\},$$

where

$$d_1 = \frac{\gamma(b^2 - a^2)\big(\frac{\pi}{\ell}\big)^4 + \gamma^2 + \lambda^2\omega_1^2}{\Delta},$$
$$d_2 = \frac{(b^2 - a^2)\big(\frac{\pi}{\ell}\big)^4\lambda\omega_1}{\Delta}, \tag{6.9}$$
$$\Delta = \lambda^2 a^2\big(\frac{\pi}{\ell}\big)^4 + \big[(b^2 - a^2)\big(\frac{\pi}{\ell}\big)^4 + \gamma\big]^2.$$

The coordinates $a_1$, $b_1$, corresponding to the lowest frequency $\omega_1$, play an important role in the oscillation of the beam. If oscillations with high frequencies are neglected,the equations (6.8) will characterize the oscillation of the beam. The stationary solutions of these equations are

$$1)\ a_1 = b_1 = 0, \tag{6.10}$$
$$2)\ \frac{9}{16}\omega_1^3 h_3 A^2 = \omega_1 h_1 - \varsigma d_2\big[b^2\big(\frac{\pi}{\ell}\big)^4 - \omega_1^2\big],\ A^2 = a_1^2 + b_1^2,$$

or

$$\frac{9}{16}\omega_1^2 h_3 A^2 = h_1 - \varsigma\frac{(b^2 - a^2)^2\big(\frac{\pi}{\ell}\big)^8\lambda}{a^2\big(\frac{\pi}{\ell}\big)^4\lambda^2 + \big[(b^2 - a^2)\big(\frac{\pi}{\ell}\big)^4 + \gamma\big]^2}. \tag{6.11}$$

The trivial solution (6.10) is stable if

$$h_1 < \varsigma\frac{(b^2 - a^2)^2\big(\frac{\pi}{\ell}\big)^8\lambda}{a^2\big(\frac{\pi}{\ell}\big)^4\lambda^2 + \big[(b^2 - a^2)\big(\frac{\pi}{\ell}\big)^4 + \gamma\big]^2}. \tag{6.12}$$

The solution (6.11) is always stable with real values of $A$.

We have met formula of type (6.11) several times before, and we have seen that the minimum of $A$ occurs with an intermediate value of $\lambda$ (see Fig. 29). It is worth mentioning that when the two lowest frequencies $\omega_1$ and $\nu_1$ of two beams

$$\omega_1^2 = a^2\big(\frac{\pi}{\ell}\big)^4,\quad \nu_1^2 = b^2\big(\frac{\pi}{\ell}\big)^4 + \gamma$$

are equal, then

$$(b^2 - a^2)\left(\frac{\pi}{\ell}\right)^4 + \gamma = 0,$$

hence,the resonance curve is of the form shown by "o" in Fig. 29. In this case, the dynamic absorber is a most effective device. Generally, to quench the oscillation of the beam at a frequency $\omega_j$, it is necessary to choose the absorber with natural frequency $\nu_j$ which is close to $\omega_j$.

For a strong absorber, when $c$, $\rho_1 S_1$, $E_1 I_1$ are not small but finite, the equations of motion are of the form:

$$\begin{aligned} \rho S\frac{\partial^2 y}{\partial t^2} + EI\frac{\partial^4 y}{\partial x^4} + c(y - z) &= \varepsilon f_1, \\ \rho_1 S_1\frac{\partial^2 z}{\partial t^2} + E_1 I_1\frac{\partial^4 y}{\partial x^4} + c(z - y) &= \varepsilon f_2, \end{aligned} \tag{6.13}$$

with the boundary conditions (6.2), here

$$\begin{aligned} f_1 &= h_1^0\frac{\partial y}{\partial t} - h_3^0\left(\frac{\partial y}{\partial t}\right)^3 - \lambda_0^*\left(\frac{\partial y}{\partial t} - \frac{\partial z}{\partial t}\right), \\ f_2 &= \lambda_0^*\left(\frac{\partial y}{\partial t} - \frac{\partial z}{\partial t}\right), \quad \lambda_0 = \varepsilon\lambda_0^*. \end{aligned} \tag{6.14}$$

We shall find the solution of the degenerate equations ($\varepsilon = 0$) of equations (6.13) in the form:

$$y(t,x) = \sum_{n=1}^{\infty} Y_n(x)T_n(t), \quad z(t,x) = \sum_{n=1}^{\infty} Z_n(x)T_n(t). \tag{6.15}$$

The unknown functions $Y_n(x)$, $Z_n(x)$ satisfy the boundary condition

$$Y_n(x) = Z_n(x) = Y_n''(x) = Z_n''(x) = 0 \text{ for } x = 0 \text{ and } x = \ell. \tag{6.16}$$

and also the orthogonality relation :

$$\int_0^\ell \left[\rho S Y_n(x)Y_k(x) + \rho_1 S_1 Z_n(x)Z_k(x)\right]dx = \begin{cases} 0 & if \quad n \neq k, \\ M_k \neq 0 & if \quad n = k. \end{cases} \tag{6.17}$$

By substituting these expressions into the degenerate equations of (6.13) and comparing the terms with index $n$, we get:

$$\ddot{T}_n + \omega_n^2 T_n = 0, \tag{6.18}$$

and

$$\begin{aligned} EIY_n^{(IV)} - \rho S\omega_n^2 Y_n + c(Y_n - Z_n) &= 0, \\ E_1 I_1 Y_n^{(IV)} - \rho_1 S_1\omega_n^2 Z_n + c(Z_n - Y_n) &= 0. \end{aligned} \tag{6.19}$$

The solution of the equations (6.19), with boundary conditions (6.16), will be found in the form:

$$Y_n(x) = \frac{2}{\ell}\sum_{j=1}^{\infty} Y_{nj} \sin\frac{j\pi}{\ell}x, \quad Z_n(x) = \frac{2}{\ell}\sum_{j=1}^{\infty} Z_{nj} \sin\frac{j\pi}{\ell}x. \tag{6.20}$$

Substituting these expressions into the degenerate equations of (6.13), multiplying them with $\sin\frac{j\pi}{\ell}x dx$, and integrating on $x$ from 0 to $\ell$, we obtain the following expressions for unknown coefficients $Z_{nj}$, $Y_{nj}$:

$$\begin{aligned} &\Big[EI\Big(\frac{j\pi}{\ell}\Big)^4 + c - \rho S\omega_n^2\Big]Y_{nj} - cZ_{nj} = 0,\\ &- cY_{nj} + \Big[E_1I_1\Big(\frac{j\pi}{\ell}\Big)^4 + c - \rho_1 S_1\omega_n^2\Big]Z_{nj} = 0,\\ &j = 1, 2, \ldots. \end{aligned} \tag{6.21}$$

The nontrivial solution of (6.21) corresponds to those values of $\omega_n$ which satisfy the frequency equations:

$$\begin{vmatrix} EI\Big(\frac{j\pi}{\ell}\Big)^4 + c - \rho S\omega_n^2 & -c \\ -c & E_1I_1\Big(\frac{j\pi}{\ell}\Big)^4 + c - \rho_1 S_1\omega_n^2 \end{vmatrix} = 0, \tag{6.22}$$

$j = 1, 2, \ldots.$ Thus, we have a set of values of frequencies $\omega_1^2, \omega_2^2, \ldots, \omega_k^2, \ldots$

For quasi-linear equations (6.13), we set

$$\begin{aligned} y(t,x) &= \sum_{n=1}^{\infty} Y_n(x)T_n(t,\varepsilon),\\ z(t,x) &= \sum_{n=1}^{\infty} Z_n(x)T_n(t,\varepsilon), \end{aligned} \tag{6.23}$$

where $Y_n(x)$, $Z_n(x)$ are of the form (6.20). Substituting (6.23) into (6.13) and taking into account the equations (6.19), we get:

$$\begin{aligned} &\sum_{n=1}^{\infty} \rho S Y_n(x)\big[\ddot{T}_n + \omega_n^2 T_n\big] = \varepsilon f_1,\\ &\sum_{n=1}^{\infty} \rho_1 S_1 Z_n(x)\big[\ddot{T}_n + \omega_n^2 T_n\big] = \varepsilon f_2. \end{aligned} \tag{6.24}$$

By multiplying the first equation of (6.24) with $Y_k(x)dx$ and the second one with $Z_k(x)dx$; adding them and integrating on $x$ from 0 to $\ell$, we have

$$\ddot{T}_k + \omega_k^2 T_k = \frac{\varepsilon}{M_k}F_k, \quad k = 1, 2, \ldots, \tag{6.25}$$

where

$$F_k = \int_0^{\ell} (f_1 Y_k + f_2 Z_k)dx. \tag{6.26}$$

Equations (6.25) can be transformed into the standard form by means of the formulae :

$$T_k = A_k \cos\psi_k, \quad \dot{T}_k = -A_k\omega_k \sin\psi_k. \tag{6.27}$$

It is easy to verify the following equations for new variables $A_k$, $\psi_k$:

$$\begin{aligned} \dot{A}_k &= -\frac{\varepsilon F_k}{M_k\omega_k}\sin\psi_k, \\ \dot{\psi}_k &= \omega_k - \frac{\varepsilon F_k}{M_k A_k \omega_k}\cos\psi_k. \end{aligned} \tag{6.28}$$

In the first approximation, we can replace the right-hand sides of (6.28) by their averaged values. If only the oscillation with the first frequency $\omega_1$ is retained, then we have:

$$\begin{aligned} \dot{A}_1 &= \frac{\varepsilon}{2M_1}\left(h_1^* - \frac{3}{4}h_3^{(1)}\omega_1^2 A_1^2\right)A_1, \\ \dot{\psi}_1 &= \omega_1, \end{aligned} \tag{6.29}$$

where

$$\begin{aligned} h_1^* &= h_1^0 \int_0^{\ell} Y_1^2(x)dx - \lambda_0^* \int_0^{\ell} (Y_1 - Z_1)^2 dx, \\ h_3^{(1)} &= h_3^0 \int_0^{\ell} Y_1^{(4)}(x)dx. \end{aligned} \tag{6.30}$$

From (6.29) it follows:

1) The equilibrium $A_1 = 0$ is stable if $h_1^* < 0$. This condition is satisfied for example with sufficiently large values of the damping coefficient $\lambda$.
2) There exists an oscillatory regime with the frequency $\omega_1$ and amplitude $A_1$ determined by

$$\frac{3}{4}h_3^{(1)}\omega_1^2 A_1^2 = h_1^* > 0. \tag{6.31}$$

If only two oscillations with the first two frequencies $\omega_1$, $\omega_2$ are retained, then we have :

$$\begin{aligned} \dot{A}_1 &= \frac{\varepsilon}{2M_1}\left(h_1^* - \frac{3}{4}h_3^{(1)}\omega_1^2 A_1^2 - \frac{3}{2}h_3^*\omega_2^2 A_2^2\right)A_1, \\ \dot{A}_2 &= \frac{\varepsilon}{2M_2}\left(h_2^* - \frac{3}{2}h_3^*\omega_1^2 A_1^2 - \frac{3}{4}h_3^{(2)}\omega_2^2 A_2^2\right)A_2, \\ \dot{\psi}_1 &= \omega_1, \quad \dot{\psi}_2 = \omega_2, \end{aligned} \tag{6.32}$$

where

$$h_2^* = h_1^0 \int_0^\ell Y_2^2 dx - \lambda_0^* \int_0^\ell (Y_2 - Z_2)^2 dx,$$
$$h_3^{(2)} = h_3^0 \int_0^\ell Y_2^4 dx, \quad h_3^* = h_3^0 \int_0^\ell Y_1^2 Y_2^2 dx. \tag{6.33}$$

From the above results it follows:

1) The equilibrium $A_1 = A_2 = 0$ is stable if $h_1^* < 0$, $h_2^* < 0$, for instance, if the damping coefficient $\lambda$ is sufficiently large. In this case, following the formulae (6.23), (6.27), the oscillation of the beam does not occur.
2) There exists an oscillatory regime with frequency $\omega_1$ $(A_2 = 0)$, determined by

$$\frac{3}{4} h_3^{(1)} \omega_1^2 A_1^2 = h_1^* > 0 \quad \text{if} \quad \frac{3}{4} h_3^{(1)} \omega_1^2 A_1^2 > \frac{h_2^* h_3^{(1)}}{2 h_3^*}. \tag{6.34}$$

3) There exists an oscillatory regime with frequency $\omega_2$ $(A_1 = 0)$, determined by

$$\frac{3}{4} h_3^{(2)} \omega_2^2 A_2^2 = h_2^* > 0 \quad \text{if} \quad \frac{3}{4} h_3^{(2)} \omega_2^2 A_2^2 > \frac{h_1^* h_3^{(2)}}{2 h_3^*}. \tag{6.35}$$

4) The oscillatory regime with two frequencies $\omega_1$ and $\omega_2$ is stable if $h_3^{(1)} h_3^{(2)} > 4 h_3^{*2}$.

It is worth mentioning that the effect of the absorber will disappear if the coefficient of $\lambda$ in the expressions of $h_1^*$, $h_2^*$ vanishes, for example, if

$$\frac{EI}{\rho S} = \frac{E_1 I_1}{\rho_1 S_1}.$$

## 7. Absorber for Self-excited Oscillations of a Plate

The problem presented in the previous paragraph may be considered for a more complicated system, namely, for a plate (Fig. 41).

Self-excited oscillation of a plate $(D, \mu, u)$ under the action of a small force

$$\varepsilon \left[ h_1^0 \frac{\partial u}{\partial t} - h_3^0 \left( \frac{\partial u}{\partial t} \right)^3 \right]$$

may be quenched with the aid of a damper which consists of another plate $(D_1, \mu_1, z)$ connected with the main plate through an elastic layer $(c, \lambda_0)$. The equations of motion for this system are [38,b]:

$$D \nabla^4 u + \mu \frac{\partial^2 u}{\partial t^2} = \varepsilon \left[ h_1^0 \frac{\partial u}{\partial t} - h_3^0 \left( \frac{\partial u}{\partial t} \right)^3 \right] - c(u - z) - \lambda_0 \left( \frac{\partial u}{\partial t} - \frac{\partial z}{\partial t} \right),$$
$$D_1 \nabla^4 z + \mu_1 \frac{\partial^2 z}{\partial t^2} = -c(z - u) - \lambda_0 \left( \frac{\partial z}{\partial t} - \frac{\partial u}{\partial t} \right), \tag{7.1}$$

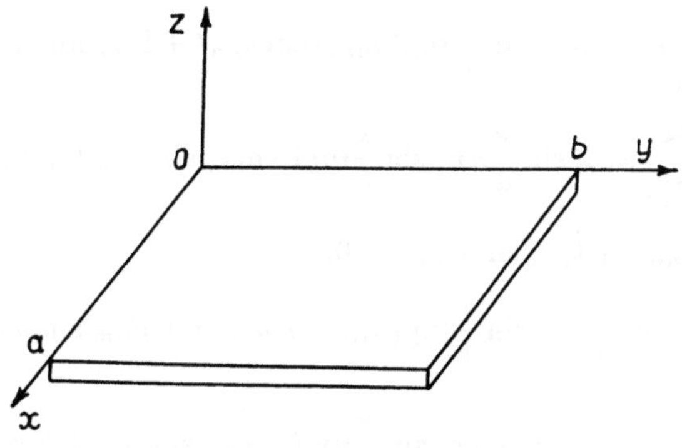


*Fig. 41.* Oscillating plate.

where

$$\nabla^4 = \frac{\partial^4}{\partial x^4} + 2\frac{\partial^4}{\partial x^2 \partial y^2} + \frac{\partial^4}{\partial y^4},$$

and $D$ and $D_1$ are the flexural rigidity of the plate, $\mu$ and $\mu_1$ are the mass density of plate materials, $u$ and $z$ are the transverse deflections of the plates; $c$ is an elastic coefficient and $\lambda_0$ - the damping coefficient of the layer between the two plates. Here, we consider rectangular plates with edge dimensions $a, b$. For hinged edges we have the following boundary conditions:

At $x = 0$, $x = a$, $y = 0$, $y = b$:

$$\begin{aligned} &u(x, y, t) = z(x, y, t) = 0, \\ &\frac{\partial^2 u}{\partial x^2} = \frac{\partial^2 u}{\partial y^2} = \frac{\partial^2 z}{\partial x^2} = \frac{\partial^2 z}{\partial y^2} = 0. \end{aligned} \tag{7.2}$$

By introducing the notation:

$$\begin{aligned} &\alpha^2 = \frac{\mu}{D}, \quad \beta^2 = \frac{\mu_1}{D_1}, \quad \sigma = \frac{D_1}{D}, \quad h_1 = \frac{h_1^0}{D}, \\ &h_3 = \frac{h_3^0}{D}, \quad k = \frac{c}{D_1}, \quad \lambda = \frac{\lambda_0}{D_1}, \end{aligned} \tag{7.3}$$

and considering the weak absorber ($c$, $\lambda_0$, $D_1$, $\mu_1$ are small), we obtain the equations

$$\begin{aligned} &\nabla^4 u + \alpha^2 \frac{\partial^2 u}{\partial t^2} = \varepsilon\left[h_1 \frac{\partial u}{\partial t} - h_3\left(\frac{\partial u}{\partial t}\right)^3\right] - \varepsilon\sigma\left(\nabla^4 z + \beta^2 \frac{\partial^2 z}{\partial t^2}\right), \\ &\nabla^4 z + \beta^2 \frac{\partial^2 z}{\partial t^2} + kz + \lambda \frac{\partial z}{\partial t} = ku + \lambda \frac{\partial u}{\partial t}. \end{aligned} \tag{7.4}$$

We shall find the solution of (7.4) with boundary conditions (7.2) in the form:

$$\begin{aligned}
&u = \sum_{n,m=1}^{\infty} \sin\frac{\pi}{a}nx \cdot \sin\frac{\pi}{b}my(a_{nm}\cos\omega_{nm}t + b_{nm}\sin\omega_{nm}t),\\
&\frac{\partial u}{\partial t} = \sum_{n,m=1}^{\infty} \omega_{nm}\sin\frac{\pi}{a}nx \cdot \sin\frac{\pi}{b}my(-a_{nm}\sin\omega_{nm}t + b_{nm}\cos\omega_{nm}t),\\
&\dot{a}_{nm}\cos\omega_{nm}t + \dot{b}_{nm}\sin\omega_{nm}t = 0, \qquad (7.5)\\
&z = \sum_{n,m=1}^{\infty} \sin\frac{\pi}{a}nx \cdot \sin\frac{\pi}{b}my(\alpha_{nm}\cos\omega_{nm}t + \beta_{nm}\sin\omega_{nm}t),\\
&\frac{\partial z}{\partial t} = \sum_{n,m=1}^{\infty} \omega_{nm}\sin\frac{\pi}{a}nx \cdot \sin\frac{\pi}{b}my(-\alpha_{nm}\sin\omega_{nm}t + \beta_{nm}\cos\omega_{nm}t),
\end{aligned}$$

where $\omega_{nm}$, $\nu_{nm}$ are natural frequencies of the plates

$$\begin{aligned}
\omega_{nm}^2 &= \frac{\left[\left(\frac{\pi}{a}n\right)^2 + \left(\frac{\pi}{b}m\right)^2\right]^2}{\alpha^2},\\
\nu_{nm}^2 &= \frac{\left[\left(\frac{\pi}{a}n\right)^2 + \left(\frac{\pi}{b}m\right)^2\right]^2 + k}{\beta^2},\\
\alpha_{nm} &= p_{nm}a_{nm} + q_{nm}b_{nm},\\
\beta_{nm} &= -q_{nm}a_{nm} + p_{nm}b_{nm}, \qquad (7.6)\\
p_{nm} &= \frac{\lambda^2\omega_{nm}^2 + k\beta^2(\nu_{nm}^2 - \omega_{nm}^2)}{\lambda^2\omega_{nm}^2 + \beta^4(\nu_{nm}^2 - \omega_{nm}^2)^2},\\
q_{nm} &= \frac{\lambda\omega_{nm}\left[\beta^2(\nu_{nm}^2 - \omega_{nm}^2) - k\right]}{\lambda^2\omega_{nm}^2 + \beta^4(\nu_{nm}^2 - \omega_{nm}^2)^2}.
\end{aligned}$$

The expressions (7.5) with constant values of $a_{nm}$, $b_{nm}$ are the solution of degenerate equations ($\varepsilon = 0$) corresponding to (7.4). Substituting (7.5) into (7.4) and comparing the coefficients of $\sin\frac{\pi}{a}nx \cdot \sin\frac{\pi}{b}my$, we get:

$$\begin{aligned}
&\alpha^2\omega_{nm}\left(-\dot{a}_{nm}\sin\omega_{nm}t + \dot{b}_{nm}\cos\omega_{nm}t\right) =\\
&= \varepsilon\Big\{h_1\omega_{nm}\left(-a_{nm}\sin\omega_{nm}t + b_{nm}\cos\omega_{nm}t\right) -\\
&- \frac{9}{16}h_3\omega_{nm}^3\left(-a_{nm}\sin\omega_{nm}t + b_{nm}\cos\omega_{nm}t\right)^3 -\\
&- \sigma\left(\nabla^4 z + \beta^2\frac{\partial^2 z}{\partial t^2}\right)\Big|_{\sin\frac{\pi}{a}nx\cdot\sin\frac{\pi}{b}my} + \ldots\Big\}. \qquad (7.7)
\end{aligned}$$

The most important oscillation of the plate is that with the lowest frequency $\omega_{11}$. Neglecting the oscillatory components with higher frequencies and solving the equations (7.5) and (7.7) with respect to $a_{11}$ and $b_{11}$, we have:

$$\begin{aligned}\omega_{11}\alpha^2\dot{a}_{11} &= \frac{\varepsilon}{2}\big[\omega_{11}h_1a_{11} - \frac{27}{64}\omega_{11}^3h_3a_{11}(a_{11}^2+b_{11}^2)+\\ &\quad +\sigma S(-q_{11}a_{11}+p_{11}b_{11})\big],\\ \omega_{11}\alpha^2\dot{b}_{11} &= \frac{\varepsilon}{2}\big[\omega_{11}h_1b_{11} - \frac{27}{64}\omega_{11}^3h_3b_{11}(a_{11}^2+b_{11}^2)-\\ &\quad -\sigma S(p_{11}a_{11}+q_{11}b_{11})\big],\end{aligned} \tag{7.8}$$

where

$$S = \Big[\Big(\frac{\pi}{a}\Big)^2+\Big(\frac{\pi}{b}\Big)^2\Big]^2 - \beta^2\omega_{11}^2. \tag{7.9}$$

From equations (7.8), we find the stationary solution:
1) $a_{11}=b_{11}=0$, which is stable if

$$h_1 < \frac{\sigma}{\omega_{11}}Sq_{11}. \tag{7.10}$$

2) $a_{11}^2+b_{11}^2 \neq 0$, determined by

$$\frac{27}{64}\omega_{11}^2h_3(a_{11}^2+b_{11}^2) = h_1 - \frac{\sigma S^2\lambda}{\beta^2(\nu_{11}^2-\omega_{11}^2)^2+\omega_{11}^2\lambda^2}, \tag{7.11}$$

which is stable if

$$h_1 > \frac{\sigma}{\omega_{11}}S.q_{11}. \tag{7.12}$$

When $\nu_{11}=\omega_{11}$, we have

$$\frac{27}{64}\omega_{11}^2h_3(a_{11}^2+b_{11}^2) = h_1 - \frac{\sigma S^2}{\omega_{11}^2\lambda}. \tag{7.13}$$

As in the previous paragraphs, formulae (7.11) and (7.13) show the effectiveness of a dynamic absorber with an intermediate value of damping coefficient $\lambda$. In general, the effectiveness of the absorber for quenching the oscillation with frequency $\omega_{nm}$ occurs in the case when the two natural frequencies $\nu_{nm}$ and $\omega_{nm}$ of the plates coincide (see curve "o", Fig. 29).

Now, let us consider the case of a strong absorber when $c$, $D_1$ and $\mu_1$ are not small, but finite. In this case, the motion of the plates is governed by the equations:

$$\begin{aligned} D\nabla^4u+\mu\frac{\partial^2u}{\partial t^2}+c(u-z) &= \varepsilon f_1,\\ D_1\nabla^4z+\mu_1\frac{\partial^2z}{\partial t^2}+c(z-u) &= \varepsilon f_2,\end{aligned} \tag{7.14}$$

here

$$f_1 = h_1^0 \frac{\partial u}{\partial t} - h_3^0 \left(\frac{\partial u}{\partial t}\right)^3 - \lambda_0^* \left(\frac{\partial u}{\partial t} - \frac{\partial z}{\partial t}\right),$$
$$f_2 = -\lambda_0^* \left(\frac{\partial z}{\partial t} - \frac{\partial u}{\partial t}\right), \quad \lambda_0 = \varepsilon \lambda_0^*, \tag{7.15}$$

with the boundary conditions (7.2).

The degenerate equations of (7.14), corresponding to $\varepsilon = 0$, have a solution of the type

$$u(t,x,y) = \sum_{n,m=1}^{\infty} U_{nm}(x,y) T_{nm}(t),$$
$$z(t,x,y) = \sum_{n,m=1}^{\infty} Z_{nm}(x,y) T_{nm}(t). \tag{7.16}$$

The unknown functions $U_{nm}$, $Z_{nm}$ satisfy the following boundary conditions at $x = 0$, $x = a$, $y = 0$, $y = b$:

$$U_{nm}(x,y) = Z_{nm}(x,y) = 0,$$
$$\frac{\partial^2 U_{nm}}{\partial x^2} = \frac{\partial^2 U_{nm}}{\partial y^2} = \frac{\partial^2 Z_{nm}}{\partial x^2} = \frac{\partial^2 Z_{nm}}{\partial y^2} = 0\,, \tag{7.17}$$

and the orthogonality relation:

$$\int_0^a \int_0^b (\mu U_{mk} U_{ij} + \mu_1 Z_{mk} Z_{ij} dx dy) =$$
$$= \begin{cases} 0 & \text{if} \quad (m-i)^2 + (k-j)^2 \neq 0, \\ M_{mk} & \text{if} \quad m = i, \ k = j. \end{cases} \tag{7.18}$$

Substituting (7.16) into the degenerate equations of (7.14), we obtain the following equations for $T_{nm}$, $U_{nm}$, $Z_{nm}$:

$$\ddot{T}_{nm} + \omega_{nm}^2 T_{nm} = 0, \tag{7.19}$$

$$D(\nabla^4 U_{nm}) + (c - \mu \omega_{nm}^2) U_{nm} - c Z_{nm} = 0,$$
$$D_1(\nabla^4 Z_{nm}) + (c - \mu_1 \omega_{nm}^2) Z_{nm} - c U_{nm} = 0. \tag{7.20}$$

The solution of equations (7.20), satisfying the boundary conditions (7.17), is found in the form:

$$U_{nm} = \frac{4}{ab} \sum_{p,q=1}^{\infty} U_{nm}^{pq} \sin \frac{p\pi}{a} x \sin \frac{q\pi}{b} y,$$
$$Z_{nm} = \frac{4}{ab} \sum_{p,q=1}^{\infty} Z_{nm}^{pq} \sin \frac{p\pi}{a} x \sin \frac{q\pi}{b} y. \tag{7.21}$$

By multiplying the equations (7.20) with $\sin\frac{p\pi}{a}x\sin\frac{q\pi}{b}ydxdy$ and integrating on $x$ from 0 to $a$ and on $y$ from 0 to $b$, we get:

$$\begin{gathered}\Big\{D\Big[\Big(\frac{p\pi}{a}\Big)^2+\Big(\frac{q\pi}{b}\Big)^2\Big]^2+c-\mu\omega_{nm}^2\Big\}U_{nm}^{pq}-cZ_{nm}^{pq}=0,\\ \Big\{D_1\Big[\Big(\frac{p\pi}{a}\Big)^2+\Big(\frac{q\pi}{b}\Big)^2\Big]^2+c-\mu_1\omega_{nm}^2\Big\}Z_{nm}^{pq}-cU_{nm}^{pq}=0.\end{gathered}\tag{7.22}$$

The nontrivial solution of these algebraic equations corresponds to the values of $\omega_{nm}$, for which

$$\begin{vmatrix} D\Big[\Big(\frac{p\pi}{a}\Big)^2+\Big(\frac{q\pi}{b}\Big)^2\Big]^2+ & \\ +c-\mu\omega_{nm}^2 & -c \\ --c & D_1\Big[\Big(\frac{p\pi}{a}\Big)^2+\Big(\frac{q\pi}{b}\Big)^2\Big]^2+ \\ & +c-\mu_1\omega_{nm}^2\end{vmatrix}=0.\tag{7.23}$$

Thus, we have a set of frequencies $\omega_{11}^2$, $\omega_{12}^2$, ... depending on the values $p, q (p, q = 1, 2, \dots)$.

We shall find the solution of the quasi-linear equations (7.14) in the form:

$$\begin{aligned} u(t,x,y)&=\sum_{n,m=1}^{\infty}U_{nm}(x,y)T_{nm}(t,\varepsilon),\\ z(t,x,y)&=\sum_{n,m=1}^{\infty}Z_{nm}(x,y)T_{nm}(t,\varepsilon).\end{aligned}\tag{7.24}$$

Substituting these expressions into (7.14), we get

$$\begin{aligned} \sum_{n,m=1}^{\infty}\mu U_{nm}(\ddot{T}_{nm}+\omega_{nm}^2T_{nm})&=\varepsilon f_1,\\ \sum_{n,m=1}^{\infty}\mu_1 Z_{nm}(\ddot{T}_{nm}+\omega_{nm}^2T_{nm})&=\varepsilon f_2.\end{aligned}\tag{7.25}$$

By multiplying the first equation of (7.25) with $U_{ij}dxdy$ and the second one with $Z_{ij}dxdy$, then adding them and integrating on $x$ from 0 to $a$ and on $y$ from 0 to $b$, we obtain:

$$\ddot{T}_{ij}+\omega_{ij}^2T_{ij}=\frac{\varepsilon}{M_{ij}}S_{ij},\tag{7.26}$$

where

$$S_{ij}=\int_0^a\int_0^b(f_1U_{ij}+f_2Z_{ij})dxdy.\tag{7.27}$$

Equations (7.26) have the same structure as equations (6.25) and we can draw some conclusions analogous to those in paragraph 6. For example, if we retain only the first shape of oscillation of the plates with lowest frequency $\omega_{11}$, then we have the following equations:

$$\ddot{T}_{11} + \omega_{11}^2 T_{11} = \frac{\varepsilon}{M_{11}} S_{11}, \tag{7.28}$$

$$\begin{aligned}
S_{11} &= H_1 \dot{T}_{11} - H_3 \dot{T}_{11}^3,\\
H_1 &= h_1^0 \int_0^a \int_0^b U_{11}^2(x,y)\,dxdy - \lambda_0^* \int_0^a \int_0^b (U_{11} - Z_{11})^2\,dxdy,\\
H_3 &= h_3^0 \int_0^a \int_0^b U_{11}^4\,dxdy,
\end{aligned} \tag{7.29}$$

and the formulae for $T_{11}$ will be

$$T_{11} = A_{11}\cos\psi_{11}, \quad \dot{T}_{11} = -A_{11}\omega_{11}\sin\psi_{11}, \tag{7.30}$$

where $A_{11}$ and $\psi_{11}$ satisfy the equations:

$$\begin{aligned}
\dot{A}_{11} &= \frac{\varepsilon A_{11}}{2M_{11}}\left(H_1 - \frac{3}{4} H_3 \omega_{11}^2 A_{11}^2\right),\\
\dot{\psi}_{11} &= \omega_{11}.
\end{aligned} \tag{7.31}$$

From the first equation of (7.31), it follows that the equilibrium $A_{11} = 0$ will be stable if $H_1 < 0$. This is the case, for example, when the damping force ($\lambda$) is sufficiently large.

## CHAPTER 3

# FORCED OSCILLATIONS

A forced oscillation is characterized by the existence of external excitation which determines the time-law of the oscillator motion. The analysis of nonlinear forced oscillations is a very difficult task. The difficulty arises mainly from the fact that the superposition principle no longer holds for nonlinear systems, so that the total solution cannot be obtained from a synthesis of the particular solutions, and the general solution cannot be found by simple superposition of the solution of the homogeneous equation and the particular solution of the inhomogeneous equation. As we shall see later, many new effects, including some of technical importance, can appear. Among these are for example, jumps in amplitude, subharmonic oscillations, combination frequencies, entrainment of frequencies and so on.

### 1. Statement of the Problem

We consider first a system with one degree of freedom described by a differential equation of the type

$$\frac{d^2x}{dt^2} + \omega^2 x = \varepsilon f\Big(\nu t, x, \frac{dx}{dt}\Big) , \tag{1.1}$$

where $\varepsilon$ is a small positive parameter. We suppose that the function $f\big(\nu t, x, \frac{dx}{dt}\big)$ is periodic with respect to $\varphi = \nu t$ with period $2\pi$ and may be expanded in a trigonometric polynomial

$$f\Big(\nu t, x, \frac{dx}{dt}\Big) = \sum_{n=-N}^{N} e^{in\nu t} f_n\Big(x, \frac{dx}{dt}\Big) , \tag{1.2}$$

where each $f_n\big(x, \frac{dx}{dt}\big)$ is a polynomial in $x$ and $\dot{x}$. For forced oscillations, it is necessary that one of the functions $f_n(0,0)$ must not vanish; because in the equations of motion of forced oscillations, there is always a time-dependent term, which is independent from motion.

Equation (1.1) can be considered as an equation representing an oscillator with unit mass and natural frequency $\omega$, acted on by a small nonlinear excitation term $\varepsilon f$ with explicit dependence on time $t$ through a trigonometric term with argument $\nu t$. Before deriving the method of approximate asymptotic solutions of equation (1.1), we analyze these solutions from a physical point of view.

When the excitation is absent in the system described by (1.1) i.e. when $\varepsilon = 0$, we have the purely harmonic oscillations

$$x = a\cos(\omega t + \varphi), \quad \frac{dx}{dt} = -a\omega\sin(\omega t + \varphi), \tag{1.3}$$

where $a$ and $\varphi$ are arbitrary constants.

If one proceeds with the general method outlined in Chapter 1, one ultimately obtains terms containing $\sin(n\nu + m\omega)t$ and $\cos(n\nu + m\omega)t$, where $n$ and $m$ are integers. Such terms will appear also in the expressions for $u_1, u_2, \ldots$. It is obvious that if one of these combination frequencies $n\nu + m\omega$ is close to the natural frequency $\omega$ of the system, then the corresponding harmonic of the exciting force has significant influence on the character of oscillation. This influence appears even in the case when the corresponding coefficient of the exciting force is small. When this coefficient is small, the detuning between the natural frequency and the exciting frequency must also be small in order that the influence is appreciable.

Thus, in nonlinear oscillatory systems, resonance phenomena are observed not only when $\nu \simeq \omega$, as in linear systems, but also when $n\nu + m\omega \simeq \omega$, that is, when

$$\nu \simeq \frac{p}{q}\omega, \tag{1.4}$$

where $p$ and $q$ are sufficiently small relatively prime integers.

We introduce the following classification:

1) $p = q = 1$, that is, $\nu \simeq \omega$; this is the fundamental (or ordinary) resonance.
2) $q = 1$, $\nu \simeq p\omega$; this is the so-called subharmonic (or parametric or a *demultiplication*) resonance.
3) $p = 1$, $\nu \simeq \frac{\omega}{q}$; this is a superharmonic resonance.

Since the ratio $\frac{p}{q}$ is a rational fraction, and the resonances occur in the neighbourhood of $\frac{p}{q}$, they appear to form a dense set everywhere. As we shall see later, in reality, this is not the case, as resonances arise only if $p$ and $q$ are sufficiently small relatively prime integers, which are fully determined by the nature of the system under consideration. Also, there are additional conditions to be fulfilled, which eliminate some of these cases.

In the process of derivation of the asymptotic solutions for the equation (1.1), one considers three separate cases: nonresonance, resonance and the general case. In the last case, the resonance zone, and also the adjacent regions will be considered. For the first case, one has the condition

$$n\nu + m\omega \neq \omega.$$

From number theory, it follows that, given any irrational ratio $\frac{\nu}{\omega}$, one can always find such integers $n$ and $m$ that the expression

$$n\nu + (m-1)\omega \tag{1.5}$$

approaches zero as closely as we please. As we do not wish that expression (1.5) should approach zero (otherwise we would be in the case of resonance which we avoid at present), it is necessary to have a rational ratio $\frac{p}{q}$ and, moreover, to impose an additional condition, which we will describe later.

## 2. Nonresonance Case

The derivation of a recursive system of approximations follows a method similar to that used in Chapter 1. However, certain differences will appear later.

If $\varepsilon = 0$, the equation (1.1) has a harmonic solution

$$x = a\cos(\omega t + \varphi)$$

with a constant amplitude $a$ and phase $\varphi$.

If $\varepsilon \neq 0$, the influence of the exciting force is that, firstly, the overtones, as well as the harmonics of combinational frequencies of different degree of smallness, appear in oscillations. So, we look for a solution of the form

$$x = a\cos\psi + \varepsilon u_1(a, \psi, \nu t) + \varepsilon^2 u_2(a, \psi, \nu t) + \dots , \tag{2.1}$$

where, the functions $u_i(a, \psi, \nu t)$ are periodic with period $2\pi$ with respect to both angular variables $\psi$ and $\nu t$. The only difference with (1.2) of Chapter 1 is that the $u_i$ depend now on $\nu t$.

Secondly, the amplitude $a$ and the phase $\varphi$ are still not constant, but are given by (1.3) of Chapter 1; namely

$$\begin{aligned} \frac{da}{dt} &= \varepsilon A_1(a) + \varepsilon^2 A_2(a) + \dots, \\ \frac{d\varphi}{dt} &= \varepsilon B_1(a) + \varepsilon^2 B_2(a) + \dots, \qquad \varphi = \psi - \omega t. \end{aligned} \tag{2.2}$$

The right-hand sides of these equations do not depend on the phase as, in the absence of resonance there is no stationary relation between the phase of the oscillation and that of the external excitation. So, the latter does not affect the amplitude of oscillations. The situation will be different, however, in the resonance case when this phase relation becomes stationary. Thus, for the nonresonance case, the matter is still similar to that discussed in 1 of Chapter 1, with the exception that the functions $u_i$ now depend on $\nu t$ in addition to $a$ and $\psi$.

Now, the problem is to determine the functions $u_i$, $A_i$ and $B_i$, so that their substitution into (1.1) satisfies this equation with any prescribed accuracy. The fundamental requirement (1.9) of Chapter 1 remains the same as before. This guarantees the absence of *small divisors* in the functions $u_i$.

By differentiation of (2.1), we have

$$\begin{aligned} \frac{dx}{dt} &= \left\{ \cos\psi + \varepsilon \frac{\partial u_1}{\partial a} + \varepsilon^2 \frac{\partial u_2}{\partial a} + \dots \right\} \frac{da}{dt} + \\ &+ \left\{ -a\sin\psi + \varepsilon \frac{\partial u_1}{\partial \psi} + \varepsilon^2 \frac{\partial u_2}{\partial \psi} + \dots \right\} \frac{d\psi}{dt} + \varepsilon \frac{\partial u_1}{\partial t} + \varepsilon^2 \frac{\partial u_2}{\partial t} + \dots , \end{aligned}$$

$$\frac{d^2x}{dt^2} = \Big\{\cos\psi + \varepsilon\frac{\partial u_1}{\partial a} + \varepsilon^2\frac{\partial u_2}{\partial a} + \dots\Big\}\frac{d^2a}{dt^2} + \Big\{-a\sin\psi + \varepsilon\frac{\partial u_1}{\partial\psi} +$$
$$+ \varepsilon^2\frac{\partial u_2}{\partial\psi} + \dots\Big\}\frac{d^2\psi}{dt^2} + \Big\{\varepsilon\frac{\partial^2 u_1}{\partial a^2} + \varepsilon^2\frac{\partial^2 u_2}{\partial a^2} + \dots\Big\}\Big(\frac{da}{dt}\Big)^2 + \tag{2.3}$$
$$+ \Big\{-a\cos\psi + \varepsilon\frac{\partial^2 u_1}{\partial\psi^2} + \varepsilon^2\frac{\partial^2 u_2}{\partial\psi^2} + \dots\Big\}\Big(\frac{d\psi}{dt}\Big)^2 +$$
$$+ 2\Big\{\varepsilon\frac{\partial^2 u_1}{\partial a\partial t} + \varepsilon^2\frac{\partial^2 u_2}{\partial a\partial t} + \dots\Big\}\frac{da}{dt} + 2\Big\{\varepsilon\frac{\partial^2 u_1}{\partial t\partial\psi} + \varepsilon^2\frac{\partial^2 u_2}{\partial t\partial\psi} + \dots\Big\}\frac{d\psi}{dt} +$$
$$+ 2\Big\{-\sin\psi + \varepsilon\frac{\partial^2 u_1}{\partial a\partial\psi} + \varepsilon^2\frac{\partial^2 u_2}{\partial a\partial\psi} + \dots\Big\}\frac{da}{dt}\frac{d\psi}{dt} + \varepsilon\frac{\partial^2 u_1}{\partial t^2} + \varepsilon^2\frac{\partial^2 u_2}{\partial t^2} + \dots$$

**The left-hand side of (1.1) after substitutions becomes:**

$$\frac{d^2x}{dt^2} + \omega^2 x = \varepsilon\Big\{\frac{\partial^2 u_1}{\partial\psi^2}\omega^2 + \frac{\partial^2 u_1}{\partial t^2} + 2\frac{\partial^2 u_1}{\partial\psi\partial t}\omega + \omega^2 u_1 - 2a\omega B_1\cos\psi -$$
$$- 2\omega A_1\sin\psi\Big\} + \varepsilon^2\Big\{\frac{\partial^2 u_2}{\partial\psi^2}\omega^2 + \frac{\partial^2 u_2}{\partial t^2} + 2\frac{\partial^2 u_2}{\partial\psi\partial t}\omega + \omega^2 u_2 - 2a\omega B_2\cos\psi -$$
$$- 2\omega A_2\sin\psi + \Big(A_1\frac{dA_1}{da} - aB_1^2\Big)\cos\psi - \Big(aA_1\frac{dB_1}{da} + 2A_1B_1\Big)\sin\psi +$$
$$+ 2\omega A_1\frac{\partial^2 u_1}{\partial a\partial\psi} + 2\omega B_1\frac{\partial^2 u_1}{\partial\psi^2} + 2\frac{\partial^2 u_1}{\partial a\partial t}A_1 + 2\frac{\partial^2 u_1}{\partial\psi\partial t}B_1\Big\} + O(\varepsilon^3). \tag{2.4}$$

**The right-hand side of (1.1) can be written as**

$$\varepsilon f\Big(\nu t, x, \frac{dx}{dt}\Big) = \varepsilon f(\nu t, a\cos\psi, -a\omega\sin\psi) +$$
$$+ \varepsilon^2\Big[f'_x(\nu t, a\cos\psi, -a\omega\sin\psi)u_1 + f'_{\dot x}(\nu t, a\cos\psi, -a\omega\sin\psi)\times$$
$$\times\Big(A_1\cos\psi - aB_1\sin\psi + \frac{\partial u_1}{\partial\psi}\omega + \frac{\partial u_1}{\partial t}\Big)\Big] + O(\varepsilon^3). \tag{2.5}$$

**In order that the series solution (2.1) should satisfy the differential equation (1.1) with accuracy $O(\varepsilon^{m+1})$, it is necessary to equate the coefficients of like powers of $\varepsilon$ up to the order $\varepsilon^m$. This recursive system of differential equation. for determination of $u_i$, $A_i$, and $B_i$ leads to:**

$$\omega^2\frac{\partial^2 u_1}{\partial\psi^2} + 2\omega\frac{\partial^2 u_1}{\partial\psi\partial t} + \frac{\partial^2 u_1}{\partial t^2} + \omega^2 u_1 =$$
$$= f_0(a, \psi, \nu t) + 2a\omega B_1\cos\psi + 2\omega A_1\sin\psi, \tag{2.6}$$
$$\omega^2\frac{\partial^2 u_2}{\partial\psi^2} + 2\omega\frac{\partial^2 u_2}{\partial\psi\partial t} + \frac{\partial^2 u_2}{\partial t^2} + \omega^2 u_2 =$$
$$= f_1(a, \psi, \nu t) + 2a\omega B_2\cos\psi + 2\omega A_2\sin\psi,$$

$$\dots\dots\dots\dots\dots\dots\dots\dots$$

$$\omega^2 \frac{\partial^2 u_m}{\partial \psi^2} + 2\omega \frac{\partial^2 u_m}{\partial \psi \partial t} + \frac{\partial^2 u_m}{\partial t^2} + \omega^2 u_m = \tag{2.7}$$
$$= f_{m-1}(a, \psi, \nu t) + 2a\omega B_m \cos\psi + 2\omega A_m \sin\psi,$$

where

$$f_0(a, \psi, \nu t) = f(\nu t, a\cos\psi, -a\omega \sin\psi),$$
$$f_1(a, \psi, \nu t) = f'_x(\nu t, a\cos\psi, -a\omega\sin\psi)u_1 +$$
$$+ f'_{\dot{x}}(\nu t, a\cos\psi, --a\omega\sin\psi)\Big[A_1\cos\psi - aB_1\sin\psi + \frac{\partial u_1}{\partial \psi}\omega + \frac{\partial u_1}{\partial t}\Big] +$$
$$+ \big(aB_1^2 - \frac{dA_1}{da}A_1\big)\cos\psi + \big(\frac{dB_1}{da}A_1 a + 2A_1B_1\big)\sin\psi -$$
$$- 2\omega B_1 \frac{\partial^2 u_1}{\partial \psi^2} - 2\frac{\partial^2 u_1}{\partial a \partial t}A_1 - 2\frac{\partial^2 u_1}{\partial \psi \partial t}B_1 - 2\frac{\partial^2 u_1}{\partial a \partial \psi}\omega A_1, \ldots$$

The functions $f_k(a, \psi, \nu t)$ are periodic with period $2\pi$ with respect to both arguments $\psi$ and $\nu t$, and also depend on $a$. An explicit expression for these functions will be known, once the functions $A_i$, $B_i$, $u_i$ are found.

For a periodic function of period $2\pi$, one can use a double Fourier series in a complex form:

$$f(x) = \sum_{n=-\infty}^{\infty} c_n e^{inx} \tag{2.8}$$

with

$$c_n = \frac{1}{2\pi}\int_0^{2\pi} f(\xi)e^{-in\xi}d\xi, \tag{2.9}$$

which yields the following relations between $c_i$, $a_i$ and $b_i$ of the usual Fourier series in real notation:

$$f(x) = \frac{a_0}{2} + \sum_{n=1}^{\infty}(a_n \cos nx + b_n \sin nx), \tag{2.10}$$

namely

$$c_n = \frac{1}{2}(a_n - ib_n), \quad c_{-n} = \frac{1}{2}(a_n + ib_n). \tag{2.11}$$

For a function $f(x, y)$ periodic with period $2\pi$ with respect to both $x$ and $y$, one can consider formally $f(x, y)$ as a function of $x$ which gives

$$f(x, y) = \sum_{n=-\infty}^{\infty} c_n(y)e^{inx} \tag{2.12}$$

with

$$c_n(y) = \frac{1}{2\pi}\int_0^{2\pi} f(\xi, y)e^{-in\xi}d\xi. \tag{2.13}$$

The coefficients $c_n(y)$, in turn, can be expanded as a complex Fourier series

$$c_n(y) = \sum_{m=-\infty}^{\infty} c_{nm} e^{imy}, \tag{2.14}$$

where

$$c_{nm} = \frac{1}{2\pi} \int_0^{2\pi} c_n(\eta) e^{-im\eta} d\eta = \frac{1}{2\pi^2} \int_0^{2\pi} \int_0^{2\pi} f(\xi, \eta) e^{-i(n\xi + m\eta)} d\xi d\eta. \tag{2.15}$$

Substituting (2.14) into (2.12), one has

$$f(x, y) = \sum_{n=-\infty}^{\infty} \sum_{m=-\infty}^{\infty} c_{nm} e^{i(nx+my)}. \tag{2.16}$$

Generally, for a periodic function $f(x_1, x_2, \dots, x_N)$ of $N$ independent variables with period $2\pi$ relative to each variable, we have

$$f(x_1, x_2, \dots, x_N) = \sum_{n_1, n_2, \dots, n_N = -\infty}^{\infty} c_{n_1 n_2 \dots n_N} e^{i(n_1 x_1 + n_2 x_2 + \dots + n_N x_N)}, \tag{2.17}$$

where

$$c_{n_1 n_2 \dots n_N} = \frac{1}{(2\pi)^N} \int_0^{2\pi} \cdots \int_0^{2\pi} f(\xi_1, \xi_2, \dots, \xi_N) \times$$
$$\times e^{-i(n_1 \xi_1 + n_2 \xi_2 + \dots + n_N \xi_N)} d\xi_1 d\xi_2 \dots d\xi_N. \tag{2.18}$$

One can now undertake the determination of $A_1(a)$, $B_1(a)$ and $u_1(a, \psi, \nu t)$ from the differential equation (2.6). For this purpose, we develop $f_0(a, \psi, \nu t)$ into a double Fourier series

$$f_0(a, \psi, \nu t) = \sum_n \sum_m f_{nm}^{(0)}(a) e^{i(n\nu t + m\psi)}, \tag{2.19}$$

where

$$f_{nm}^{(0)}(a) = \frac{1}{4\pi^2} \int_0^{2\pi} \int_0^{2\pi} f(\theta, a\cos\psi, -a\omega\sin\psi) e^{-i(n\theta + m\psi)} d\theta d\psi.$$

The unknown function $u_1(a, \psi, \nu t)$ will be found in the form of a Fourier series:

$$u_1(a, \psi, \nu t) = \sum_n \sum_m c_{nm}(a) e^{i(n\theta + m\psi)} \tag{2.20}$$

with unknown coefficients $c_{nm}$.

Substituting (2.19) and (2.20) for $f_0$ and $u_1$ into (2.6), one gets

$$\sum_n \sum_m \left[\omega^2 - (n\nu + m\omega)^2\right] c_{nm}(a) e^{i(n\nu t + m\psi)} =$$
$$= 2\omega a B_1 \cos\psi + 2\omega A_1 \sin\psi + \sum_n \sum_m f_{nm}^{(0)}(a) e^{i(n\nu t + m\psi)}. \tag{2.21}$$

It is necessary to determine $c_{nm}$, $A_1$ and $B_1$ so that $u_1$ has no resonance terms. This condition is fulfilled if one has

$$2\omega a B_1 \cos\psi + 2\omega A_1 \sin\psi = - \sum_{\substack{n,m \\ [\omega^2 - (n\nu + m\omega)^2 = 0]}} f_{nm}^{(0)}(a) e^{i(n\nu t + m\psi)}. \tag{2.22}$$

Equating the coefficients of like harmonics in (2.21), one has

$$c_{nm}(a) = \frac{f_{nm}^{(0)}(a)}{\omega^2 - (n\nu + m\omega)^2} \tag{2.23}$$

for all values of $n$ and $m$ satisfying the condition

$$\omega^2 - (n\nu + m\omega)^2 \neq 0.$$

For the nonresonance case, this condition is equivalent to

$$n^2 + (m^2 - 1)^2 \neq 0 \quad (\text{that is } n \neq 0,\ m \neq \pm 1).$$

If one substitutes (2.23) into (2.20) and sets $\nu t = \theta$, one has

$$u_1(a, \psi, \theta) = \frac{1}{4\pi^2} \sum_{\substack{n,m \\ [n^2 + (m^2 - 1)^2 \neq 0]}} \frac{e^{i(n\theta + m\psi)}}{\omega^2 - (n\nu + m\omega)^2} \times$$
$$\times \int_0^{2\pi} \int_0^{2\pi} f_0(a, \psi, \theta) e^{-i(n\theta + m\psi)} d\theta d\psi. \tag{2.24}$$

Returning to the trigonometric functions, one gets

$$u_1(a, \psi, \theta) = \frac{1}{2\pi^2} \sum_{\substack{n,m \\ [n^2 + (m^2 - 1)^2 \neq 0]}} \left\{ \frac{\cos(n\theta + m\psi)}{\omega^2 - (n\nu + m\omega)^2} \times \right.$$
$$\times \int_0^{2\pi} \int_0^{2\pi} f_0(a, \psi, \theta) \cos(n\theta + m\psi) d\theta d\psi +$$
$$\left. + \frac{\sin(n\theta + m\psi)}{\omega^2 - (n\nu + m\omega)^2} \int_0^{2\pi} \int_0^{2\pi} f_0(a, \psi, \theta) \sin(n\theta + m\psi) d\theta d\psi \right\}. \tag{2.25}$$

Equating the coefficients of like harmonics in (2.22), one has

$$A_1(a) = -\frac{1}{4\pi^2\omega}\int_0^{2\pi}\int_0^{2\pi} f_0(a,\psi,\theta)\sin\psi d\theta d\psi,$$
$$B_1(a) = -\frac{1}{4\pi^2\omega a}\int_0^{2\pi}\int_0^{2\pi} f_0(a,\psi,\theta)\cos\psi d\theta d\psi. \tag{2.26}$$

By a similar way, one can determine $u_2(a,\psi,\theta)$, $A_2(a)$ and $B_2(a)$ which are necessary for the equations of second approximations. If this is done, one has

$$u_2(a,\psi,\theta) = \frac{1}{4\pi^2} \sum_{\substack{n,m \\ [n^2+(m^2-1)^2\neq 0]}} \frac{e^{i(n\theta+m\psi)}}{\omega^2-(n\nu+m\omega)^2}\times$$
$$\times\int_0^{2\pi}\int_0^{2\pi} f_1(a,\psi,\theta)e^{-i(n\theta+m\psi)}d\theta d\psi, \tag{2.27}$$

$$A_2(a) = -\frac{1}{2\omega}\left[\frac{dB_1}{da}aA_1 + 2A_1B_1\right]-$$
$$-\frac{1}{4\pi^2\omega}\int_0^{2\pi}\int_0^{2\pi}\Big\{f'_x(\theta, a\cos\psi, -a\omega\sin\psi)u_1+$$
$$+f'_{\dot x}(\theta, a\cos\psi, -a\omega\sin\psi)\Big(A_1\cos\psi-$$
$$-aB_1\sin\psi + \frac{\partial u_1}{\partial\psi}\omega + \frac{\partial u_1}{\partial\theta}\nu\Big)\Big\}\sin\psi\, d\theta d\psi,$$
$$B_2(a) = \frac{1}{2a\omega}\left[\frac{dA_1}{da}A_1 - aB_1^2\right]- \tag{2.28}$$
$$-\frac{1}{4\pi^2\omega a}\int_0^{2\pi}\int_0^{2\pi}\Big\{f'_x(\theta, a\cos\psi, -a\omega\sin\psi)u_1+$$
$$+f'_{\dot x}(\theta, a\cos\psi, -a\omega\sin\psi)\Big(A_1\cos\psi-$$
$$-aB_1\sin\psi + \frac{\partial u_1}{\partial\psi}\omega + \frac{\partial u_1}{\partial\theta}\nu\Big)\Big\}\cos\psi\, d\theta d\psi.$$

Continuing this process, one will obtain the solution of equation (1.1) with any desired accuracy.

From (2.26), it follows that in the equations of the first approximation only the free term $f_0\left(x, \frac{dx}{dt}\right)$ of the expansion (1.2) of the exciting force $f\left(\theta, x, \frac{dx}{dt}\right)$ appears, so that in view of (1.2), one has

$$f_0\left(x,\frac{dx}{dt}\right) = \lim_{T\to\infty}\frac{1}{T}\int_0^T f\left(\tau, x, \frac{dx}{dt}\right)d\tau. \tag{2.29}$$

Therefore, for the equations of the first approximation, it is sufficient to average the exciting term $\varepsilon f$ with respect to the time appearing explicitly. One thus falls back on the determination of $A_1(a)$ and $B_1(a)$ in Chapter 1.

It follows that the investigation of the nonresonance case leads to the same procedure as for the free systems (Chapter 1), except that a multiple Fourier series is used instead of an ordinary Fourier series.

The second approximation is

$$x = a\cos\psi + \varepsilon u_1(a, \psi, \theta), \tag{2.30}$$

where $u_1(a, \psi, \theta)$ is given by (2.25). Equations (2.30) and (2.25) show that the effect of the external periodic excitation in this nonresonance case is felt only in the second approximation. This means that the combination frequencies appear only in the second approximation, and that the amplitudes of these combination harmonics are $O(\varepsilon)$.

Now consider formula (2.30) for the case of stationary oscillations:

$$a = \text{const}, \quad \psi = \omega(a)t + \theta, \quad \theta = \text{const}. \tag{2.31}$$

In this case, the oscillation of $x$ consists of a *free oscillation* with frequency $\omega(a)$, *forced oscillation* with frequencies $n\nu$ $(n = 1, 2, 3, \dots)$, and a *combination oscillation* with frequencies $n\nu \pm m\omega$ $(n, m = 1, 2, 3, \dots)$. The amplitudes of the latter increase, approaching to a resonance when the divisor $\omega^2 - (n\nu \pm m\omega)^2$ becomes small.

For a particular case when free oscillations are absent $(a = 0)$, (2.30) becomes

$$x = \varepsilon \sum_{n=1}^{\infty} \frac{A_n \cos n\theta + B_n \sin n\theta}{\omega^2 - n^2\nu^2}, \tag{2.32}$$

where

$$\begin{aligned} A_n &= \frac{1}{\pi} \int_0^{2\pi} f_0(\theta, 0, 0) \cos n\theta d\theta, \\ B_n &= \frac{1}{\pi} \int_0^{2\pi} f_0(\theta, 0, 0) \sin n\theta d\theta. \end{aligned} \tag{2.33}$$

Here, one has a purely forced oscillation with the frequencies of external excitation $n\nu$ $(n = 1, 2, 3, \dots)$. The oscillatory regime corresponding to the formula (2.32) is called heteroperiodic, because the periods of all harmonics are imposed from outside.

If the oscillatory system is such that the equivalent decrement is positive:

$$\lambda_e^*(a) > 0, \tag{2.34}$$

then $a(t) \underset{t\to\infty}{\longrightarrow} 0$. This decrement is given by the expression

$$\lambda_e^*(a) = \frac{1}{4\pi^2\omega}\int_0^{2\pi}\int_0^{2\pi} f(\theta, \cos\psi, -a\omega\sin\psi)\sin\psi d\theta d\psi =$$

$$= \frac{1}{2\pi\omega}\int_0^{2\pi} f_0(a\cos\psi, -a\omega\sin\psi)\sin\psi d\psi. \tag{2.35}$$

Therefore, if the inequality (2.34) is satisfied, then all oscillations approach the heteroperiodic one, so that the heteroperiodic regime will be the only possible stationary regime.

In general, the condition (2.34) for damping free oscillations depends on the amplitude of the external periodic force. If the external periodic excitation is absent (that is, the function $f$ in (1.1) does not depend on $t$ explicitly), one has the usual condition of self-excitation:

$$\lambda_e(a) < 0 \tag{2.36}$$

and also the condition of damping

$$\lambda_e(a) > 0, \tag{2.37}$$

where

$$\lambda_e(a) = \frac{1}{2\pi\omega}\int_0^{2\pi} f(0, a\cos\psi, -a\omega\sin\psi)\sin\psi d\psi. \tag{2.38}$$

Here, we denote the exciting force $f(\theta, x, \frac{dx}{dt})$ by $f(0, a\cos\psi, -a\omega\sin\psi)$ in which one sets $x = a\cos\psi$ , $\frac{dx}{dt} = -a\omega\sin\psi$, and the amplitudes of the external periodic excitations are zero.

Depending exclusively on the form of the nonlinear function $f(\theta, x, dx/dt)$, it may happen that the conditions (2.34) and (2.36) are satisfied simultaneously. In this case, one has a system, which, being self-excited in the absence of the external periodic excitation, loses its self-excitation as soon as the external periodic excitation is applied. This corresponds to the so-called nonresonance or asynchronous quenching. Similarly, there may be also an opposite effect of asynchronous excitation.

**Example 18**

Consider the generalized Van der Pol equation

$$\frac{d^2y}{dt^2} + y - \varepsilon(1 - y^2)\frac{dy}{dt} = E\sin\nu t, \tag{2.39}$$

where $\nu \neq p/q$; $p$, $q$ are relatively prime (nonresonance case). By setting

$$y = x + U\sin\nu t, \quad U = \frac{E}{1-\nu^2}, \tag{2.40}$$

we find that the equation (2.39) becomes

$$\frac{d^2x}{dt^2} + x = \varepsilon\big[1 - (x + U\sin\nu t)^2\big]\Big(\frac{dx}{dt} + U\nu\cos\nu t\Big). \tag{2.41}$$

Using formulae (2.26), we obtain the solution of the equation (2.41) in the first approximation as:

$$x = a\cos(t+\varphi), \quad \varphi = \text{const}, \tag{2.42}$$

where $a$ is given by the differential equation

$$\frac{da}{dt} = -\varepsilon\Big\langle[1 - (x + U\sin\nu t)^2]\Big(\frac{dx}{dt} + U\nu\cos\nu t\Big)\sin(t+\varphi)\Big\rangle \tag{2.43}$$

and $\langle \; - \; \rangle$ is an averaging operator on time. It is easy to verify that equation (2.43) has the form

$$\frac{da}{dt} = \Phi(a) = \frac{\varepsilon a}{2}\Big(1 - \frac{a^2}{4} - \frac{U^2}{2}\Big). \tag{2.44}$$

If $U^2 < 2$, then $\Phi'(0) < 0$ and the origin is unstable. The system is self-excited and there exists a stable stationary state $\Big(\frac{da}{dt} = 0\Big)$ with the amplitude

$$a^2 = 4 - 2U^2. \tag{2.45}$$

For $U^2 > 2$, $a(t) \underset{t\to\infty}{\longrightarrow} 0$, which is the phenomenon of asynchronous quenching.

Using the formulae (2.30) and (2.28), we have the solution of equation (2.41) in the second approximation:

$$\begin{aligned} x = a\cos\psi &+ \frac{\varepsilon U\nu(4 - U^2 - 2a^2)}{4(1-\nu^2)}\cos\theta + \frac{\varepsilon U^3\nu}{4(1-9\nu^2)}\cos 3\theta + \\ &+ \frac{\varepsilon U a^2(2+\nu)}{4(1+\nu)(3+\nu)}\cos(\theta + 2\psi) - \frac{\varepsilon U a^2(2-\nu)}{4(1-\nu)(3-\nu)}\cos(\theta - 2\psi) + \\ &+ \frac{\varepsilon U^2 a(1+2\nu)}{16\nu(1+\nu)}\sin(2\theta + \psi) + \frac{\varepsilon U^2 a(1-2\nu)}{16\nu(1-\nu)}\sin(2\theta - \psi) - \frac{\varepsilon a^3}{32}\sin 3\psi, \end{aligned} \tag{2.46}$$

where $a$ and $\psi$ are determined from the equations

$$\begin{aligned} \frac{da}{dt} &= \frac{\varepsilon a}{2}\Big(1 - \frac{a^2}{4} - \frac{U^2}{2}\Big), \\ \frac{d\psi}{dt} &= 1 - \varepsilon^2\Big(\frac{1}{8} - \frac{a^2}{8} + \frac{7a^4}{256}\Big) + \frac{\varepsilon^2 U^2}{8} - \\ &\quad - \frac{\varepsilon^2 U^2 a^2(\nu^4 - 14\nu^2 + 21)}{16(9-\nu^2)(1-\nu^2)} + \frac{\varepsilon^2 U^4(2\nu^2 - 3)}{64(1-\nu^2)}. \end{aligned} \tag{2.47}$$

In the second approximation, in addition to the forced oscillations with frequencies $\nu$ and $3\nu$, there are also oscillations with frequency $3\omega$ as well as the combination frequencies $\nu \pm 2\omega$; $2\nu \pm \omega$. Besides, if $U^2 < 2$, the heteroperiodic oscillation becomes unstable; if $U^2 > 2$, the heteroperiodic oscillation is the only one possible in the stationary state (since the free oscillation vanishes), so that, the following heteroperiodic oscillations will be established.

$$x = \frac{\varepsilon U\nu(4-U^2)}{4(1-\nu^2)}\cos\theta + \frac{\varepsilon U^3\nu}{4(1-9\nu^2)}\cos 3\theta. \tag{2.48}$$

## 3. Resonance Case

We now assume that

$$\omega \simeq \frac{p}{q}\nu, \tag{3.1}$$

where $p$ and $q$ are relatively prime. Two cases will be considered: 1) when (3.1) is an equality, that is, the exact resonance; and 2) when (3.1) corresponds to an approximate equality, that is, the neighbourhood of the resonance.

We begin from the first case, which is simple. Denoting by $\varepsilon\Delta$ the deviation between $\omega^2$ and $\left(\frac{p}{q}\nu\right)^2$, i.e.

$$\varepsilon\Delta = \omega^2 - \left(\frac{p}{q}\nu\right)^2, \tag{3.2}$$

we can write the basic differential equation (1.1) as

$$\frac{d^2x}{dt^2} + \left(\frac{p}{q}\nu\right)^2 x = \varepsilon\left\{f\left(\nu t, x, \frac{dx}{dt}\right) - \Delta x\right\}. \tag{3.3}$$

We again look for the solution of this equation in the form

$$x = a\cos\psi + \varepsilon u_1(a, \nu t, \psi) + \varepsilon^2 u_2(a, \nu t, \psi) + \dots, \tag{3.4}$$

where $u_i(a, \nu t, \psi)$ are periodic functions with period $2\pi$ with respect to both angular variables $\nu t$ and $\psi$; and $a$ and $\psi$ are functions of time which will be determined from the corresponding differential equation.

For convenience, it is necessary to introduce a difference of phase

$$\theta = \psi - \frac{p}{q}\nu t. \tag{3.5}$$

As was mentioned in the beginning of the previous paragraph, in the resonance case, the difference of phases between free oscillation and external excitation has significant influence on the amplitude and frequency of forced oscillation. This means that in the differential equations for $a$ and $\psi$ the terms $A_1, A_2, \ldots,$ $B_1, B_2, \ldots$ will depend not only on $a$, as previously, but also on $\theta$, so that we have

$$\begin{aligned} \frac{da}{dt} &= \varepsilon A_1(a,\theta) + \varepsilon^2 A_2(a,\theta) + \ldots, \\ \frac{d\psi}{dt} &= \frac{p}{q}\nu + \varepsilon B_1(a,\theta) + \varepsilon^2 B_2(a,\theta) + \ldots, \end{aligned} \tag{3.6}$$

where $A_i(a,\theta)$, $B_i(a,\theta)$ are periodic functions of the angular variable $\theta$ with period $2\pi$.

Inasmuch as on the right-hand side of (3.6), the phase angle $\theta$ and not the total phase $\psi$ appears, it is useful to keep the variable $\theta$ everywhere so that instead of (3.4),(3.6), we can write

$$x = a\cos\left(\frac{p}{q}\nu t + \theta\right) + \varepsilon u_1\left(a,\theta,\frac{\nu}{q}t\right) + \varepsilon^2 u_2\left(a,\theta,\frac{\nu}{q}t\right) + \ldots, \tag{3.7}$$

$$\begin{aligned} \frac{da}{dt} &= \varepsilon A_1(a,\theta) + \varepsilon^2 A_2(a,\theta) + \ldots, \\ \frac{d\theta}{dt} &= \varepsilon B_1(a,\theta) + \varepsilon^2 B_2(a,\theta) + \ldots \end{aligned} \tag{3.8}$$

Now, the problem is to determine the functions $A_i(a,\theta)$, $B_i(a,\theta)$ and $u_i\left(a,\theta,\frac{\nu}{q}t\right)$, in such a manner that their substitution into (3.3) satisfies this differential equation. Differentiating (3.7), we obtain

$$\begin{aligned} \frac{dx}{dt} = &\left\{\cos\psi + \varepsilon\frac{\partial u_1}{\partial a} + \varepsilon^2\frac{\partial u_2}{\partial a} + \ldots\right\}\frac{da}{dt} + \left\{-a\sin\psi + \varepsilon\frac{\partial u_1}{\partial \theta} + \right. \\ &\left. + \varepsilon^2\frac{\partial u_2}{\partial \theta} + \ldots\right\}\frac{d\theta}{dt} + \left\{-a\frac{p}{q}\nu\sin\psi + \varepsilon\frac{\partial u_1}{\partial t} + \varepsilon^2\frac{\partial u_2}{\partial t} + \ldots\right\}, \end{aligned} \tag{3.9}$$

$$\begin{aligned} \frac{d^2x}{dt^2} = &\left\{\cos\psi + \varepsilon\frac{\partial u_1}{\partial a} + \varepsilon^2\frac{\partial u_2}{\partial a} + \ldots\right\}\frac{d^2a}{dt^2} + \\ &+ \left\{\varepsilon\frac{\partial^2 u_1}{\partial a^2} + \varepsilon^2\frac{\partial^2 u_2}{\partial a^2} + \ldots\right\}\left(\frac{da}{dt}\right)^2 + \\ &+ 2\left\{-\sin\psi + \varepsilon\frac{\partial^2 u_1}{\partial a\partial\theta} + \varepsilon^2\frac{\partial^2 u_2}{\partial a\partial\theta} + \ldots\right\}\frac{da}{dt}\frac{d\theta}{dt} + \\ &+ 2\left\{-\frac{p}{q}\nu\sin\psi + \varepsilon\frac{\partial^2 u_1}{\partial a\partial t} + \varepsilon^2\frac{\partial^2 u_2}{\partial a\partial t} + \ldots\right\}\frac{da}{dt} + \end{aligned}$$

$$+\Big\{-a\sin\psi+\varepsilon\frac{\partial u_1}{\partial\theta}+\varepsilon^2\frac{\partial u_2}{\partial\theta}+\dots\Big\}\frac{d^2\theta}{dt^2}+$$
$$+\Big\{-a\cos\psi+\varepsilon\frac{\partial^2 u_1}{\partial\theta^2}+\varepsilon^2\frac{\partial^2 u_2}{\partial\theta^2}+\dots\Big\}\Big(\frac{d\theta}{dt}\Big)^2+$$
$$+2\Big\{-a\frac{p}{q}\nu\cos\psi+\varepsilon\frac{\partial^2 u_1}{\partial\theta\partial t}+\varepsilon^2\frac{\partial^2 u_2}{\partial\theta\partial t}+\dots\Big\}\frac{d\theta}{dt}+$$
$$+\Big\{-a\Big(\frac{p}{q}\nu\Big)^2\cos\psi+\varepsilon\frac{\partial^2 u_1}{\partial t^2}+\varepsilon^2\frac{\partial^2 u_2}{\partial t^2}+\dots\Big\}. \tag{3.10}$$

Likewise from (3.8), we have

$$\begin{aligned}\frac{d^2a}{dt^2}&=\varepsilon^2\Big\{\frac{\partial A_1}{\partial a}A_1+\frac{\partial A_1}{\partial\theta}B_1\Big\}+O(\varepsilon^3),\\ \Big(\frac{da}{dt}\Big)^2&=\varepsilon^2A_1^2+O(\varepsilon^3),\quad \Big(\frac{d\theta}{dt}\Big)^2=\varepsilon^2B_1^2+O(\varepsilon^3),\\ \frac{d^2\theta}{dt^2}&=\varepsilon^2\Big\{\frac{\partial B_1}{\partial a}A_1+\frac{\partial B_1}{\partial\theta}B_1\Big\}+O(\varepsilon^3),\\ \frac{da}{dt}\frac{d\theta}{dt}&=\varepsilon^2A_1B_1+O(\varepsilon^3).\end{aligned} \tag{3.11}$$

Substituting (3.7) and (3.10) into the left-hand side of (3.3), and taking into account (3.8) and (3.11), we obtain the development in the powers of $\varepsilon$:

$$\begin{aligned}&\frac{d^2x}{dt^2}+\Big(\frac{p}{q}\nu\Big)^2x=\\ &=\varepsilon\Big\{-2\frac{p}{q}\nu\sin\psi A_1-2a\frac{p}{q}\nu\cos\psi B_1+\frac{\partial^2u_1}{\partial t^2}+\Big(\frac{p}{q}\nu\Big)^2u_1\Big\}+\\ &+\varepsilon^2\Big\{\Big[\frac{\partial A_1}{\partial a}A_1+\frac{\partial A_1}{\partial\theta}B_1-aB_1^2-2a\frac{p}{q}\nu B_2\Big]\cos\psi+\\ &+\Big[-2A_1B_1-2\frac{p}{q}\nu A_2-a\frac{\partial B_1}{\partial a}A_1-a\frac{\partial B_1}{\partial\theta}B_1\Big]\sin\psi+\\ &+2\frac{\partial^2u_1}{\partial a\partial t}A_1+2\frac{\partial^2u_1}{\partial\theta\partial t}B_1+\frac{\partial^2u_2}{\partial t^2}+\Big(\frac{p}{q}\nu\Big)^2u_2\Big\}+O(\varepsilon^3).\end{aligned} \tag{3.12}$$

For the right-hand side of (3.3), we have

$$\begin{aligned}&\varepsilon\Big\{f\Big(\nu t,x,\frac{dx}{dt}\Big)-\Delta x\Big\}=\varepsilon\Big\{-\Delta a\cos\psi+f\Big(\nu t,a\cos\psi,-a\frac{p}{q}\nu\sin\psi\Big)\Big\}+\\ &+\varepsilon^2\Big\{f'_x\Big(\nu t,a\cos\psi,-a\frac{p}{q}\nu\sin\psi\Big)u_1+f'_{\dot x}\Big(\nu t,a\cos\psi,-a\frac{p}{q}\nu\sin\psi\Big)\times\\ &\times\Big(A_1\cos\psi-aB_1\sin\psi+\frac{\partial u_1}{\partial t}\Big)-\Delta u_1\Big\}+O(\varepsilon^3).\end{aligned} \tag{3.13}$$

Equating the coefficients of like powers of $\varepsilon$ in (3.3), one has a recursive system of the differential equation of successive approximations:

$$\frac{\partial^2 u_1}{\partial t^2} + \left(\frac{p}{q}\nu\right)^2 u_1 = f_0(a, \nu t, \psi) + 2\frac{p}{q}\nu A_1 \sin\psi + 2a\frac{p}{q}\nu B_1 \cos\psi - \Delta a \cos\psi, \tag{3.14}$$

$$\frac{\partial^2 u_2}{\partial t^2} + \left(\frac{p}{q}\nu\right)^2 u_2 = f_1(a, \nu t, \psi) + \Big[2\frac{p}{q}\nu A_2 + a\frac{\partial B_1}{\partial a}A_1 + a\frac{\partial B_1}{\partial \theta}B_1 +$$

$$+ 2A_1B_1\Big] \sin\psi + \Big[2a\frac{p}{q}\nu B_2 - \frac{\partial A_1}{\partial a}A_1 - \frac{\partial A_1}{\partial \theta}B_1 + aB_1^2\Big]\cos\psi, \tag{3.15}$$

where

$$f_0(a, \nu t, \psi) = f\big(\nu t, a\cos\psi, -a\frac{p}{q}\nu\sin\psi\big), \tag{3.16}$$

$$f_1(a, \nu t, \psi) = f'_x\big(\nu t, a\cos\psi, -a\frac{p}{q}\nu\sin\psi\big)u_1 +$$

$$+ f'_{\dot{x}}\big(\nu t, a\cos\psi, -a\frac{p}{q}\nu\sin\psi\big)\big(A_1\cos\psi - aB_1\sin\psi + \frac{\partial u_1}{\partial t}\big) -$$

$$- \Delta u_1 - 2A_1\frac{\partial^2 u_1}{\partial a \partial t} - 2B_1\frac{\partial^2 u_1}{\partial \theta \partial t} \cdot \tag{3.17}$$

As previously, $f_k(a, \nu t, \psi)$ are periodic with period $2\pi$ with respect to both angular variables $\nu t$ and $\psi$; and $A_i(a,\theta)$, $B_i(a,\theta)$ are periodic in $\theta$ with period $2\pi$.

One can now undertake the determination of $u_1(a, \nu t, \psi)$, $A_1(a,\theta)$ and $B_1(a,\theta)$ from equation (3.14), so that $u_1$ has no resonance terms. Representing $u_1$ and $f_0$ in the form of a Fourier series, we have

$$u_1\big(a, \nu t, \frac{p}{q}\nu t + \theta\big) = \sum_{n,m} u_{nm}^{(1)}(a) e^{i\left\{n\nu t + m\left(\frac{p}{q}\nu t + \theta\right)\right\}},$$

$$f_0\big(a, \nu t, \frac{p}{q}\nu t + \theta\big) = \sum_{n,m} f_{nm}^{(0)}(a) e^{i\left\{n\nu t + m\left(\frac{p}{q}\nu t + \theta\right)\right\}}, \tag{3.18}$$

$$f_{nm}^{(0)}(a) = \frac{1}{4\pi^2}\int_0^{2\pi}\int_0^{2\pi} f_0(a, \theta, \psi) e^{-i(n\theta + m\psi)}\, d\theta d\psi.$$

Substituting (3.18) into (3.14) gives

$$\sum_{n,m}\Big\{\big(\frac{p}{q}\nu\big)^2 - \big(n\nu + m\frac{p}{q}\nu\big)^2 e^{i\left\{n\nu t + m\left(\frac{p}{q}\nu t + \theta\right)\right\}} u_{nm}^{(1)}(a) =$$

$$= \sum_{n,m} f_{nm}^{(0)}(a) e^{i\left\{n\nu t + m\left(\frac{p}{q}\nu t + \theta\right)\right\}} + 2\frac{p}{q}\nu A_1 \sin\big(\frac{p}{q}\nu t + \theta\big) +$$

$$+ 2a\frac{p}{q}\nu B_1 \cos\big(\frac{p}{q}\nu t + \theta\big) - \Delta a\cos\big(\frac{p}{q}\nu t + \theta\big). \tag{3.19}$$

Identifying the terms corresponding to the same harmonics, one has

$$u_{nm}^{(1)}(a) = \frac{f_{nm}^{(0)}(a)}{\left(\frac{p}{q}\nu\right)^2 - \left(n\nu + m\frac{p}{q}\nu\right)^2} \tag{3.20}$$

for all $n$ and $m$ for which the denominator is different from zero, which is equivalent to the condition

$$nq + (m \pm 1)p \neq 0. \tag{3.21}$$

We also obtain a relation for determining $A_1(a, \theta)$ and $B_1(a, \theta)$

$$\begin{aligned} &2\frac{p}{q}\nu A_1 \sin\left(\frac{p}{q}\nu t + \theta\right) + \left(2a\frac{p}{q}\nu B_1 - \Delta a\right)\cos\left(\frac{p}{q}\nu t + \theta\right) + \\ &+ \sum_{\substack{n,m \\ [nq+(m\pm 1)p=0]}} e^{i\left\{n\nu t + m\left(\frac{p}{q}\nu t + \theta\right)\right\}} f_{nm}^{(0)}(a) = 0, \end{aligned} \tag{3.22}$$

where the double summation extends for all $m$ and $n$ for which

$$nq + (m \pm 1)p = 0. \tag{3.23}$$

Substituting (3.20) into the first equation of (3.18), we obtain the explicit expression for $u_1(a, \nu t, \psi)$:

$$u_1\left(a, \nu t, \frac{p}{q}\nu t + \theta\right) = \sum_{\substack{n,m \\ [nq+(m\pm 1)p\neq 0]}} \frac{f_{nm}^{(0)}(a) e^{i\left\{n\nu t + m\left(\frac{p}{q}\nu t + \theta\right)\right\}}}{\left(\frac{p}{q}\nu\right)^2 - \left(n\nu + m\frac{p}{q}\nu\right)^2}, \tag{3.24}$$

where in the double summation, the numbers $n$ and $m$ must be taken for all values for which $nq + (m \pm 1)p \neq 0$. Hence, in (3.24), the following complex exponential functions appear

$$\begin{aligned} &e^{i\{(n+m\frac{p}{q})\nu t + m\theta\}} = e^{i\{(nq+mp)\frac{\nu}{q}t + m\theta\}} = e^{i\{\mp\frac{p}{q}\nu t + m\theta\}} = \\ &e^{i\{\mp(\frac{p}{q}\nu t + \theta) + (m\pm 1)\theta\}} = \left\{\cos\left(\frac{p}{q}\nu t + \theta\right) \mp i \sin\left(\frac{p}{q}\nu t + \theta\right)\right\} e^{i(m\pm 1)\theta}. \end{aligned}$$

It is noted that, by virtue of (3.23), $m \pm 1$ is divisible by $q$, so that this term is designated by $\sigma q$ $(-\infty < \sigma < +\infty)$. Equating the coefficients of $\cos\psi$ and $\sin\psi$ in (3.22), one gets:

$$\begin{aligned} A_1(a, \theta) &= \frac{-q}{4\pi^2 \nu p} \sum_\sigma e^{iq\sigma\theta} \int_0^{2\pi}\int_0^{2\pi} f_0(a, \theta, \psi) e^{-iq\sigma\theta'} \sin\psi \, d\theta \, d\psi, \\ B_1(a, \theta) &= \frac{\Delta}{2}\frac{q}{p\nu} - \frac{q}{4\pi^2 a p \nu} \sum_\sigma e^{iq\sigma\theta} \int_0^{2\pi}\int_0^{2\pi} f_0(a, \theta, \psi) e^{-iq\sigma\theta'} \cos\psi \, d\theta \, d\psi, \\ &\left(\theta' = \psi - \frac{p}{q}\theta\right). \end{aligned} \tag{3.25}$$

In the formulae (3.25), the summation extends for all values of $\sigma$ for which the integrals are different from zero.

Thus, for the first approximation of the resonance case, the solution of the differential equation (1.1) is

$$x = a\cos\left(\frac{p}{q}\nu t + \theta\right), \tag{3.26}$$

where $a$ and $\theta$ are given by the differential equation

$$\frac{da}{dt} = -\frac{\varepsilon q}{4\pi^2 \nu p}\sum_{\sigma} e^{iq\sigma\theta}\int_0^{2\pi}\int_0^{2\pi} f_0(a,\theta,\psi)e^{-iq\sigma\theta'}\sin\psi d\theta d\psi,$$
$$\frac{d\theta}{dt} = \frac{\varepsilon\Delta q}{2p\nu} - \frac{\varepsilon q}{4\pi^2 a\nu p}\sum_{\sigma} e^{iq\sigma\theta}\int_0^{2\pi}\int_0^{2\pi} f_0(a,\theta,\psi)e^{-iq\sigma\theta'}\cos\psi d\theta d\psi. \tag{3.27}$$

Inasmuch as in the resonance case $\varepsilon\Delta$ is of the first order of smallness, one can write the differential equation (3.27) with the same accuracy in the form

$$\frac{da}{dt} = -\frac{\varepsilon q}{4\pi^2 \nu p}\sum_{\sigma} e^{iq\sigma\theta}\int_0^{2\pi}\int_0^{2\pi} f_0(a,\theta,\psi)e^{-iq\sigma\theta'}\sin\psi d\theta d\psi,$$
$$\frac{d\theta}{dt} = \omega - \frac{p}{q}\nu - \frac{\varepsilon q}{4\pi^2 a\nu p}\sum_{\sigma} e^{iq\sigma\theta}\int_0^{2\pi}\int_0^{2\pi} f_0(a,\theta,\psi)e^{-iq\sigma\theta'}\cos\psi d\theta d\psi. \tag{3.28}$$

Once $u_1$, $A_1$ and $B_1$ are determined, one can calculate $f_1(a,\nu t,\psi)$ and therefore $A_2(a,\theta)$ and $B_2(a,\theta)$:

$$A_2(a,\theta) = -\frac{q}{2\nu p}\left[a\frac{\partial B_1}{\partial a}A_1 + a\frac{\partial B_1}{\partial \theta}B_1 + 2A_1B_1\right] -$$
$$- \frac{q}{4\pi^2\nu p}\sum_{\sigma} e^{iq\sigma\theta}\int_0^{2\pi}\int_0^{2\pi} f_1(a,\theta,\psi)e^{-iq\sigma\theta'}\sin\psi d\theta d\psi,$$
$$B_2(a,\theta) = \frac{q}{2a\nu p}\left[\frac{\partial A_1}{\partial a}A_1 + \frac{\partial A_1}{\partial \theta}B_1 - aB_1^2\right] -$$
$$- \frac{q}{4\pi^2 a\nu p}\sum_{\sigma} e^{iq\sigma\theta}\int_0^{2\pi}\int_0^{2\pi} f_1(a,\theta,\psi)e^{-iq\sigma\theta'}\cos\psi d\theta d\psi. \tag{3.29}$$

This analysis can also be extended to the case when the deviation $\Delta$ is not too small. This allows us to investigate the behaviour of the system between the resonance zone and the nonresonance zone. This is the most general case from which both the nonresonance and the resonance ases appear as special cases. In this case, the solution of (1.1) is taken in the form

$$x = a\cos\left(\frac{p}{q}\nu t + \theta\right) + \varepsilon u_1(a, \nu t, \psi) + \varepsilon^2 u_2(a, \nu t, \psi) + \dots, \tag{3.30}$$

where $a$ and $\theta$ must be determined from the following differential equation

$$\begin{aligned} \frac{da}{dt} &= \varepsilon A_1(a, \theta) + \varepsilon^2 A_2(a, \theta) + \dots, \\ \frac{d\theta}{dt} &= \omega - \frac{p}{q}\nu + \varepsilon B_1(a, \theta) + \varepsilon^2 B_2(a, \theta) + \dots, \end{aligned} \tag{3.31}$$

where $\omega - \frac{p}{q}\nu$ is not necessarily small, $u_i(a, \nu t, \psi)$ are periodic functions with period $2\pi$ with respect to both angular variables $\psi$ and $\nu t$, and $A_i(a, \theta)$, $B_i(a, \theta)$ are periodic with period $2\pi$ with respect to the angular variable $\theta$.

For the determination of these functions, we could use the procedure of direct differentiations and substitutions into the fundamental differential equation (1.1), and subsequently equating terms with equal powers of $\varepsilon$ as was done previously. However, it is more convenient to use the method of harmonic balance.

In order to obtain the first approximation, we consider the fundamental harmonic

$$x = a\cos\psi, \quad \psi = \frac{p}{q}\nu t + \theta. \tag{3.32}$$

We replace this expression into equation (1.1), taking into account (3.31) and equating the fundamental harmonics on both sides of (1.1) after these substitutions.

For the determination of fundamental harmonics on the left-hand side of (1.1) in the second approximation, one must take into account the terms with $\varepsilon^2$ and for $f\left(\nu t, x, \frac{dx}{dt}\right)$, one must introduce the terms with $\varepsilon u_1(a, \nu t, \psi)$. Thus, in the second approximation for the fundamental harmonic of the left-hand side of (1.1), we obtain:

$$\begin{aligned} &\text{fundamental harmonics}\left\{\frac{d^2x}{dt^2} + \omega^2 x\right\} = \\ &= \Big\{\varepsilon\Big[\Big(\omega - \frac{p}{q}\nu\Big)\frac{\partial A_1}{\partial\theta} - 2\omega a B_1\Big] + \\ &+ \varepsilon^2\Big[\Big(\omega - \frac{p}{q}\nu\Big)\frac{\partial A_2}{\partial\theta} - 2a\omega B_2 + \\ &+ \frac{\partial A_1}{\partial a}A_1 + \frac{\partial A_1}{\partial\theta}B_1 - B_1^2 a\Big]\Big\}\cos\psi - \\ &- \Big\{\varepsilon\Big[\Big(\omega - \frac{p}{q}\nu\Big)a\frac{\partial B_1}{\partial\theta} + 2\omega A_1\Big] + \varepsilon^2\Big[\Big(\omega - \frac{p}{q}\nu\Big)a\frac{\partial B_2}{\partial\theta} + \\ &+ 2\omega A_2 + a\frac{\partial B_1}{\partial a}A_1 + a\frac{\partial B_1}{\partial\theta}B_1 + 2A_1B_1\Big]\Big\}\sin\psi. \end{aligned} \tag{3.33}$$

Substituting $x = a\cos\psi + \varepsilon u_1(a, \nu t, \psi)$ into the right-hand side of (1.1), we obtain the following expression for the fundamental harmonics with second order accuracy

$$\text{fundamental harmonics}\left\{\varepsilon f\left(\nu t, x, \frac{dx}{dt}\right)\right\} =$$

$$= \varepsilon \sum_{\substack{n,m \\ [nq+(m\pm1)p=0]}} f^{(0)}_{nm}(a) e^{i\{n\nu t + m(\frac{p}{q}\nu t+\theta)\}} +$$

$$\varepsilon^2 \sum_{\substack{n,m \\ [nq+(m\pm1)p=0]}} f^{(1)}_{nm}(a) e^{i\{n\nu t + m(\frac{p}{q}\nu t+\theta)\}} =$$

$$= \varepsilon\left\{\cos\psi \frac{1}{2\pi^2}\sum_\sigma e^{i\sigma q\theta}\int_0^{2\pi}\int_0^{2\pi} f_0(a,\theta,\psi)e^{-i\sigma q\theta'}\cos\psi d\theta d\psi + \right.$$

$$\left. + \sin\psi \frac{1}{2\pi^2}\sum_\sigma e^{i\sigma q\theta}\int_0^{2\pi}\int_0^{2\pi} f_0(a,\theta,\psi)e^{-i\sigma q\theta'}\sin\psi d\theta d\psi \right\} +$$

$$+ \varepsilon^2\left\{\cos\psi \frac{1}{2\pi^2}\sum_\sigma e^{i\sigma q\theta}\int_0^{2\pi}\int_0^{2\pi} f_1(a,\theta,\psi)e^{-i\sigma q\theta'}\cos\psi d\theta d\psi + \right.$$

$$\left. + \sin\psi \frac{1}{2\pi^2}\sum_\sigma e^{i\sigma q\theta}\int_0^{2\pi}\int_0^{2\pi} f_1(a,\theta,\psi)e^{-i\sigma q\theta'}\sin\psi d\theta d\psi \right\}, \tag{3.34}$$

where

$$f_0(a,\theta,\psi) = f(\theta, a\cos\psi, -a\omega\sin\psi),$$

$$f_1(a,\theta,\psi) = f'_x(\theta, a\cos\psi, -a\omega\sin\psi)u_1 +$$
$$+ f'_{\dot{x}}(\theta, a\cos\psi, -a\omega\sin\psi)\Big[A_1\cos\psi - aB_1\sin\psi +$$
$$+ \frac{\partial u_1}{\partial\theta}\left(\omega - \frac{p}{q}\nu\right)\Big] - \frac{\partial u_1}{\partial a}\left(\omega - \frac{p}{q}\nu\right)\frac{\partial A_1}{\partial\theta} - \frac{\partial u_1}{\partial\theta}\left(\omega - \frac{p}{q}\nu\right)\frac{\partial B_1}{\partial\theta} -$$
$$- 2\frac{\partial^2 u_1}{\partial\theta^2}B_1\left(\omega - \frac{p}{q}\nu\right) - 2\frac{\partial^2 u_1}{\partial a\partial t}A_1 - 2\frac{\partial^2 u_1}{\partial\theta\partial t}B_1 - 2\frac{\partial^2 u_1}{\partial a\partial\theta}\left(\omega - \frac{p}{q}\nu\right)A_1. \tag{3.35}$$

Equating the coefficients of like harmonics in (3.33) and (3.34), one obtains in the first approximation:

$$\left(\omega - \frac{p}{q}\nu\right)\frac{\partial A_1}{\partial\theta} - 2a\omega B_1 = \frac{1}{2\pi^2}\sum_\sigma e^{i\sigma q\theta}\int_0^{2\pi}\int_0^{2\pi} f_0(a,\theta,\psi)e^{-iq\sigma\theta'}\cos\psi d\theta d\psi,$$

$$\left(\omega - \frac{p}{q}\nu\right)a\frac{\partial B_1}{\partial\theta} + 2\omega A_1 = -\frac{1}{2\pi^2}\sum_\sigma e^{i\sigma q\theta}\int_0^{2\pi}\int_0^{2\pi} f_0(a,\theta,\psi)e^{-iq\sigma\theta'}\sin\psi d\theta d\psi, \tag{3.36}$$

and in the second approximation

$$\left(\omega-\frac{p}{q}\nu\right)\frac{\partial A_2}{\partial\theta}-2a\omega B_2=-\left\{\frac{\partial A_1}{\partial a}A_1+\frac{\partial A_1}{\partial\theta}B_1-aB_1^2\right\}+$$

$$+\frac{1}{2\pi^2}\sum_\sigma e^{i\sigma q\theta}\int_0^{2\pi}\int_0^{2\pi}f_1(a,\theta,\psi)e^{-iq\sigma\theta'}\cos\psi\,d\theta d\psi,$$

$$\left(\omega-\frac{p}{q}\nu\right)a\frac{\partial B_2}{\partial\theta}+2\omega A_2=-\left\{a\frac{\partial B_1}{\partial a}A_1+a\frac{\partial B_1}{\partial\theta}B_1+2A_1B_1\right\}-$$

$$-\frac{1}{2\pi^2}\sum_\sigma e^{i\sigma q\theta}\int_0^{2\pi}\int_0^{2\pi}f_1(a,\theta,\psi)e^{-iq\sigma\theta'}\sin\psi\,d\theta d\psi, \tag{3.37}$$

It is noted that the right-hand sides of the equations (3.36) and (3.37) are periodic in $\theta$ and can be represented as $\sum k_n(a)e^{in\theta}$. In view of this, for the determination of $A_i(a,\theta)$ and $B_i(a,\theta)$, one can use expressions of a similar form.

The expression for $\varepsilon u_1(a,\nu t,\psi)$ is determined as forced oscillations excited in $x$ by the action of higher harmonics of the external force $\varepsilon f\left(\nu t, x, \frac{dx}{dt}\right)$ in the regime of harmonic oscillations ($x=a\cos\psi$, $\frac{dx}{dt}=-a\omega\sin\psi$)

$$\varepsilon u_1(a,\nu t,\psi)=\sum_{\substack{n,m\\ [nq+(m\pm1)p\neq0]}}\frac{e^{i(n\nu t+m\psi)}}{\omega^2-(n\nu+m\omega)^2}f_{nm}^{(0)}(a), \tag{3.38}$$

where

$$f_{nm}^{(0)}(a)=\frac{1}{4\pi^2}\int_0^{2\pi}\int_0^{2\pi}f_0(a,\theta,\psi)e^{-i(n\theta+m\psi)}\,d\theta d\psi. \tag{3.39}$$

The formulae, found in the previous paragraphs for investigation of the resonance zone and adjacent region, may be obtained from the above formulae. For example, assuming that in the equations (3.36), the quantity $\omega-\frac{p}{q}\nu$ is small of the first order: $\omega-\frac{p}{q}\nu=\varepsilon\Delta$, we obtain the expressions $A_1(a,\theta)$ and $B_1(a,\theta)$ (3.25) for the resonance case with the same accuracy.

In summary, we have in the first approximation

$$x=a\cos\left(\frac{p}{q}\nu t+\theta\right), \tag{3.40}$$

where $a$ and $\theta$ must be determined from the following equations

$$\begin{aligned}\frac{da}{dt}&=\varepsilon A_1(a,\theta),\\ \frac{d\theta}{dt}&=\omega-\frac{p}{q}\nu+\varepsilon B_1(a,\theta),\end{aligned} \tag{3.41}$$

in which $A_1(a,\theta)$ and $B_1(a,\theta)$ are partial periodic solutions of the equations (3.36).

In the second approximation, one has

$$x = a\cos\left(\frac{p}{q}\nu t + \theta\right) + \varepsilon u_1\left(a, \nu t, \frac{p}{q}\nu t + \theta\right), \tag{3.42}$$

where $a$ and $\theta$ are determined by the equations

$$\begin{aligned} \frac{da}{dt} &= \varepsilon A_1(a,\theta) + \varepsilon^2 A_2(a,\theta), \\ \frac{d\theta}{dt} &= \omega - \frac{p}{q}\nu + \varepsilon B_1(a,\theta) + \varepsilon^2 B_2(a,\theta), \end{aligned} \tag{3.43}$$

in which $A_1(a,\theta)$, $B_1(a,\theta)$, $A_2(a,\theta)$ and $B_2(a,\theta)$ must be found from the equations (3.36) and (3.37); and $u_1(a, \nu t, \frac{p}{q}\nu t + \theta)$ by the formula (3.38).

We consider now, in more detail, the equations of the first approximation. Contrary to the nonresonance case, in the equations of the first approximation (3.41), the variables are not separate, and we have a system of two connected equations for determination of the unknowns $a$ and $\theta$. These equations generally cannot be integrated in closed form. However, the qualitative aspect of the solution can be established from the Poincaré theory, because we have two equations of first order (see Chapter 2). The basic theory shows that any solution of equations (3.41) approaches either a periodic solution, or the constant solution

$$a = a_0, \qquad \theta = \theta_0$$

determined from equation $\frac{da}{dt} = 0$, $\frac{d\theta}{dt} = 0$:

$$A_1(a_0,\theta_0) = 0, \quad \omega - \frac{p}{q}\nu + \varepsilon B_1(a_0,\theta_0) = 0. \tag{3.44}$$

This leads to two principal forms of stationary oscillations: (a) those which correspond to the constant solution (equilibrium points) and (b) oscillations corresponding to the periodic solution.

In the first case (a), for the first approximation, oscillations occur with frequency exactly equal to $\frac{p}{q}\nu$; which is in a simple rational relation with the frequency of excitation. Hence, this oscillatory regime is called the synchronous one.

For the higher-order approximations in the expression for $u_1(a, \nu t, \psi)$ (3.24), other overtones of the *divided* frequency $\frac{\nu}{q}$ are present in addition to the fundamental frequency $\frac{p}{q}\nu$.

If the system has a constant solution of type $a = 0$, which corresponds to the absence of free oscillations, the expression for $u_1(a, \nu t, \psi)$ (3.24) will be the same as in the nonresonance case (2.25). This may be regarded as a heteroperiodic solution.

We now study the stability of the stationary synchronous state $(a_0, \theta_0)$ determined by equations (3.44). Introducing a perturbation of $a$ and $\theta$, namely,

$$a = a_0 + \delta a, \quad \theta = \theta_0 + \delta\theta,$$

linearizing equations (3.41) in $\delta a$, $\delta\theta$ and taking into account the relations (3.44), we obtain the following set of first order differential equations (variational equations) with constant coefficients:

$$\begin{aligned} \frac{d\delta a}{dt} &= \varepsilon A_{1a}(a_0, \theta_0)\delta a + \varepsilon A_{1\theta}(a_0, \theta_0)\delta\theta, \\ \frac{d\delta\theta}{dt} &= \varepsilon B_{1a}(a_0, \theta_0)\delta a + \varepsilon B_{1\theta}(a_0, \theta_0)\delta\theta, \end{aligned} \tag{3.45}$$

where subscripts $a$ and $\theta$ are used to designate the partial derivatives with respect to these variables:

$$A_{1a} = \frac{\partial A_1(a,\theta)}{\partial a}, \quad A_{1\theta} = \frac{\partial A_1(a,\theta)}{\partial \theta}, \dots$$

The characteristic equation of the system (3.45) is

$$\begin{vmatrix} \varepsilon A_{1a} - \lambda & \varepsilon A_{1\theta} \\ \varepsilon B_{1a} & \varepsilon B_{1\theta} - \lambda \end{vmatrix} = 0$$

or

$$\lambda^2 - \varepsilon(A_{1a} + B_{1\theta})\lambda + \varepsilon^2(A_{1a}B_{1\theta} - B_{1a}A_{1\theta}) = 0. \tag{3.46}$$

The roots of (3.46) determine stable and unstable solutions(see Chapter 2). The conditions of stability of stationary synchronous solutions are

$$A_{1a}(a_0, \theta_0) + B_{1\theta}(a_0, \theta_0) < 0 \tag{3.47}$$

$$A_{1a}(a_0, \theta_0)B_{1\theta}(a_0, \theta_0) - A_{1\theta}(a_0, \theta_0)B_{1a}(a_0, \theta_0) > 0. \tag{3.48}$$

For the second case (b), which corresponds to the periodic solution of (3.41), in the first approximation, there are beats of two oscillations with frequencies $\omega$ and $\frac{p}{q}\nu + \Delta\omega$ and beat frequency $\Delta\omega$, where $\Delta\omega = \frac{2\pi}{T}$ ($T$ is the period of given periodic solution). These oscillations are called asynchronous ones.

## 4. External Harmonic Excitation of a Nonlinear Oscillator. Duffing's Equation. Jump Phenomenon

We consider an oscillatory system specified by the differential equation:

$$\ddot{x} + \omega^2 x = \varepsilon f(x, \dot{x}) + \varepsilon E \sin \nu t, \tag{4.1}$$

where $\omega$, $E$ and $\nu$ are constants. It is worth mentioning that the differential equation

$$m\ddot{y} + ky = \varepsilon f(y, \dot{y}) + E \sin \nu t$$

can be reduced to the type (4.1) by the change of variable $y = x + \dfrac{E \sin \nu t}{k - \nu^2 m}$.

It is assumed that between $\nu$ and $\omega$ there is a resonance relation

$$\nu^2 = \omega^2 + \epsilon\Delta,$$

where $\Delta$ is a detuning parameter. We look for a solution of the equation (4.1) in the form:

$$\begin{aligned} x &= a\cos(\nu t + \theta), \\ \dot{x} &= -a\nu \sin(\nu t + \theta), \end{aligned} \tag{4.2}$$

where $a$ and $\theta$ are unknown functions of time, which satisfy the additional condition

$$\dot{a}\cos\varphi - a\dot{\theta}\sin\varphi = 0, \quad \varphi = \nu t + \theta. \tag{4.3}$$

Substituting (4.2) into (4.1) gives

$$-\nu\dot{a}\sin\varphi - a\nu\dot{\theta}\cos\varphi = \varepsilon[\Delta x + f(x, \dot{x}) + E\sin\nu t]. \tag{4.4}$$

Solving (4.3) and (4.4) with respect to the derivatives $\dot{a}$ and $\dot{\theta}$, we obtain

$$\begin{aligned} \nu\dot{a} &= -\varepsilon\big[\Delta x + f(x, \dot{x}) + E\sin\nu t\big]\sin\varphi, \\ \nu a\dot{\theta} &= -\varepsilon\big[\Delta x + f(x, \dot{x}) + E\sin\nu t\big]\cos\varphi. \end{aligned} \tag{4.5}$$

Equations (4.5) are fully equivalent to (4.1), because only a change of variables was made. The approximation, which is now to be introduced, consists of averaging the right sides of the equations (4.5) over $2\pi$ in $\varphi$. In the process of averaging, the amplitude $a$ and phase $\theta$ will be considered as constants, and the relations $x = a\cos\varphi$, $\dot{x} = -a\nu\sin\varphi$, $\nu t = \varphi - \theta$, $\sin\nu t = \sin\varphi\cos\theta - \cos\varphi\sin\theta$ should be taken into account. If this is done, we get in the first approximation the following averaged equations

$$\begin{aligned} \dot{a} &= -\varepsilon\Big[a\Phi(\nu, a) + \frac{E}{2\nu}\cos\theta\Big], \\ \dot{\theta} &= -\varepsilon\Big[\frac{\Delta}{2\nu} + \Psi(\nu, a) - \frac{E}{2\nu a}\sin\theta\Big], \end{aligned} \tag{4.6}$$

where

$$\Phi(\nu,a) = \frac{1}{\nu a}\langle f(x,\dot{x})\sin\varphi\rangle = \frac{1}{2\pi\nu a}\int_0^{2\pi} f(a\cos\varphi, -a\nu\sin\varphi)\sin\varphi d\varphi,$$
$$\Psi(\nu,a) = \frac{1}{\nu a}\langle f(x,\dot{x})\cos\varphi\rangle = \frac{1}{2\pi\nu a}\int_0^{2\pi} f(a\cos\varphi, -a\nu\sin\varphi)\cos\varphi d\varphi, \tag{4.7}$$

and $\langle g\rangle$ denotes an averaging operator on time of the function $g$.

The stationary solutions $a_0$, $\theta_0$ are given by the equations: $\dot{a} = \dot{\theta} = 0$ or

$$a_0\Phi(\nu,a_0) + \frac{E}{2\nu}\cos\theta_0 = 0,$$
$$a_0\Big[\frac{\Delta}{2\nu} + \Psi(\nu,a_0)\Big] - \frac{E}{2\nu}\sin\theta_0 = 0. \tag{4.8}$$

Eliminating $\theta_0$ from these equations, one has:

$$W(\nu,a_0) = 0, \tag{4.9}$$

where

$$W(\nu,a_0) = a_0^2\Big\{\Phi^2(\nu,a_0) + \big[\frac{\Delta}{2\nu} + \Psi(\nu,a_0)\big]^2\Big\} - \frac{E^2}{4\nu^2}\cdot \tag{4.10}$$

To study the stability of stationary solutions, one substitutes the expressions $a = a_0 + \delta a$ and $\theta = \theta_0 + \delta\theta$ into the equations (4.6), then expands them in a Taylor's series and uses the equations (4.8), to obtain

$$\frac{d(\delta a)}{dt} = -\varepsilon\frac{\partial}{\partial a_0}[a_0\Phi(\nu,a_0)]\delta a + \varepsilon a_0\big[\frac{\Delta}{2\nu} + \Psi(\nu,a_0)\big]\delta\theta,$$
$$\frac{d(\delta\theta)}{dt} = -\varepsilon\Big\{\frac{\partial}{\partial a_0}\Psi(\nu,a_0) + \frac{1}{a_0}\big[\frac{\Delta}{2\nu} + \Psi(\nu,a_0)\big]\Big\}\delta_a - \varepsilon\Phi(\nu,a_0).\delta\theta\}. \tag{4.11}$$

These differential equations are linear with respect to the two unknown functions $\delta a$ and $\delta\theta$, and have constant coefficients. One would seek a solution of (4.11) in the form

$$\delta a = M_1 e^{\lambda t}, \quad \delta\theta = M_2 e^{\lambda t}, \tag{4.12}$$

where $M_1$ and $M_2$ are constants which are not simultaneously zero. Substituting (4.12) into (4.11), we obtain two homogeneous algebraic equations for $M_1$, $M_2$:

$$-\Big\{\lambda + \varepsilon\frac{\partial}{\partial a_0}[a_0\Phi(\nu,a_0)]\Big\}M_1 + \varepsilon a_0\big[\frac{\Delta}{2\nu} + \Psi(\nu,a_0)\big]M_2 = 0,$$
$$-\varepsilon\Big\{\frac{\partial}{\partial a_0}\Psi(\nu,a_0) -- \frac{1}{a_0}\big[\frac{\Delta}{2\nu} + \Psi(\nu,a_0)\big]\Big\}M_1 - [\lambda + \varepsilon\Phi(\nu,a_0)]M_2 = 0,$$

with the determinant

$$\Delta(\lambda) = \lambda^2 + \varepsilon P(\nu, a_0)\lambda + \varepsilon^2 Q(\nu, a_0),$$

where

$$\begin{aligned} P(a, \nu_0) &= \Phi(\nu, a_0) + \frac{\partial}{\partial a_0}\left[a_0 \Phi(\nu, a_0)\right], \\ Q(\nu, a_0) &= \Phi(\nu, a_0) \cdot \frac{\partial}{\partial a_0}\left[a_0 \Phi(\nu, a_0)\right] + \\ &+ a_0\left[\frac{\Delta}{2\nu} + \Phi(\nu, a_0)\right]\left\{\frac{\partial}{\partial a_0}\Psi(\nu, a_0) + \frac{1}{a_0}\left[\frac{\Delta}{2\nu} + \Psi(\nu, a_0)\right]\right\}. \end{aligned} \tag{4.13}$$

The constants $M_1$ and $M_2$ will be zero simultaneously unless the determinant $\Delta(\lambda)$ vanishes. Hence, we obtain a quadratic equation for determination of $\lambda$:

$$\lambda^2 + \varepsilon P(\nu, a_0)\lambda + \varepsilon^2 Q(\nu, a_0) = 0. \tag{4.14}$$

If none of the $\lambda$'s has a positive real part, the solution $a_0$, $\theta_0$ of (4.6) is asymptotically stable. If the real part of any $\lambda$ is positive, the solution is unstable. As we have seen in Chapter 2, the conditions of stability are:

$$P(\nu, a_0) > 0, \tag{4.15}$$

$$Q(\nu, a_0) > 0. \tag{4.16}$$

The dependence of the amplitude of forced oscillation $a_0$ on the excitation frequency $\nu$ will be represented graphically in the plane $(\nu, a)$ by means of the resonance curve $W(\nu, a_0) = 0$, where $W(\nu, a_0)$ is determined by the formula (4.10). The stability condition $Q(\nu, a_0) > 0$ (4.16) is connected with this resonance curve. In fact, from (4.10) and (4.13), we have

$$\begin{aligned} \frac{\partial W(\nu, a_0)}{\partial a_0} &= 2a_0\left\{\Phi^2(\nu, a_0) + \left[\frac{\Delta}{2\nu} + \Psi(\nu, a_0)\right]^2\right\} + \\ &+ 2a_0^2\left\{\Phi(\nu, a_0) \cdot \frac{\partial \Phi(\nu, a_0)}{\partial a_0} + \left[\frac{\Delta}{2\nu} + \Psi(\nu, a_0)\right]\frac{\partial \Psi}{\partial a_0}\right\}. \end{aligned}$$

Hence,

$$Q(\nu, a_0) = \frac{1}{2a_0} \cdot \frac{\partial W(\nu, a_0)}{\partial a_0}$$

and the stability condition (4.16) is equivalent to

$$\frac{\partial W(\nu, a_0)}{\partial a_0} > 0. \tag{4.17}$$

In the plane $(\nu, a_0)$, the resonance curve separates the zone in which the function $W(\nu, a_0)$ is negative from that where $W(\nu, a_0)$ is positive (Fig. 42). Moving upwards along a straight line parallel to the ordinate axis which cuts the resonance curve, one goes from the first zone ($W < 0$) to the second one ($W > 0$). At the intersection point of the straight line with the resonance curve the derivative $\partial W/\partial a_0$ is positive. If one crosses the resonance curve from $w >$ to $w < 0$, then $\partial W/\partial a_0$ is negative. The inequality (4.17) shows that the stability condition (4.16) is satisfied in the first case. The change of the sign of $Q$ happens only with vanishing $\dfrac{\partial W}{\partial a_0}$, and since

$$\frac{\partial W}{\partial \nu} + \frac{\partial W}{\partial a_0} \cdot \frac{\partial a_0}{\partial \nu} = 0,$$

this change takes place either at the points of vertical tangents with the resonance curve $\left(\dfrac{\partial W}{\partial \nu} \neq 0, \dfrac{\partial W}{\partial a_0} = 0, \dfrac{\partial a}{\partial \nu} = \infty\right)$ or at the singular points $\left(\dfrac{\partial W}{\partial \nu} = \dfrac{\partial W}{\partial a_0} = 0\right)$.

The arguments presented are usually called the investigation of the stability. However, the averaged equations are only approximate, and we consider the stability not of the original process, but of some solution of approximate equations. Nevertheless, those arguments may be very useful for studying the tendency of the change of $x$ in some finite (large enough) interval of time. Bogoliubov and his collaborators have shown that in certain cases, the stability of solutions of averaged equations defines the stability of the solutions of the original equations.

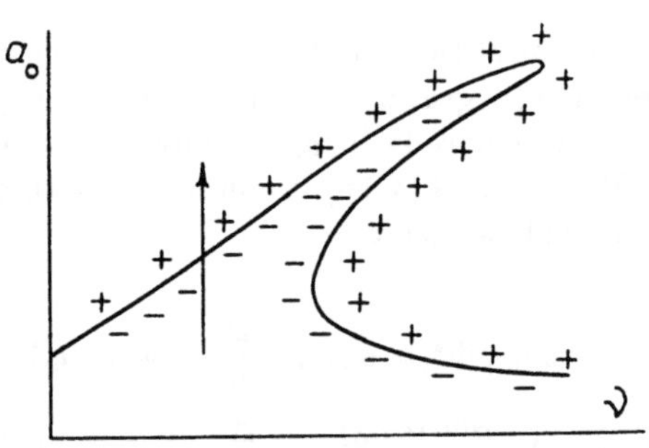


*Fig. 42.* Resonance curve.

**Example 19. Duffing's equation without damping**

Let us consider the Duffing's equation

$$\ddot{x} + \omega^2 x + \varepsilon\beta x^3 = \varepsilon E \sin \nu t, \tag{4.18}$$

where $\beta$ is a constant. The function $f(x, \dot{x})$ is now

$$f(x, \dot{x}) = -\beta x^3.$$

The expressions (4.7) take the form

$$\Phi(\nu, a) = -\frac{\beta a^2}{\nu}\langle\cos^3\varphi\sin\varphi\rangle = 0,$$

$$\Psi(\nu, a) = -\frac{\beta a^2}{\nu}\langle\cos^4\varphi\rangle = -\frac{3}{8\nu}\beta a^3.$$

So, in the first approximation, one has the following averaged equations (4.6):

$$\dot{a} = -\frac{\varepsilon}{2\nu}E\cos\theta, \qquad \dot{\theta} = -\frac{\varepsilon}{2\nu}\Big(\Delta - \frac{3}{4}\beta a^2 - \frac{E}{a}\sin\theta\Big). \tag{4.19}$$

The stationary solution of these equations $a_0$, $\theta_0$ satisfies the relations

$$\cos\theta_0 = 0, \qquad \Delta - \frac{3}{4}\beta a_0^2 - \frac{E}{a}\sin\theta_0 = 0. \tag{4.20}$$

It then follows from (4.20) that $\cos\theta_0$ must vanish, so that $\theta_0 = \pm\frac{\pi}{2}$. Hence, we obtain the following relationship between the frequency $\nu$ and the amplitude $a_0$:

$$\nu^2 = \omega^2 + \frac{3}{4}\varepsilon\beta a_0^2 \pm \varepsilon\frac{E}{a_0}\cdot \tag{4.21}$$

From (4.21), we deduce that there are two values of $\nu^2$, symmetric with respect to the skeleton line

$$\nu^2 = \omega^2 + \frac{3}{4}\varepsilon\beta a_0^2,$$

and that these values become $\pm\infty$ when $a_0 \to 0$. These indications allow us to predict resonance curves of the types illustrated in Figures 43 and 44. We have plotted $a_0$ against $\nu^2$ for the case $\beta > 0$ (Fig. 43) and for $\beta < 0$ (Fig. 44). These cases are not strictly similar, because the resonance curve which is bent to the left ($\beta < 0$, Fig. 44) has three intersection points with the axis $a_0$, whereas the one which is bent to the right ($\beta > 0$) has only one intersection point (Fig. 43). It is interesting to compare the resonance curves of Figs. 43 and 44 with the one obtained in the linear case where $\beta = 0$ (Fig. 45)

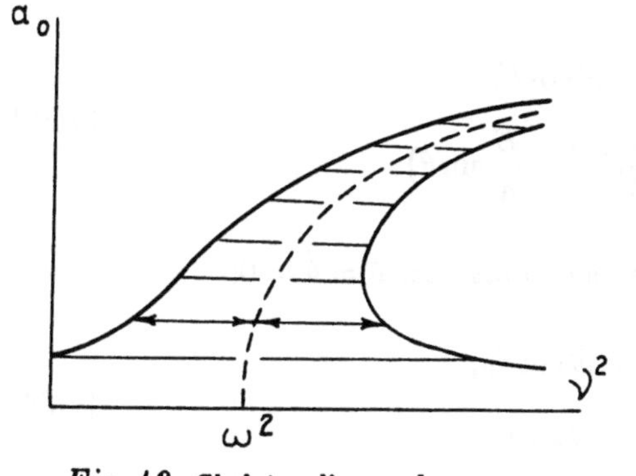


*Fig. 43.* Skeleton line and resonance curve for $\beta > 0$, $h = 0$.

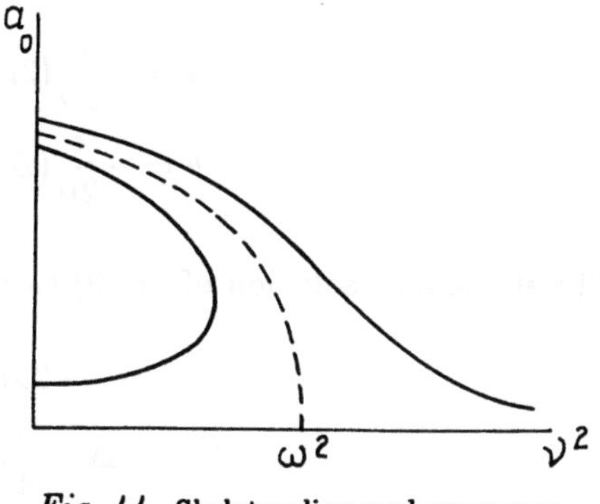


*Fig. 44.* Skeleton line and resonance curve for $\beta < 0$, $h = 0$.

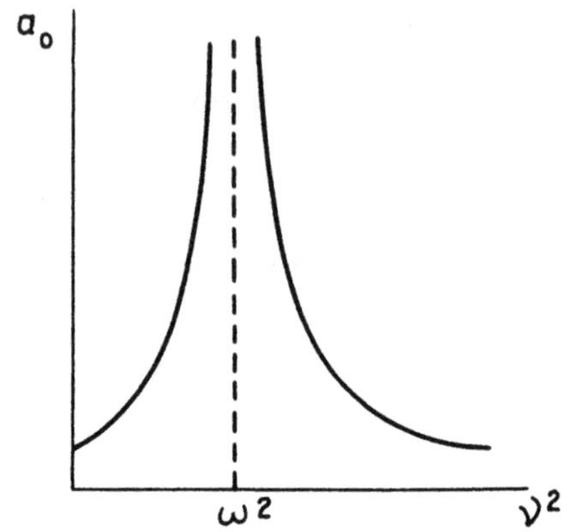


*Fig. 45.* Resonance curve for $\beta = 0$, $h = 0$.

**Example 20. Duffing equation with damping**

Consider the Duffing equation of the type

$$\ddot{x} + \omega^2 x + \varepsilon(2h\dot{x} + \beta x^3) = \varepsilon E \sin \nu t, \tag{4.22}$$

where $h$ and $\beta$ are constants and $h > 0$. Comparing with (4.1), we have

$$f(x, \dot{x}) = -2h\dot{x} - \beta x^3.$$

Hence, the expressions (4.7) will be

$$\Phi(\nu, a) = \frac{1}{\nu a}\langle [2ha\nu \sin\varphi - \beta a^3 \cos^3\varphi] \sin\varphi \rangle = h,$$
$$\Psi(\nu, a) = \frac{1}{\nu a}\langle [2ha\nu \sin\varphi - \beta a^3 \cos^3\varphi] \cos\varphi \rangle = -\frac{3\beta}{8\nu} a^2.$$

The averaged equations (4.6) now take the form:

$$\begin{aligned} \dot{a} &= -\frac{\varepsilon}{2\nu}(2\nu h a + E\cos\theta), \\ \dot{\theta} &= -\frac{\varepsilon}{2\nu}\Big(\Delta - \frac{3}{4}\beta a^2 - \frac{E}{a}\sin\theta\Big). \end{aligned} \tag{4.23}$$

The stationary solution of (4.19) satisfies the equations: $\dot{a} = \dot{\theta} = 0$:

$$\begin{aligned} -2\nu h a_0 &= E\cos\theta_0, \\ \Delta - \frac{3}{4}\beta a_0^2 &= E\sin\theta_0. \end{aligned} \tag{4.24}$$

From here, it follows that:

$$\cot\theta_0 = \frac{2\varepsilon\nu h a_0}{\frac{3}{4}\varepsilon\beta a_0^2 + \omega^2 - \nu^2}. \tag{4.25}$$

Eliminating the phase $\theta_0$ gives

$$W(\nu, a_0) = 0, \tag{4.26}$$

where

$$W(\nu, a_0) = a_0^2\Big[4\nu^2 h^2 + \Big(\frac{\nu^2 - \omega^2}{\varepsilon} - \frac{3}{4}\beta a_0^2\Big)^2\Big] - E^2. \tag{4.27}$$

Equation (4.26) is the relation between the amplitude $a_0$ and the frequency $\nu$. The phase lag $\theta_0$ between the exciting force and the oscillation is given by (4.25).

Equation (4.26) is quadratic with respect to $\nu^2$. Solving this equation for $\nu^2$, we obtain for the first approximation:

$$\nu^2 = \omega^2 + \frac{3}{4}\varepsilon\beta a_0^2 \pm \varepsilon\sqrt{\frac{E^2}{a_0^2} - 4h^2\omega^2}\,. \tag{4.28}$$

Equation (4.28) gives two values of $\nu^2$ for every value of $a_0$. In the plane $(a_0, \nu^2)$, the resonance curve runs on both sides of the skeleton line

$$\nu^2 = \omega^2 + \frac{3}{4}\varepsilon\beta a_0^2. \tag{4.29}$$

The horizontal distances from this line to the resonance curve are identical and equal to:

$$\delta(a) = \varepsilon\sqrt{\frac{E^2}{a_0^2} - 4h^2\omega^2}\,.$$

When the expression under the radical vanishes, the resonance curve cuts the skeleton curve and has a horizontal tangent.

In Figs 46 - 48, we have plotted $a_0$ against $\nu^2$ for the three cases corresponding, respectively, to $\beta > 0$, $\beta < 0$ and $\beta = 0$. These curves should be compared with the ones of Figs 43 - 45, sketched for $h = 0$.

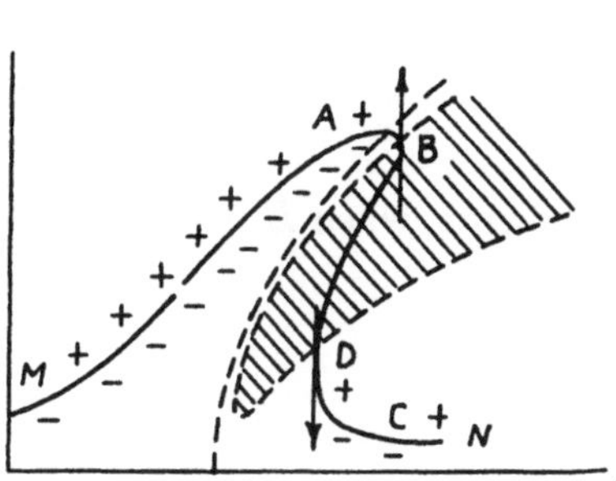


*Fig. 46.* Resonance curve for $\beta > 0$, $h > 0$.

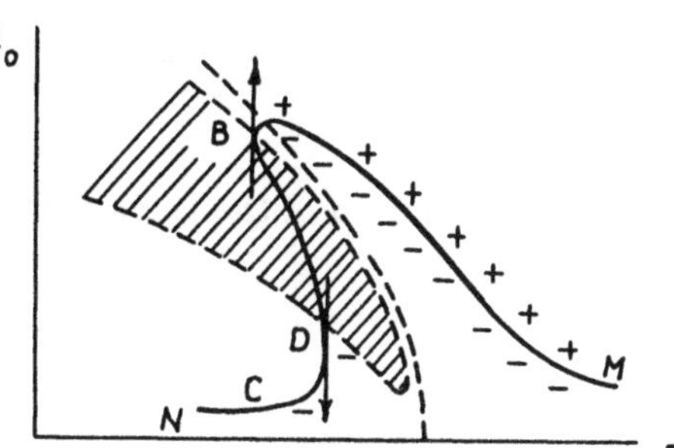


*Fig. 47.* Resonance curve for $\beta < 0$, $h > 0$.

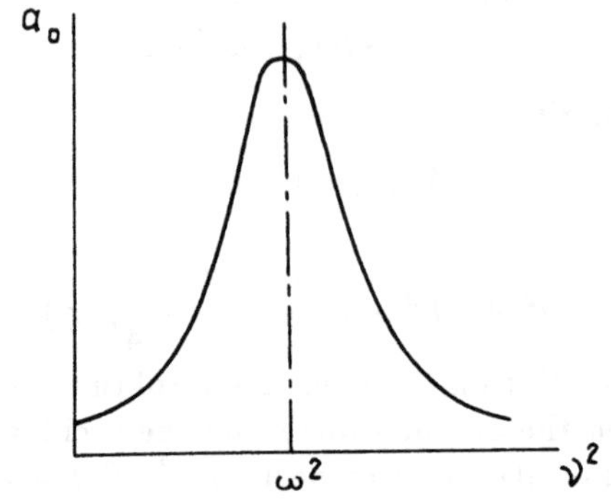


*Fig. 48.* Resonance curve for $\beta = 0$, $h > 0$.

The first condition for asymptotic stability (4.15) is then

$$P(\nu, a_0) = h + \frac{\partial}{\partial a_0}[h a_0] = 2h > 0,$$

which is always satisfied, because $h > 0$. The second stability condition (4.16) is easy to interpret by using (4.27). When $a = 0$, one has $W(\nu, 0) = -E^2 < 0$. This region is marked in Figs 46 and 47 by the minus sign. By the rule indicated when analyzing equation (4.13), the condition (4.16) shows that the arcs MAB and DCN correspond to the stable stationary regimes.

These diagrams give us a simple explanation of the physical phenomenon of the jump in amplitude observed when the excitation frequency is varied (sufficiently slowly however, to allow us to assume that the stationary state is reached at each point in time). See, for example, Fig. 46. As $\nu^2$ increases from $-\infty$ to $+\infty$, the image point describes the arc MAB, jumps to C and then describes the arc CN. If, on the contrary, $\nu^2$ decreases from $+\infty$ to $-\infty$, the image point describes the arc NCD, jumps to A, and then describes the arc AM.

Differentiating (4.9) implicitly with respect to $a_0$, we have

$$\frac{\partial W(\nu, a_0)}{\partial a_0} + \frac{\partial W}{\partial \nu} \cdot \frac{d\nu}{da_0} = 0. \tag{4.30}$$

Hence, the equation of the locus of the vertical tangents, when $E$ is varied, will be $\dfrac{d\nu}{da_0} = 0$, or

$$\frac{\partial W(\nu, a_0)}{\partial a_0} = 0.$$

If the expression (4.27) is substituted into (4.30), one obtains:

$$4\varepsilon^2\nu^2h^2 + \left(\nu^2 - \omega^2 - \frac{3}{4}\varepsilon\beta a_0^2\right)\left(\nu^2 - \omega^2 - \frac{9}{4}\varepsilon\beta a_0^2\right) = 0. \tag{4.31}$$

If there is no damping, that is if $h = 0$, the equation (4.31) yields the following pair of equations:

$$\nu^2 = \omega^2 + \frac{3}{4}\varepsilon\beta a_0^2,$$
$$\nu^2 = \omega^2 + \frac{9}{4}\varepsilon\beta a_0^2.$$

The first equation is the equation of the skeleton line (4.21), and the second one is the locus of the vertical tangents of the resonance curves for the case of undamped Duffing's equation (see Figs. 43 and 44). Hence, if $h$ is not equal to zero, but small, the relation (4.31) yields a curve near these two curves as shown by the dashed curve in Fig. 49. The shaded area of Fig. 49 corresponds to the instability of the stationary solutions of (4.23). Accordingly, either one or two stable solutions is associated with each value of $\nu$. The third solution, when it exists inside the shaded area, is an unstable one.

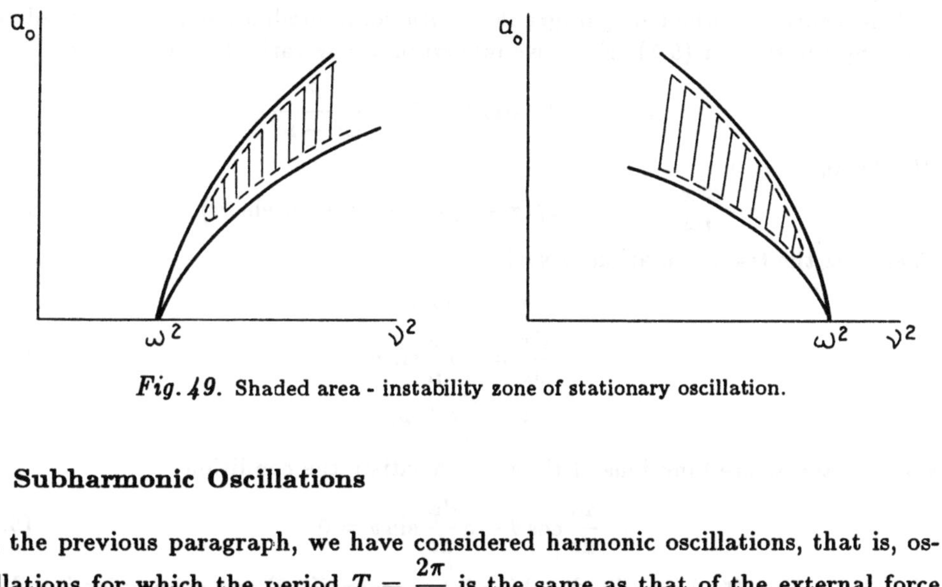


*Fig. 49.* Shaded area - instability zone of stationary oscillation.

## 5. Subharmonic Oscillations

In the previous paragraph, we have considered harmonic oscillations, that is, oscillations for which the period $T = \dfrac{2\pi}{\nu}$ is the same as that of the external force $E \sin \nu t$. However, oscillations, whose period is $nT$, where $n$ is an integer, can occur in nonlinear systems. These oscillations are called subharmonic oscillations of $1/n$ order and the corresponding resonance is called the resonance of $n$ kind. This phenomenon appears only in nonlinear systems. For a linear system

$$\frac{d^2x}{dt^2} + \omega^2 x + 2h\frac{dx}{dt} = E \sin \nu t,$$

if its frequency $\omega$ is equal to $\nu/n$ ($n$ is an integer), then a periodic external force of frequency $\nu$ can excite the free oscillation of frequency $\nu/n$ in addition to the forced oscillation of frequency $\nu$. But, since some damping is always present in real systems, the free oscillation is damped out, so that the eventual steady state consists solely of the oscillation of frequency $\nu$. For a nonlinear system, any free oscillation contains higher harmonics, and hence it is possible that an external force with frequency identical to one of the higher harmonics might be able to excite and sustain the harmonic of lowest frequency.

Consider now an oscillatory system, described by a differential equation of the type

$$\frac{d^2z}{dt^2} + \omega^2 z = \varepsilon f\Big(z, \frac{dz}{dt}\Big) + F\cos\nu t, \tag{5.1}$$

where $\omega$, $\nu$ and $F$ are constants. It is supposed that the natural frequency $\omega$ is close to $\nu/n$, so that there is a relation

$$\omega^2 = \frac{\nu^2}{n^2} - \varepsilon\sigma, \tag{5.2}$$

between $\omega$ and $\nu$; $\sigma$ is a detuning parameter. Comparing with (3.1), we have $p = 1$, $q = n$.

The general method of paragraph 3 with some modification can be used for studying the system (5.1). First, we introduce a new variable $x$ as follows:

$$x = z - f_0\cos\nu t, \quad f_0 = \frac{F}{\omega^2 - \nu^2}. \tag{5.3}$$

We obtain

$$\frac{d^2x}{dt^2} + \omega^2 x = \varepsilon f(x + f_0\cos\nu t, \dot{x} - \nu f_0 \sin\nu t). \tag{5.4}$$

Then, use the transformation of variables

$$\begin{aligned} x &= a\cos\theta, \\ \frac{dx}{dt} &= -a\frac{\nu}{n}\sin\theta, \\ \theta &= \frac{\nu}{n}t + \psi, \end{aligned} \tag{5.5}$$

where $a$ and $\psi$ are functions of time which satisfy the condition

$$\frac{da}{dt}\cos\theta - a\frac{d\psi}{dt}\sin\theta = 0. \tag{5.6}$$

Substituting (5.5) into (5.4), taking into account (5.6) and solving for $\dot{a}$ and $\dot{\psi}$, we obtain:

$$\begin{aligned} \frac{\nu}{n}\frac{da}{dt} &= -\varepsilon\Big\{\sigma x + f\Big(x + f_0\cos\nu t, \frac{dx}{dt} - \nu f_0\sin\nu t\Big)\Big\}\sin\theta, \\ \frac{\nu}{n}a\frac{d\psi}{dt} &= -\varepsilon\Big\{\sigma x + f\Big(x + f_0\cos\nu t, \frac{dx}{dt} - \nu f_0\sin\nu t\Big)\Big\}\cos\theta. \end{aligned} \tag{5.7}$$

Since the amplitude $a$ and phase $\psi$ are slowly varying functions of time, one can, in the first approximation, replace the right-hand sides of (5.7) by their averaged values on time. If this is done, taking into account (5.2) and $\cos\nu t = \cos n(\theta - \psi)$; $\sin\nu t = \sin n(\theta - \psi)$, one has

$$\frac{\nu}{n}\frac{da}{dt} = \varepsilon S(a,\psi,\nu),$$
$$\frac{\nu}{n}a\frac{d\psi}{dt} = \frac{1}{2}(\omega^2 - \frac{\nu^2}{n^2})a_0 + \varepsilon C(a,\psi,\nu), \tag{5.8}$$

where

$$S(a,\psi,\nu) = \frac{-1}{2\pi}\int_0^{2\pi} f\Big[a\cos\theta + f_0\cos n(\theta-\psi), - \frac{\nu}{n}a\sin\theta - \nu f_0 \sin n(\theta-\psi)\Big]\sin\theta d\theta,$$
$$C(a,\psi,\nu) = \frac{-1}{2\pi}\int_0^{2\pi} f\Big[a\cos\theta + f_0\cos n(\theta-\psi), - \frac{\nu}{n}a\sin\theta - \nu f_0 \sin n(\theta-\psi)\Big]\cos\theta d\theta. \tag{5.9}$$

The stationary solution $(a_0, \psi_0)$ exists if (5.8) has a singular point, that is, if

$$S(a_0,\psi_0,\nu) = 0,$$
$$\frac{1}{2}(\omega^2 - \frac{\nu^2}{n^2})a_0 + \varepsilon C(a_0,\psi_0,\nu) = 0. \tag{5.10}$$

By introducing a perturbation of $a$ and $\psi$, and holding $\nu$ constant, namely,

$$a = a_0 + \delta a, \quad \psi = \psi_0 + \delta\psi$$

and linearizing equations (5.8) in $\delta a$, $\delta\psi$, we have

$$\frac{\nu}{n}\frac{d\delta a}{dt} = \varepsilon\big[S_a(a_0,\psi_0,\nu)\delta a + S_\psi(a_0,\psi_0,\nu)\delta\psi\big], \tag{5.11}$$
$$\frac{\nu}{n}a_0\frac{d\delta\psi}{dt} = \frac{1}{2}(\omega^2 - \frac{\nu^2}{n^2})\delta a + \varepsilon\big[C_a(a_0,\psi_0,\nu)\delta a + C_\psi(a_0,\psi_0,\nu)\delta\psi\big],$$

where the subscripts $a$ and $\psi$ designate the partial derivatives with respect to these variables, and $a_0$, $\psi_0$ are the roots common to equations (5.10). We seek a solution of (5.11) in the form

$$\delta a = Me^{\lambda t}, \quad \delta\psi = Ne^{\lambda t}$$

and proceed to determine the eigenvalues $\lambda$. Substituting these equations into (5.11) gives the characteristic equation, governing $\lambda$:

$$\frac{\nu^2}{n^2}a_0\lambda^2 - \frac{\nu}{n}\varepsilon H(a_0,\psi_0,\nu)\lambda + \varepsilon K(a_0,\psi_0,\nu) = 0, \tag{5.12}$$

where

$$H(a_0,\psi_0,\nu) = a_0 S_a(a_0,\psi_0,\nu) + C_\psi(a_0,\psi_0,\nu), \tag{5.13}$$
$$K(a_0,\psi_0,\nu) = \varepsilon\big[S_a(a_0,\psi_0,\nu)\cdot C_\psi(a_0,\psi_0,\nu) - $$
$$- S_\psi(a_0,\psi_0,\nu)C_a(a_0,\psi_0,\nu)\big] - \frac{1}{2}\Big(\omega^2 - \frac{\nu^2}{n^2}\Big)S_\psi(a_0,\psi_0,\nu).$$

For asymptotic stability of the stationary solution $(a_0,\ \psi_0)$, we require all of the eigenvalues to have a negative real part, that is:

$$H(a_0,\psi_0,\nu) < 0, \tag{5.14}$$
$$K(a_0,\psi_0,\nu) > 0. \tag{5.15}$$

**Example 21**

Consider again the Duffing equation

$$\ddot z + \omega^2 z + \varepsilon(h\dot z + \beta z^3) = F\cos\nu t \tag{5.16}$$

where $h$ and $\beta$ are constants and $h > 0$. Comparing with (5.1), we have

$$f(z,\dot z) = -h\dot z - \beta z^3. \tag{5.17}$$

Our interest is focused on the case $n = 3$. The formulae (5.9) are now

$$S(a,\psi,\nu) = \Big\langle\Big\{h\Big[-\frac{\nu}{3}a\sin\theta - \nu f_0\sin 3(\theta-\psi)\Big] +$$
$$+ \beta\big[a\cos\theta + f_0\cos 3(\theta-\psi)\big]^3\Big\}\sin\theta\Big\rangle,$$
$$C(a,\psi,\nu) = \Big\langle\Big\{h\Big[-\frac{\nu}{3}a\sin\theta - \nu f_0\sin 3(\theta-\psi)\Big] +$$
$$+ \beta\big[a\cos\theta + f_0\cos 3(\theta-\psi)\big]^3\Big\}\cos\theta\Big\rangle,$$

where $\langle\ \ \rangle$ is an averaging operator on $\theta$ over the interval $[0,2\pi]$. Taking into account the expressions:

$$\cos^3 3(\theta-\psi) = \frac{1}{4}\big[3\cos 3(\theta-\psi) + \cos 9(\theta-\psi)\big],$$
$$\cos^2 3(\theta-\psi) = \frac{1}{2}\big[1 + \cos 6(\theta-\psi)\big],$$
$$\cos 3(\theta-\psi).\cos^2\theta = \frac{1}{4}\big(\cos\theta\cos 3\psi + \sin\theta\sin 3\psi\big) + \frac{1}{4}\cos(5\theta - 3\psi),$$

and invoking the averaging operator, we obtain

$$S(a,\psi,\nu) = \frac{3}{8}a\Big(-\frac{4}{9}h\nu + af_0\beta\sin 3\psi\Big),$$
$$C(a,\psi,\nu) = \frac{3}{8}\beta a\big(a^2 + 2f_0^2 + af_0\cos 3\psi\big). \tag{5.18}$$

Equations (5.10) for stationary values $a_0$, $\psi_0$ give:

$$a_0\left(-\frac{4}{9}h\nu + a_0 f_0\beta\sin 3\psi_0\right) = 0,$$
$$\left(\omega^2 - \frac{\nu^2}{9}\right)a_0 + \frac{3}{4}\varepsilon\beta a_0\left(a_0^2 + 2f_0^2 + a_0 f_0\cos 3\psi_0\right) = 0.$$

From here, we obtain a trivial solution $a_0 = 0$, which is always stable, and also nontrivial ones which satisfy the equations:

$$\begin{aligned}\frac{1}{3}\varepsilon h\nu &= \frac{3}{4}\varepsilon a_0 f_0\beta\sin 3\psi_0,\\ \omega^2 - \frac{\nu^2}{9} + \frac{3}{4}\varepsilon\beta(a_0^2 + 2f_0^2) &= -\frac{3}{4}\varepsilon\beta a_0 f_0\cos 3\psi_0.\end{aligned} \tag{5.19}$$

If one eliminates $3\psi_0$ from the equations (5.19), one has

$$W(a_0, \nu) = 0,$$

where

$$W(a_0, \nu) = \left[\omega^2 - \frac{\nu^2}{9} + \frac{3}{4}\varepsilon\beta(a_0^2 + 2f_0^2)\right]^2 + \frac{1}{9}\varepsilon^2 h^2\nu^2 - \frac{9}{16}\varepsilon^2\beta^2 a_0^2 f_0^2. \tag{5.20}$$

From this, we obtain:

$$\begin{aligned}a_0^2 =& \frac{1}{4}\Big\{f_0^2 + \left[\frac{16}{27\varepsilon\beta}(\nu^2 - 9\omega^2) - 7f_0^2\right]\pm\\ &\pm 2f_0\sqrt{\frac{16}{27\varepsilon\beta}(\nu^2 - 9\omega^2) - 7f_0^2 - \left(\frac{8}{9\beta}\frac{h\nu}{f_0}\right)^2}\Big\}.\end{aligned} \tag{5.21}$$

This relation gives the dependence of the amplitude $a_0$ on the frequency of the external force $\nu$. The resonance curve has the form shown in Fig. 50, where $\xi$ is proportional to the excitation amplitude.

Thus, in the first approximation, one has the solution of the equation (5.16) in the form (see(5.3), (5.5)):

$$z = a_0\cos\left(\frac{\nu}{3}t + \psi_0\right) + f_0\cos\nu t. \tag{5.22}$$

This solution has a period $\frac{6\pi}{\nu}$, which is triple that of the exciting force. The corresponding oscillation is called subharmonic oscillation of $\frac{1}{3}$ order.

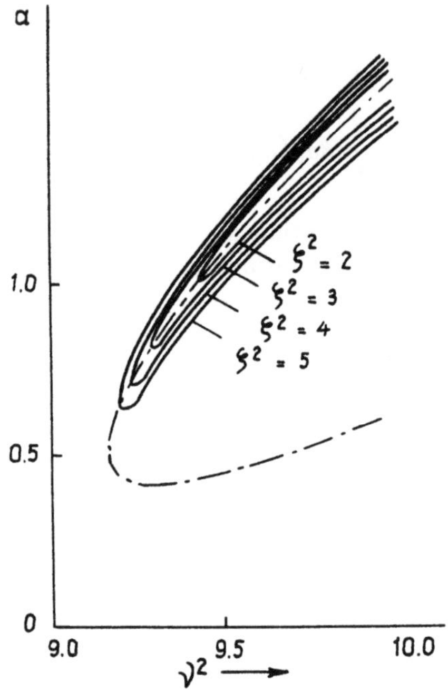


*Fig. 50.* Resonance curves for subharmonic oscillation.

The stability of the stationary oscillations is considered by using the conditions (5.14) and (5.15). From (5.18), we have:

$$S_a(a_0, \psi_0, \nu) = -\frac{h\nu}{6} + \frac{3}{4} a_0 f_0 \beta \sin 3\psi_0,$$
$$S_\psi(a_0, \psi_0, \nu) = \frac{9}{8} a_0^2 f_0 \beta \cos 3\psi_0,$$
$$C_a(a_0, \psi_0, \nu) = \frac{3}{4}\beta\Big(\frac{3}{2}a_0^2 + f_0^2 + a_0 f_0 \cos 3\psi_0\Big),$$
$$C_\psi(a_0, \psi_0, \nu) = -\frac{9}{8}\beta a_0^2 f_0 \sin 3\psi_0.$$

Eliminating $\sin 3\psi_0$ and $\cos 3\psi_0$ by the help of (5.19), we get:

$$S_a = \frac{1}{6}h\nu, \quad S_\psi = \Big[\frac{1}{6\varepsilon}(\nu^2 - 9\omega^2) - \frac{9}{8}\beta(a_0^2 + 2f_0^2)\Big]a_0,$$
$$C_a = \frac{1}{9\varepsilon}(\nu^2 - 9\omega^2) + \frac{3}{8}\beta(a_0^2 - 2f_0^2), \quad C_\psi = -\frac{1}{2}h\nu a_0. \tag{5.23}$$

Substituting these expressions into (5.13), we obtain

$$H(a_0, \psi_0, \nu) = -\frac{1}{3}h\nu a_0.$$

Hence, the stability condition (5.14) is always satisfied ($h > 0$, $\nu > 0$, $a_0 > 0$). The problem of stability depends on the fulfillment of the second condition (5.15). We have

$$K = \frac{1}{4}\varepsilon a_0\Big[\frac{27}{16}\beta^2(a_0^4 - 4f_0^4) + \frac{\nu^2 - 9\omega^2}{\varepsilon}\beta f_0^2 - \frac{1}{27}\frac{(\nu^2 - 9\omega^2)^2}{\varepsilon^2} - \frac{1}{3}h^2\nu^2\Big].$$

Taking into account the relation $W = 0$ (5.20), one can write $K$ as

$$K = \frac{3}{8}a_0^3\Big[\frac{\beta}{3}(9\omega^2 - \nu^2) + \frac{9}{8}\varepsilon\beta^2(3f_0^2 + 2a_0^2)\Big]. \tag{5.24}$$

Hence, the condition $K > 0$ is equivalent to

$$\frac{\partial W}{\partial a_0} > 0,$$

where $W$ is determined by (5.20).

When $a_0 = 0$, we have from (5.20)

$$W(0, \nu) > 0.$$

that is, $W$ is positive on the $\nu$ - axis and therefore one can easily identify the signs of W in the plane $(\nu, a_0)$. Then, using the rule formulated in paragraph 4, one can easily define the branch of the resonance curve, which corresponds to the stable stationary oscillation. So, the stable part is the upper portion of the resonance curve, bounded by its vertical tangent.

**Example 22**

We consider the resonance of the second kind in a Van der Pol oscillator:

$$\ddot{x} + \omega^2 x = \varepsilon(\alpha + \beta x - \gamma x^2)\dot{x} + e\sin 2t, \tag{5.25}$$

where $\omega$, $\alpha$, $\beta$, $\gamma$ and $e$ are constants; $\alpha$ and $\gamma$ are positive. It is supposed that $\omega$ is close to unity, so that

$$\omega^2 = 1 - \varepsilon\Delta, \tag{5.26}$$

where $\Delta$ is a detuning parameter. We are interested in the solution of (5.25) with a period $T = 2\pi$, which is twice the period of the exciting force.

The solution of (5.25) will be found in the form:

$$\begin{aligned} x &= M\cos t + N\sin t - \frac{e}{3}\sin 2t, \\ \dot{x} &= -M\sin t + N\cos t - \frac{2}{3}e\cos 2t, \end{aligned} \tag{5.27}$$

where $M$ and $N$ are functions of time, which satisfy the condition:

$$\dot{M}\cos t + \dot{N}\sin t = 0. \tag{5.28}$$

Substituting (5.27) into (5.25) gives:

$$-\dot{M}\sin t + \dot{N}\cos t = \varepsilon\big[\Delta x + (\alpha + \beta x - \gamma x^2)\dot{x}\big].$$

Solving this equation and (5.28) with respect to the derivatives $\dot{M}$ and $\dot{N}$, we obtain

$$\begin{aligned} \dot{M} &= -\varepsilon\big[\Delta x + (\alpha + \beta x - \gamma x^2)\dot{x}\big]\sin t, \\ \dot{N} &= \varepsilon\big[\Delta x + (\alpha + \beta x - \gamma x^2)\dot{x}\big]\cos t. \end{aligned} \tag{5.29}$$

Since $M$ and $N$ are slowly varying functions of time (their derivatives are proportional to $\varepsilon$), we may replace, in the first approximation, the right-hand sides of (5.29) by their averaged values over time:

$$\begin{aligned} \dot{M} &= \frac{\varepsilon}{2}\Big\{M\big[\alpha - \frac{\gamma}{4}(M^2 + N^2 + \frac{2}{9}e^2)\big] - N(\Delta + \frac{1}{6}e\beta)\Big\}, \\ \dot{N} &= \frac{\varepsilon}{2}\Big\{M(\Delta - \frac{1}{6}e\beta) + N\big[\alpha - \frac{\gamma}{4}(M^2 + N^2 + \frac{2}{9}e^2)\big]\Big\}. \end{aligned} \tag{5.30}$$

The stationary solution $M = M_0 = \text{const}$, $N = N_0 = \text{const}$ of these equations satisfies the relations:

$$\begin{aligned} M_0\big[\alpha - \frac{\gamma}{4}(M_0^2 + N_0^2 + \frac{2}{9}e^2)\big] - N_0(\Delta + \frac{1}{6}e\beta) &= 0, \\ M_0(\Delta - \frac{1}{6}e\beta) + N_0\big[\alpha - \frac{\gamma}{4}(M_0^2 + N_0^2 + \frac{2}{9}e^2)\big] &= 0. \end{aligned} \tag{5.31}$$

There is a trivial solution, namely

$$M_0 = N_0 = 0. \tag{5.32}$$

The nontrivial solutions are determined by

$$\begin{aligned} M_0^2 + N_0^2 &= \frac{4\alpha}{\gamma} - \frac{2}{9}e^2 \pm \frac{2}{3\gamma}\sqrt{e^2\beta^2 - 36\Delta^2}\,, \\ \frac{M_0}{N_0} &= \pm\sqrt{\frac{e\beta + 6\Delta}{e\beta - 6\Delta}}\,. \end{aligned} \tag{5.33}$$

For small enough values of $\Delta$, these solutions are real and they actually correspond to the subharmonic oscillation of the system described by (5.25).

## 6. Nonstationary Oscillations

The asymptotic method has been applied very successfully by Mitropolskii to the investigation of nonstationary oscillations of dynamical systems. The term *nonstationary* oscillations refers to systems in which the parameters vary slowly in time, assuming that the rate of variation of parameters is much slower than the frequency of the motion considered at a fixed value of these parameters [29,a].

One can consider a mechanical system specified by a differential equation of the form

$$\frac{d}{dt}\left[m(\tau)\frac{dx}{dt}\right] + c(\tau)x = \varepsilon F\left(\tau, \theta, x, \frac{dx}{dt}\right), \tag{6.1}$$

where $m(\tau)$ is a mass, $c(\tau)$ is the coefficient of the restoring force, $\tau = \varepsilon t$ is *slow* time, and $t$ is ordinary or *rapid* time. It is obvious that the functions $m(\tau)$ and $c(\tau)$ are always positive; otherwise the problem would have no physical meaning. $F\left(\tau, \theta, x, \frac{dx}{dt}\right)$ is a periodic function of $\theta$ with the period $2\pi$ which can be expanded in the form

$$F\left(\tau, \theta, x, \frac{dx}{dt}\right) = \sum_{n=-N}^{N} e^{in\theta} F_n\left(\tau, x, \frac{dx}{dt}\right), \tag{6.2}$$

where $F_n\left(\tau, x, \frac{dx}{dt}\right)$ are polynomials of $x$ and $\frac{dx}{dt}$. It is supposed also that $\frac{d\theta}{dt} = \nu(\tau)$, so that the *frequency* of external force is slowly varying in time.

In order to apply the asymptotic methods for construction of the approximate solution of equation (7.1), it is necessary to assume that, for the interval ($0 \le \tau \le L$, $L = T/\varepsilon$, $T$ is some finite value), the functions $m(\tau)$, $c(\tau)$, $\frac{d\theta}{dt} = \nu(\tau)$ and $F\left(\tau, \theta, x, \frac{dx}{dt}\right)$ have sufficiently many continuous derivatives with respect to $\tau$ for all finite values of their arguments.

It is noted that the phenomena occurring in the system are not necessarily periodic; even if $m(\tau)$ and $c(\tau)$ vary periodically. The instantaneous frequency $\omega(\tau)$, defined in the conventional manner as $\omega(\tau) = \sqrt{c(\tau)/m(\tau)}$, does not now remain constant, and all familiar concepts of the theory of oscillations governed by differential equation with fixed parameters cease to hold here.

As we know, for a system with fixed parameters, we consider the resonance case and the nonresonance case separately. The approximate solutions for these two cases are quite different. But for the system with slowly varying parameters, the *natural* frequency $\omega(\tau)$ and the exciting frequency $\nu(\tau)$ both depend on the slow time $\tau$. Hence, the system may change from the nonresonance zone to the resonance one and vice versa.

In connection with this, we try to work out a method for the construction of solutions of the equation (6.1) in a general form, which is suitable for both the resonance and nonresonance zones.

When the excitation is absent ($\varepsilon = 0$) and when $\tau$ is fixed, we have, instead of

(6.1), the differential equation

$$m\frac{d^2x}{dt^2} + cx = 0,$$

whose solution is

$$x = a\cos(\omega t + \varphi),$$

where $\omega = \sqrt{c/m}$ and $a$ and $\varphi$ are arbitrary constants determined by initial conditions.

We follow the usual argument (see paragraph 1) with the difference that, the calculation is more complicated in view of the presence of the additional variable $\tau$.

We look for a solution of (6.1) of the form:

$$x = a\cos\left(\frac{p}{q}\theta + \psi\right) + \varepsilon u_1\left(\tau, a, \theta, \frac{p}{q}\theta + \psi\right) + \varepsilon^2 u_2\left(\tau, a, \theta, \frac{p}{q}\theta + \psi\right) + \ldots, \tag{6.3}$$

where $u_1, u_2, \ldots$ are periodic functions of $\theta$ and $\frac{p}{q}\theta + \psi$ with a period $2\pi$, and $p$ and $q$ are small relatively prime integers. The choice of $p$ and $q$ depends on which resonance case is under consideration.

As previously, we assume that the amplitude $a$ and the phase $\psi$ are given by the differential equation

$$\begin{aligned} \frac{da}{dt} &= \varepsilon A_1(\tau, a, \psi) + \varepsilon^2 A_2(\tau, a, \psi) + \ldots, \\ \frac{d\psi}{dt} &= \omega(\tau) - \frac{p}{q}\nu(\tau) + \varepsilon B_1(\tau, a, \psi) + \varepsilon^2 B_2(\tau, a, \psi) + \ldots, \end{aligned} \tag{6.4}$$

where $\omega(\tau) = \sqrt{c(\tau)/m(\tau)}$, $\nu(\tau) = \frac{d\theta}{dt}$ and $\omega(\tau) - \frac{p}{q}\nu(\tau) = \Delta(\tau)$ is the difference between the instantaneous frequency of the system $\omega(\tau)$ and the frequency of the external excitation $\nu(\tau)$.

The problem is to find such functions $u_1, u_2, \ldots, A_1, A_2, \ldots, B_1, B_2, \ldots$ that the series (6.3) after the substitution of $a$ and $\psi$, as given by the series (6.4) should be the solution of (6.1) with any desired accuracy. The applicability of the method does not depend on the actual convergence of the series (6.3) and (6.4) (which may even be divergent), but depends on their asymptotic properties for $\varepsilon \to 0$. One can impose, however, an additional condition, as we did in paragraph 1, which consists of the requirement that the fundamental harmonics should be absent in all functions $u_1, u_2, \ldots$. This requirement is expressed by the conditions

$$\begin{aligned} &\int_0^{2\pi} u_i(\tau, a, \theta, \xi)\cos\xi\,d\xi = 0, \\ &\int_0^{2\pi} u_i(\tau, a, \theta, \xi)\sin\xi\,d\xi = 0, \end{aligned} \tag{6.5}$$

where $\xi = \frac{p}{q}\theta + \psi$. Physically, this means that we take as amplitude $a$ the full amplitude of the fundamental harmonics, so that no resonance terms appear in $u_1, u_2, \ldots$ for any $\tau$ in the interval $0 \leq \tau \leq L$.

In order to determine $u_1, u_2, \ldots, A_1, A_2, \ldots, B_1, B_2, \ldots$ under the conditions (6.5), one has to determine $\frac{d^2a}{dt^2}, \frac{d^2\psi}{dt^2}$ and also the function F in (6.1). We have:

$$\frac{d^2a}{dt^2} = \varepsilon\left[\omega(\tau) - \frac{p}{q}\nu(\tau)\right]\frac{\partial A_1}{\partial\psi} + \\ + \varepsilon^2\left\{\frac{\partial A_1}{\partial a}A_1 + \frac{\partial A_1}{\partial\psi}B_1 + \frac{\partial A_1}{\partial\tau} + \left[\omega(\tau) - \frac{p}{q}\nu(\tau)\right]\frac{\partial A_2}{\partial\psi}\right\} + O(\varepsilon^3),$$

$$\frac{d^2\psi}{dt^2} = \varepsilon\left\{\frac{d\omega}{d\tau} - \frac{p}{q}\frac{d\nu}{d\tau} + \left[\omega(\tau) - \frac{p}{q}\nu(\tau)\right]\frac{\partial B_1}{\partial\psi}\right\} + \\ + \varepsilon^2\left\{\frac{\partial B_1}{\partial a}A_1 + \frac{\partial B_1}{\partial\psi}B_1 + \frac{\partial B_1}{\partial\tau} + \left[\omega(\tau) - \frac{p}{q}\nu(\tau)\right]\frac{\partial B_2}{\partial\psi}\right\} + O(\varepsilon^3),$$

$$\frac{dx}{dt} = -a\omega(\tau)\sin(p\varphi + \psi) + \varepsilon\Big\{A_1\cos(p\varphi + \psi) - B_1 a\sin(p\varphi + \psi) + \\ + \frac{\partial u_1}{\partial\theta}\nu(\tau) + \frac{\partial u_1}{\partial(p\varphi + \psi)}\omega(\tau)\Big\} + \varepsilon^2\Big\{A_2\cos(p\varphi + \psi) - B_2 a\sin(p\varphi + \psi) + \\ + \frac{\partial u_1}{\partial\tau} + \frac{\partial u_1}{\partial a}A_1 + \frac{\partial u_1}{\partial(p\varphi + \psi)}B_1 + \frac{\partial u_2}{\partial\theta}\nu(\tau) + \frac{\partial u_2}{\partial(p\varphi + \psi)}\omega(\tau)\Big\} + O(\varepsilon^3),$$

$$\frac{d^2x}{dt^2} = -a\omega^2(\tau)\cos(p\varphi + \psi) + \varepsilon\Big\{\Big[\left[\omega(\tau) - \frac{p}{q}\nu(\tau)\right]\frac{\partial A_1}{\partial\psi} - \\ - 2a\omega(\tau)B_1\Big]\cos(p\varphi + \psi) - \Big[\left[\omega(\tau) - \frac{p}{q}\nu(\tau)\right]a\frac{\partial B_1}{\partial\psi} + \\ + 2\omega(\tau)A_1\Big]\sin(p\varphi + \psi) - \frac{d\omega(\tau)}{d\tau}a\sin(p\varphi + \psi) + \frac{\partial^2 u_1}{\partial\theta^2}\nu^2(\tau) + \\ + 2\frac{\partial^2 u_1}{\partial\theta\,\partial(p\varphi + \psi)}\nu(\tau)\omega(\tau) + \frac{\partial^2 u_1}{\partial(p\varphi + \psi)^2}\omega^2(\tau)\Big\} + \\ + \varepsilon^2\Big\{\Big[\left[\omega(\tau) - \frac{p}{q}\nu(\tau)\right]\frac{\partial A_2}{\partial\psi} - 2a\omega(\tau)B_2\Big]\cos(p\varphi + \psi) - \\ - \Big[\left[\omega(\tau) - \frac{p}{q}\nu(\tau)\right]a\frac{\partial B_2}{\partial\psi} + 2\omega(\tau)A_2\Big]\sin(p\varphi + \psi) + \\ + \left[\frac{\partial A_1}{\partial a}A_1 + \frac{\partial A_1}{\partial\psi}B_1 + \frac{\partial A_1}{\partial\tau} - aB_1^2\right]\cos(p\varphi + \psi) - \\ - \left[2A_1B_1 + a\frac{\partial B_1}{\partial a}A_1 + a\frac{\partial B_1}{\partial\psi}B_1 + a\frac{\partial B_1}{\partial\tau}\right]\sin(p\varphi + \psi) + \tag{6.6}$$

$$+2\frac{\partial^2 u_1}{\partial\tau\partial(p\varphi+\psi)}\omega(\tau)+2\frac{\partial^2 u_1}{\partial\tau\partial\theta}\nu(\tau)+2\frac{\partial^2 u_1}{\partial a\partial\theta}\nu(\tau)A_1+$$
$$+2\frac{\partial^2 u_1}{\partial a\partial(p\varphi+\psi)}\omega(\tau)A_1+2\frac{\partial^2 u_1}{\partial\theta\partial(p\varphi+\psi)}\nu(\tau)B_1+$$
$$+2\frac{\partial^2 u_1}{\partial(p\varphi+\psi)^2}\omega(\tau)B_1+\frac{\partial u_1}{\partial(p\varphi+\psi)}\left[\omega(\tau)-\frac{p}{q}\nu(\tau)\right]\frac{\partial B_1}{\partial\psi}+$$
$$+\frac{\partial u_1}{\partial a}\left[\omega(\tau)-\frac{p}{q}\nu(\tau)\right]\frac{\partial A_1}{\partial\psi}+\frac{\partial u_1}{\partial\theta}\frac{d\nu(\tau)}{d\tau}+\frac{\partial u_1}{\partial(p\varphi+\psi)}\frac{d\omega(\tau)}{d\tau}+$$
$$+\frac{\partial^2 u_2}{\partial\theta^2}\nu^2(\tau)+2\frac{\partial^2 u_2}{\partial\theta\partial(p\varphi+\psi)}\nu(\tau)\omega(\tau)+\frac{\partial^2 u_2}{\partial(p\varphi+\psi)^2}\omega^2(\tau)\Big\}+O(\varepsilon^3).$$

Substituting (6.3), (6.6) into equation (6.1), we get

$$\frac{d}{dt}\left[m(\tau)\frac{dx}{dt}\right]+c(\tau)x=$$
$$=\varepsilon\Big\{m(\tau)\Big[\left[\omega(\tau)-\frac{p}{q}\nu(\tau)\right]\frac{\partial A_1}{\partial\psi}-2a\omega(\tau)B_1\Big]\cos(p\varphi+\psi)-$$
$$-m(\tau)\Big[\left[\omega(\tau)-\frac{p}{q}\nu(\tau)\right]a\frac{\partial B_1}{\partial\psi}+2\omega(\tau)A_1\Big]\sin(p\varphi+\psi)-$$
$$-\frac{d[m(\tau)\omega(\tau)]}{d\tau}a\sin(p\varphi+\psi)+m(\tau)\Big[\frac{\partial^2 u_1}{\partial\theta^2}\nu^2(\tau)+ \tag{6.7}$$
$$+2\frac{\partial^2 u_1}{\partial\theta\partial(p\varphi+\psi)}\nu(\tau)\omega(\tau)+\frac{\partial^2 u_1}{\partial(p\varphi+\psi)^2}\omega^2(\tau)+\omega^2(\tau)u_1\Big]\Big\}+O(\varepsilon^2),$$

$$\varepsilon F\left(\tau,\theta,x,\frac{dx}{dt}\right)=\varepsilon F\big(\tau,\theta,a\cos(p\varphi+\psi),-a\omega\sin(p\varphi+\psi)\big)+$$
$$\varepsilon^2\Big\{F'_x u_1+F'_{\dot{x}}\Big[A_1\cos(p\varphi+\psi)-B_1 a\sin(p\varphi+\psi)+$$
$$+\frac{\partial u_1}{\partial\theta}\nu(\tau)+\frac{\partial u_1}{\partial(p\varphi+\psi)}\omega(\tau)\Big]\Big\}+O(\varepsilon^3). \tag{6.8}$$

Comparing the coefficients of like powers of $\varepsilon$ in the equation (6.1), we obtain

$$m(\tau)\Big[\frac{\partial^2 u_1}{\partial\theta^2}\nu^2(\tau)+2\frac{\partial^2 u_1}{\partial\theta\partial(p\varphi+\psi)}\nu(\tau)\omega(\tau)+\frac{\partial^2 u_1}{\partial(p\varphi+\psi)^2}\omega^2(\tau)+$$
$$+\omega^2(\tau)u_1\Big]=F_0(\tau,a,\theta,p\varphi+\psi)-m(\tau)\Big[\left[\omega(\tau)-\frac{p}{q}\nu(\tau)\right]\frac{\partial A_1}{\partial\psi}-$$
$$-2a\omega(\tau)B_1\Big]\cos(p\varphi+\psi)+m(\tau)\Big[\left[\omega(\tau)-\frac{p}{q}\nu(\tau)\right]a\frac{\partial B_1}{\partial\psi}+$$
$$+2\omega(\tau)A_1+\frac{1}{m(\tau)}\frac{d[m(\tau)\omega(\tau)]}{d\tau}a\Big]\sin(p\varphi+\psi), \tag{6.9}$$

$$m(\tau)\Big[\frac{\partial^2 u_2}{\partial\theta^2}\nu^2(\tau) + 2\frac{\partial^2 u_2}{\partial\theta\,\partial(p\varphi+\psi)}\nu(\tau)\omega(\tau)+$$
$$+\frac{\partial^2 u_2}{\partial(p\varphi+\psi)^2}\omega^2(\tau) + \omega^2(\tau)u_2\Big] = F_1(\tau,a,\theta,p\varphi+\psi)-$$
$$-m(\tau)\Big[\Big[\omega(\tau)-\frac{p}{q}\nu(\tau)\Big]\frac{\partial A_2}{\partial\psi} - 2a\omega(\tau)B_2 + \frac{\partial A_1}{\partial a}A_1+$$
$$+\frac{\partial A_1}{\partial\psi}B_1 + \frac{\partial A_1}{\partial\tau} - aB_1^2 + \frac{dm(\tau)}{d\tau}\frac{A_1}{m(\tau)}\Big]\cos(p\varphi+\psi)+$$
$$+m(\tau)\Big[\Big[\omega(\tau)-\frac{p}{q}\nu(\tau)\Big]a\frac{\partial B_2}{\partial\psi} + 2\omega(\tau)A_2 + 2A_1B_1 + a\frac{\partial B_1}{\partial a}A_1+$$
$$+a\frac{\partial B_1}{\partial\psi}B_1 + a\frac{\partial B_1}{\partial\tau} + \frac{dm(\tau)}{d\tau}\frac{a}{m(\tau)}B_1\Big]\sin(p\varphi+\psi), \tag{6.10}$$

where

$$F_0(\tau,a,\theta,p\varphi+\psi) = F[\tau,\theta,a\cos(p\varphi+\psi),-a\omega\sin(p\varphi+\psi)], \tag{6.11}$$

$$F_1(\tau,a,\theta,p\varphi+\psi) = F'_x u_1 + F'_{\dot{x}}\big[A_1\cos(p\varphi+\psi)-$$
$$-B_1 a\sin(p\varphi+\psi) + \frac{\partial u_1}{\partial\theta}\nu(\tau) + \frac{\partial u_1}{\partial(p\varphi+\psi)}\omega(\tau)\big]-$$
$$-m(\tau)\Big[2\frac{\partial^2 u_1}{\partial\tau\,\partial(p\varphi+\psi)}\omega(\tau) + 2\frac{\partial^2 u_1}{\partial\tau\,\partial\theta}\nu(\tau) + 2\frac{\partial^2 u_1}{\partial a\,\partial\theta}\nu(\tau)A_1+$$
$$+2\frac{\partial^2 u_1}{\partial a\,\partial(p\varphi+\psi)}\omega(\tau)A_1 + 2\frac{\partial^2 u_1}{\partial\theta\,\partial(p\varphi+\psi)}\nu(\tau)B_1+$$
$$+2\frac{\partial^2 u_1}{\partial(p\varphi+\psi)^2}\omega(\tau)B_1 + \frac{\partial u_1}{\partial(p\varphi+\psi)}\Big[\omega(\tau)-\frac{p}{q}\nu(\tau)\Big]\frac{\partial B_1}{\partial\psi}+$$
$$+\frac{\partial u_1}{\partial a}\Big[\omega(\tau)-\frac{p}{q}\nu(\tau)\Big]\frac{\partial A_1}{\partial\psi} + \frac{\partial u_1}{\partial\theta}\frac{d\nu(\tau)}{d\tau} + \frac{\partial u_1}{\partial(p\varphi+\psi)}\frac{d\omega(\tau)}{d\tau}+$$
$$+\frac{1}{m(\tau)}\frac{\partial u_1}{\partial\theta}\nu(\tau)\frac{dm(\tau)}{d\tau} + \frac{\partial u_1}{\partial(p\varphi+\psi)}\frac{\omega(\tau)}{m(\tau)}\frac{dm(\tau)}{d\tau}\Big]. \tag{6.12}$$

In order to determine $A_1$, $B_1$ and $u_1$ from the equation (6.9), we represent the function (6.11) in the form of a double trigonometrical sum:

$$F_0(\tau,a,\theta,p\varphi+\psi) = \sum_n\sum_m F_{0,n,m}(\tau,a)e^{i\{n\theta+m(p\varphi+\psi)\}}, \tag{6.13}$$

where

$$F_{0,n,m}(\tau,a) = \frac{1}{4\pi^2}\int_0^{2\pi}\int_0^{2\pi} F_0(\tau,a,\theta,p\varphi+\psi)\,e^{-i\{n\theta+m(p\varphi+\psi)\}}d\theta\,d(p\varphi+\psi).$$

The unknown function $u_1(\tau, a, \theta, p\varphi+\psi)$, which is a periodic function of $\theta$ and $p\varphi+\psi$ with period $2\pi$, is also found in the form of a Fourier sum:

$$u_1(\tau, a, \theta, p\varphi+\psi) = \sum_n \sum_m g_{n,m}(\tau, a) e^{i\{n\theta+m(p\varphi+\psi)\}}, \tag{6.14}$$

where the coefficients $g_{n,m}(\tau, a)$ are to be determined.

Substituting (6.13) and (6.14) into (6.9) gives:

$$\begin{aligned} & m(\tau) \sum_n \sum_m \{\omega^2(\tau) - [m\omega(\tau) + n\nu(\tau)]^2\} \times \\ & \times g_{n,m}(\tau, a) e^{i\{n\theta+m(p\varphi+\psi)\}} = \sum_n \sum_m e^{i\{n\theta+m(p\varphi+\psi)\}} \times \\ & \times \frac{1}{4\pi^2} \int_0^{2\pi}\int_0^{2\pi} F_0(\tau, a, \theta, p\varphi+\psi) e^{-i\{n\theta+m(p\varphi+\psi)\}} d\theta d(p\varphi+\psi) - \\ & - m(\tau)\Big\{\Big[\omega(\tau) - \frac{p}{q}\nu(\tau)\Big]\frac{\partial A_1}{\partial\psi} - 2a\omega(\tau)B_1\Big\}\cos(p\varphi+\psi) + m(\tau)\Big\{\Big[\omega(\tau) - \\ & - \frac{p}{q}\nu(\tau)\Big]a\frac{\partial B_1}{\partial\psi} + 2\omega(\tau)A_1 + \frac{1}{m(\tau)}\frac{d[m(\tau)\omega(\tau)]}{d\tau}a\Big\}\sin(p\varphi+\psi). \end{aligned} \tag{6.15}$$

It is noted that the function $u_1$ (6.14) does not contain the fundamental harmonics $\sin(p\varphi+\psi)$, $\cos(p\varphi+\psi)$, which correspond to the values $m, n$ satisfying the relations

$$n\theta + m(p\varphi+\psi) = \pm(p\varphi+\psi) + p_1\psi,$$

where $p_1$ is an integer. Because $\theta = p\varphi$, this relation can be rewritten as:

$$\Big(\frac{nq}{p} + m\Big)(p\varphi+\psi) - \frac{nq}{p}\psi = \pm(p\varphi+\psi) + p_1\psi.$$

From here, it follows that

$$nq + p(m \pm 1) = 0. \tag{6.16}$$

Comparing now the coefficients of like harmonics in (6.15), we obtain:

$$\begin{aligned} g_{n,m}(\tau, a) &= \frac{\frac{1}{4\pi^2 m(\tau)}}{\omega^2(\tau) - [m\omega(\tau) + n\nu(\tau)]^2} \times \\ & \times \int_0^{2\pi}\int_0^{2\pi} F_0(\tau, a, \theta, p\varphi+\psi) e^{-i\{n\theta+m(p\varphi+\psi)\}} d\theta d(p\varphi+\psi) \\ & (nq + p(m \pm 1) \neq 0), \end{aligned} \tag{6.17}$$

$$
\begin{aligned}
& m(\tau)\Big\{\big[\omega(\tau)-\frac{p}{q}\nu(\tau)\big]a\frac{\partial B_1}{\partial\psi}+2\omega(\tau)A_1+ \\
& +\frac{1}{m(\tau)}\frac{d[m(\tau)\omega(\tau)]}{d\tau}a\Big\}\sin(p\varphi+\psi)- \\
& -m(\tau)\Big\{\big[\omega(\tau)-\frac{p}{q}\nu(\tau)\big]\frac{\partial A_1}{\partial\psi}-2a\omega(\tau)B_1\Big\}\cos(p\varphi+\psi)+ \\
& +\sum_{\substack{n,m \\ [nq+p(m\pm1)=0]}}\frac{1}{4\pi^2}e^{i\{n\theta+m(p\varphi+\psi)\}}\int_0^{2\pi}\int_0^{2\pi}F_0(\tau,a,\theta,p\varphi+\psi)\times \\
& \times e^{-i\{n\theta+m(p\varphi+\psi)\}}d\theta d(p\varphi+\psi)=0,
\end{aligned}
\tag{6.18}
$$

where the sum is taken for the values $n$ and $m$ satisfying (6.16). Hence, in the double sum there are complex terms of the form:

$$
\begin{aligned}
& e^{i\{n\theta+m(\frac{p}{q}\theta+\psi)\}}=e^{i\{(n+m\frac{p}{q})\theta+m\psi\}}=e^{i\{\mp\frac{p}{q}\theta+m\psi\}}= \\
& e^{i\{\mp(\frac{p}{q}\theta+\psi)+(m\pm1)\psi\}}=\Big\{\cos(p\varphi+\psi)\mp i\sin(p\varphi+\psi)\Big\}e^{i(m\pm1)\psi}.
\end{aligned}
\tag{6.19}
$$

Moreover, from (6.16), it follows that $m\pm1$ is divisible by $q$, so that $m\pm1=\sigma q$ $(-\infty<\sigma<\infty)$.

Comparing the coefficients of the fundamental harmonics in (6.8), we have

$$
\begin{aligned}
& \big[\omega(\tau)-\frac{p}{q}\nu(\tau)\big]\frac{\partial A_1}{\partial\psi}-2a\omega(\tau)B_1= \\
& =\frac{1}{2\pi^2 m(\tau)}\sum_{\sigma}e^{i\sigma q\psi}\int_0^{2\pi}\int_0^{2\pi}F_0(\tau,a,\theta,\vartheta)e^{-i\sigma q\psi_1}\cos\vartheta d\theta d\vartheta, \\
& \big[\omega(\tau)-\frac{p}{q}\nu(\tau)\big]a\frac{\partial B_1}{\partial\psi}+2\omega(\tau)A_1=-\frac{1}{m(\tau)}\frac{d[m(\tau)\omega(\tau)]}{d\tau}a- \\
& -\frac{1}{2\pi^2 m(\tau)}\sum_{\sigma}e^{i\sigma q\psi}\int_0^{2\pi}\int_0^{2\pi}F_0(\tau,a,\theta,\vartheta)e^{-i\sigma q\psi_1}\sin\vartheta d\theta d\vartheta, \\
& (\psi_1=\psi-p\varphi),
\end{aligned}
\tag{6.20}
$$

where the sum is taken for all positive as well as negative values of $\sigma$.

From (6.20), we obtain

$$
\begin{aligned}
A_1(\tau, a, \psi) = & -\frac{a}{2m(\tau)\omega(\tau)}\frac{d[m(\tau)\omega(\tau)]}{d\tau} + \\
& + \frac{1}{2\pi^2 m(\tau)}\sum_{\sigma} e^{i\sigma q\psi}\frac{[q\omega(\tau) - p\nu(\tau)]\sigma i}{4\omega^2(\tau) - [q\omega(\tau) - p\nu(\tau)]^2\sigma^2}\times \\
& \times \int_0^{2\pi}\int_0^{2\pi} F_0(\tau, a, \theta, \vartheta)e^{-i\sigma q\psi_1}\cos\vartheta \, d\theta d\vartheta - \\
& - \frac{2\omega(\tau)\int_0^{2\pi}\int_0^{2\pi} F_0(\tau, a, \theta, \vartheta)e^{-i\sigma q\psi_1}\sin\vartheta \, d\theta d\vartheta}{4\omega^2(\tau) - [q\omega(\tau) - p\nu(\tau)]^2\sigma^2},
\end{aligned}
\tag{6.21}
$$

$$
\begin{aligned}
B_1(\tau, a, \psi) = & -\frac{1}{2\pi^2 m(\tau)a}\sum_{\sigma} e^{i\sigma q\psi}\left\{\frac{[q\omega(\tau) - p\nu(\tau)]\sigma i}{4\omega^2(\tau) - [q\omega(\tau) - p\nu(\tau)]^2\sigma^2}\times \right. \\
& \times \int_0^{2\pi}\int_0^{2\pi} F_0(\tau, a, \theta, \vartheta)e^{-i\sigma q\psi_1}\sin\vartheta \, d\theta d\vartheta + \\
& \left. + \frac{2\omega(\tau)\int_0^{2\pi}\int_0^{2\pi} F_0(\tau, a, \theta, \vartheta)e^{-i\sigma q\psi_1}\cos\vartheta \, d\theta d\vartheta}{4\omega^2(\tau) - [q\omega(\tau) - p\nu(\tau)]^2\sigma^2}\right\}.
\end{aligned}
$$

Thus, in the first approximation, we have

$$
x = a\cos(p\varphi + \psi), \tag{6.22}
$$

where $a$ and $\psi$ are determined from the equations

$$
\begin{aligned}
\frac{da}{dt} &= \varepsilon A_1(\tau, a, \psi), \\
\frac{d\psi}{dt} &= \omega(\tau) - \frac{p}{q}\nu(\tau) + \varepsilon B_1(\tau, a, \psi),
\end{aligned}
\tag{6.23}
$$

with $A_1$ and $B_1$ given by (6.21).

Similarly, it is not difficult to find the higher approximations. For example, from the equation (6.10), one can find the following equations for $A_2$ and $B_2$:

$$
\begin{aligned}
&\left[\omega(\tau) - \frac{p}{q}\nu(\tau)\right]\frac{\partial A_2}{\partial \psi} - 2\omega(\tau)aB_2 = \\
&= -\left[\frac{\partial A_1}{\partial a}A_1 + \frac{\partial A_1}{\partial \psi}B_1 + \frac{\partial A_1}{\partial \tau} - aB_1^2 + \frac{dm(\tau)}{d\tau}\frac{A_1}{m(\tau)}\right] + \\
&+ \frac{1}{2\pi^2 m(\tau)}\sum_{\sigma} e^{i\sigma q\psi}\int_0^{2\pi}\int_0^{2\pi} F_1(\tau, a, \theta, \vartheta)e^{-i\sigma q\psi_1}\cos\vartheta\, d\theta\, d\vartheta, \\
&\left[\omega(\tau) - \frac{p}{q}\nu(\tau)\right]a\frac{\partial B_2}{\partial \psi} + 2\omega(\tau)A_2 = \\
&= -\left[\frac{\partial B_1}{\partial a}aA_1 + \frac{\partial B_1}{\partial \psi}aB_1 + \frac{\partial B_1}{\partial \tau}a + 2A_1B_1 + \frac{dm(\tau)}{d\tau}\frac{aB_1}{m(\tau)}\right] - \\
&- \frac{1}{2\pi^2 m(\tau)}\sum_{\sigma} e^{i\sigma q\psi}\int_0^{2\pi}\int_0^{2\pi} F_1(\tau, a, \theta, \vartheta)e^{-i\sigma q\psi_1}\sin\vartheta\, d\theta\, d\vartheta.
\end{aligned}
\tag{6.24}
$$

Hence, in the second approximation, we have

$$
x = a\cos(p\varphi + \psi) + \varepsilon u_1(\tau, a, \theta, p\varphi + \psi), \tag{6.25}
$$

where

$$
\begin{aligned}
u_1(\tau, a, \theta, p\varphi + \psi) &= \frac{1}{4\pi^2 m(\tau)} \sum_{\substack{n,m \\ [nq+p(m\pm 1)\neq 0]}} \frac{e^{i\{n\theta + m(p\varphi+\psi)\}}}{\omega^2 - [m\omega(\tau) + n\nu(\tau)]^2} \times \\
&\times \int_0^{2\pi}\int_0^{2\pi} F_0(\tau, a, \theta, p\varphi + \psi)e^{-i\{n\theta + m(p\varphi+\psi)\}}\, d\theta\, d(p\varphi + \psi),
\end{aligned}
\tag{6.26}
$$

and $a$ and $\psi$ are determined from the equations

$$
\begin{aligned}
\frac{da}{dt} &= \varepsilon A_1(\tau, a, \psi) + \varepsilon^2 A_2(\tau, a, \psi), \\
\frac{d\psi}{dt} &= \omega(\tau) - \frac{p}{q}\nu(\tau) + \varepsilon B_1(\tau, a, \psi) + \varepsilon^2 B_2(\tau, a, \psi),
\end{aligned}
\tag{6.27}
$$

with $A_1$, $A_2$, $B_1$, $B_2$ given by (6.21) and (6.24).

A particular case of equation (6.1) is

$$
m\frac{d^2x}{dt^2} + cx = \varepsilon F\left(x, \frac{dx}{dt}\right) + \varepsilon E(\tau)\sin\theta, \tag{6.28}
$$

where $\frac{d\theta}{dt} = \nu(\tau)$, $\tau = \varepsilon t$; $m$ and $c$ are constants. This equation is very often confronted in the problems of physics and engineering. For the system specified by the equation (6.28), the most interesting resonance is the fundamental one, so that $p = q = 1$; Under this assumption, $\frac{p}{q}\theta + \psi = \theta + \psi$.

In the first approximation, we have

$$x = a\cos(\theta + \psi), \tag{6.29}$$

where $a$ and $\psi$ are determined by the following equations of the first approximation

$$\begin{aligned} \frac{da}{dt} &= \varepsilon A_1(\tau, a, \psi), \\ \frac{d\psi}{dt} &= \omega - \nu(\tau) + \varepsilon B_1(\tau, a, \psi), \end{aligned} \tag{6.30}$$

in which $\omega = \sqrt{\frac{c}{m}}$. It is easy to verify that, in this case:

$$\begin{aligned} m\frac{d^2x}{dt^2} + cx = \varepsilon m\Big\{[\omega - \nu(\tau)]\frac{\partial A_1}{\partial\psi} - 2a\omega B_1\Big\}\cos(\theta + \psi) - \\ - \varepsilon m\Big\{[\omega - \nu(\tau)]a\frac{\partial B_1}{\partial\psi} + 2\omega A_1\Big\}\sin(\theta + \psi) + \dots, \end{aligned} \tag{6.31}$$

$$\begin{aligned} \varepsilon F\Big(x, \frac{dx}{dt}\Big) + \varepsilon E(\tau)\sin\theta = \frac{\varepsilon}{\pi}\cos(\theta + \psi)\int_0^{2\pi} F_0(a, \xi)\cos\xi d\xi + \\ + \frac{\varepsilon}{\pi}\sin(\theta + \psi)\int_0^{2\pi} F_0(a, \xi)\sin\xi d\xi + \\ + \varepsilon E(\tau)\big[\cos\psi\sin(\theta + \psi) - \sin\psi\cos(\theta + \psi)\big] + \dots, \end{aligned} \tag{6.32}$$

where

$$F_0(a, \xi) = F(a\cos\xi, -a\omega\sin\xi), \quad \xi = \theta + \psi.$$

Equating the coefficients of like powers of $\varepsilon$ and like harmonics in the right-hand sides of (6.31) and (6.32), we obtain the following equations for $A_1$ and $B_1$:

$$\begin{aligned} m\Big\{[\omega - \nu(\tau)]\frac{\partial A_1}{\partial\psi} - 2a\omega B_1\Big\} &= \frac{1}{\pi}\int_0^{2\pi} F_0(a, \xi)\cos\xi d\xi - E(\tau)\sin\psi, \\ m\Big\{[\omega - \nu(\tau)]a\frac{\partial B_1}{\partial\psi} + 2\omega A_1\Big\} &= -\frac{1}{\pi}\int_0^{2\pi} F_0(a, \xi)\sin\xi d\xi - E(\tau)\cos\psi. \end{aligned} \tag{6.33}$$

From these equations, one can find approximately

$$
\begin{aligned}
A_1(\tau, a, \psi) &= \frac{-1}{2\pi m\omega}\int_0^{2\pi} F_0(a,\xi)\sin\xi d\xi - \frac{E(\tau)}{m[\omega+\nu(\tau)]}\cos\psi, \\
B_1(\tau, a, \psi) &= \frac{-1}{2\pi m\omega a}\int_0^{2\pi} F_0(a,\xi)\cos\xi d\xi + \frac{E(\tau)}{ma[\omega+\nu(\tau)]}\sin\psi.
\end{aligned}
\tag{6.34}
$$

**Example 23**

We consider a nonlinear oscillator passing through resonance, whose motion is specified by the differential equation

$$
\frac{d^2x}{dt^2} + x = -\left(2\delta\frac{dx}{dt} + x^3\right) + E\sin\theta; \quad \delta \ll 1,\ E \ll 1,\ x^2 \le 1. \tag{6.35}
$$

Using the formulae (6.29) and (6.30), we obtain in the first approximation the solution of (6.35) in the form

$$
x = a\cos(\theta + \psi),
$$

where $a$ and $\psi$ are determined from the system of the following differential equation in the first approximation

$$
\begin{aligned}
\frac{da}{dt} &= -\delta a - \frac{E}{1+\nu(\tau)}\cos\psi, \\
\frac{d\psi}{dt} &= 1 - \nu(\tau) + \frac{3}{8}a^2 + \frac{E}{a(1+\nu(\tau))}\sin\psi, \\
\nu(\tau) &= \frac{d\theta}{dt}, \quad \tau = \varepsilon t.
\end{aligned}
\tag{6.36}
$$

These equations cannot be solved in the closed form. With the aid of numerical integration or by computer, one can find the dependence of the amplitude $a$ on the frequency of external excitation $\nu(\tau)$.

We are specially interested in the process of passing through resonance. The following form of $\nu(\tau)$ is assumed :

$$
\nu(\tau) = \nu_0 + \beta\tau, \tag{6.37}
$$

where $\nu_0$ and $\beta$ are constants. The resonance curves in Fig. 51 correspond to the parameters $\delta = 0.01$, $E = 0.02$, $\beta = 0.001$ (curve $ML_1T_1$) and $\beta = -0.001$ (curve $NF_1G_1$).

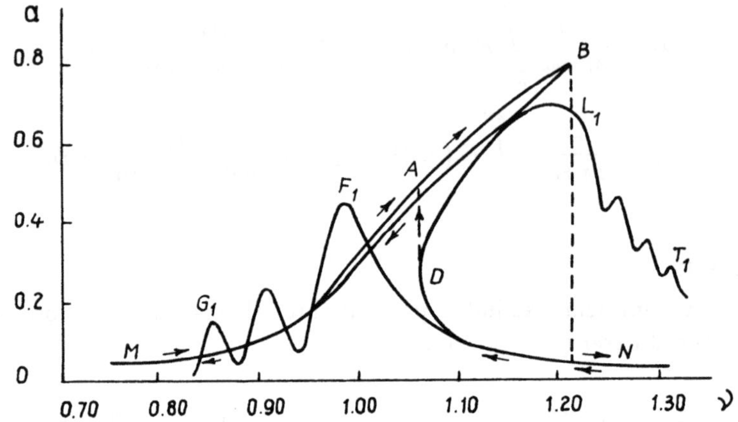


*Fig. 51.* Amplitude curves for oscillating system passing through resonance.

In the second approximation, following (6.25), (6.27), we have

$$x = a\cos(\theta + \psi) + \frac{a^3}{32}\cos 3(\theta + \psi),$$

where $a$ and $\psi$ satisfy the equations:

$$\begin{aligned}\frac{da}{dt} &= -\delta a + \frac{3a^2\delta}{8} - E\Big[\frac{1}{1+\nu} - \frac{3a^2(7-\nu)}{8(3-\nu)(1+\nu)^2}\Big]\cos\psi + \\ &\quad + E\Big[\frac{1}{(1+\nu)^3}\frac{d\nu}{dt} - \frac{\delta}{(1+\nu)^2}\Big]\sin\psi,\end{aligned} \tag{6.38}$$

$$\begin{aligned}\frac{d\psi}{dt} &= 1 - \nu + \frac{3a^2}{8} - \frac{\delta^2}{2} - \frac{15a^4}{256} + \frac{E}{a}\Big[\frac{1}{1+\nu} - \frac{3a^2(5-3\nu)}{8(3-\nu)(1+\nu)^2}\Big]\sin\psi + \\ &\quad + \frac{E}{a}\Big[\frac{1}{(1+\nu)^3}\frac{d\nu}{dt} - \frac{\delta}{(1+\nu)^2}\Big]\cos\psi.\end{aligned}$$

## 7. Multi-frequency Oscillations in Systems with One Degree of Freedom

In this paragraph, we investigate the oscillations of a system with one degree of freedom, which is subjected to a multi-frequency force, by means of the asymptotic method.

Let us consider a differential equation of the general form [38,c]:

$$\frac{d}{dt}\Big[m(\tau)\frac{dx}{dt}\Big] + b(\tau)x = \varepsilon F\Big(\tau, \theta_1, \ldots, \theta_r, x, \frac{dx}{dt}, \varepsilon\Big)\ , \tag{7.1}$$

where $\tau = \varepsilon t$ is slow time, $\varepsilon$ is a small parameter, $\frac{d\theta_k}{dt} = \nu_k(\tau)$ is the frequency of the external force and $F$ is the periodic functions of $\theta_i$ with period $2\pi$. It is supposed that $m(\tau)$, $b(\tau)$ and $F$ have enough derivatives with respect to $\tau$, $\theta_k$, $x$, $\frac{dx}{dt}$. Moreover, for any $\tau$ in the interval $0 \le \tau \le L$, $m(\tau)$ and $b(\tau)$ are strictly positive. The function $F$ is assumed to be expanded in series as:

$$F\Big(\tau,\theta_1,\dots,\theta_r,x,\frac{dx}{dt},\varepsilon\Big) = F_1\Big(\tau,\theta_1,\dots,\theta_r,x,\frac{dx}{dt}\Big)+ \\ +\varepsilon F_2\Big(\tau,\theta_1,\dots,\theta_r,x,\frac{dx}{dt},\Big) + \varepsilon^2\dots, \tag{7.2}$$

$$F_j\Big(\tau,\theta_1,\dots,\theta_r,x,\frac{dx}{dt},\Big) = \sum_c F_{jc_1,\dots,c_r}\Big(\tau,x,\frac{dx}{dt}\Big)e^{i(c_1\theta_1+\dots+c_r\theta_r)}, \tag{7.3}$$

where the coefficients $F_{jc_1\dots c_r}$ are polynomials of $x$ and $\frac{dx}{dt}$ .

It is supposed that between the *natural* frequency $\omega(\tau) = \sqrt{b(\tau)/m(\tau)}$ and the frequencies of external forces $\nu_1(\tau),\dots,\ \nu_r(\tau)$ there exists a resonance relation of the type

$$\begin{aligned} &p^*\Omega(\tau) + c_1^*\nu_1(\tau) + \dots + c_r^*\nu_r(\tau) \equiv 0, \\ &\Omega(\tau) \simeq \omega(\tau), \end{aligned} \tag{7.4}$$

where $p^*, c_1^*, \dots, c_r^*$ are small relatively prime integers, $p^*$ is different from zero, but some of the $c_i^*$ may vanish.

The solution of the equation (7.1) is found in the form

$$x = a\cos(\xi+\psi) + \varepsilon u_1(\tau,a,\xi+\psi,\theta_1,\dots,\theta_r)+ \\ +\varepsilon^2 u_2(\tau,a,\xi+\psi,\theta_1,\dots,\theta_r) + \dots, \tag{7.5}$$

where $u_i$ are periodic functions of the angular variables $\xi+\psi$, $\theta_1,\dots,\theta_r$ with period $2\pi$, and the amplitude $a$ and phase $\psi$ are determined from the differential equations

$$\begin{aligned} \frac{da}{dt} &= \varepsilon A_1(\tau,a,\psi) + \varepsilon^2 A_2(\tau,a,\psi) + \dots, \\ \frac{d\psi}{dt} &= \omega(\tau) - \Omega(\tau) + \varepsilon B_1(\tau,a,\psi) + \varepsilon^2 B_2(\tau,a,\psi) + \dots, \\ \frac{d\xi}{dt} &= \Omega(\tau), \end{aligned} \tag{7.6}$$

$$p^*\xi + c_1^*\theta_1 + \dots + c_r^*\theta_r = 0. \tag{7.7}$$

The problem is now to determine the unknown functions $u_i$, $A_i$, $B_i$ and to integrate the equations (7.6). One can impose an additional condition, as we did in the paragraph 1, which consists of the requirement that the fundamental harmonics should be absent in all functions $u_i$. This condition is equivalent to the requirement of the absence of the resonance terms in $u_i$, which contain zero denominators.

In order to determine the functions $u_i$, $A_i$, $B_i$, we first calculate some derivatives:

$$\frac{d^2a}{dt^2} = \varepsilon\big[\omega(\tau) - \Omega(\tau)\big]\frac{\partial A_1}{\partial\psi} + \varepsilon^2\Big[\frac{\partial A_1}{\partial\tau} + A_1\frac{\partial A_1}{\partial a} + \\ + B_1\frac{\partial A_1}{\partial\psi} + (\omega - \Omega)\frac{\partial A_2}{\partial\psi}\Big] + O(\varepsilon^3),$$

$$\frac{d^2\psi}{\partial t^2} = \varepsilon\Big[\frac{d(\omega - \Omega)}{d\tau} + (\omega - \Omega)\frac{\partial B_1}{\partial\psi}\Big] + \varepsilon^2\Big[A_1\frac{\partial B_2}{\partial a} + \\ + \frac{\partial B_1}{\partial\tau} + B_1\frac{\partial B_1}{\partial\psi} + (\omega - \Omega)\frac{\partial B_2}{\partial\psi}\Big] + O(\varepsilon^3),$$

$$\frac{dx}{dt} = -a\omega(\tau)\sin(\xi + \psi) + \varepsilon\Big[A_1\cos(\xi + \psi) - aB_1\sin(\xi + \psi) + \\ + \omega(\tau)\frac{\partial u_1}{\partial(\xi + \psi)} + \sum_{k=1}^{r}\nu_k\frac{\partial u_1}{\partial\theta_k}\Big] + \varepsilon^2\big[A_2\cos(\xi + \psi) - aB_2\sin(\xi + \psi) + \\ + \frac{\partial u_1}{\partial\tau} + A_1\frac{\partial u_1}{\partial a} + B_1\frac{\partial u_1}{\partial(\xi + \psi)} + \omega\frac{\partial u_2}{\partial(\xi + \psi)} + \sum_{k=1}^{r}\nu_k\frac{\partial u_2}{\partial\theta_k}\Big] + O(\varepsilon^3),$$

$$\frac{d^2x}{dt^2} = -a\omega^2\cos(\xi + \psi) + \varepsilon\Big\{\big[(\omega - \Omega)\frac{\partial A_1}{\partial\psi} - 2a\omega B_1\big]\cos(\xi + \psi) - \\ - \big[(\omega - \Omega)a\frac{\partial B_1}{\partial\psi} + 2\omega A_1\big]\sin(\xi + \psi) - a\frac{d\omega}{d\tau}\sin(\xi + \psi) + \omega^2\frac{\partial^2 u_1}{\partial(\xi + \psi)^2} + \\ + 2\omega\sum_{k=1}^{r}\nu_k\frac{\partial^2 u_1}{\partial\theta_k\partial(\xi + \psi)} + \sum_{k,j=1}^{r}\nu_k\nu_j\frac{\partial^2 u_1}{\partial\theta_k\partial\theta_j}\Big\} + \varepsilon^2\Big\{\big[(\omega - \Omega)\frac{\partial A_2}{\partial\psi} - \\ - 2a\omega B_2 + A_1\frac{\partial A_1}{\partial a} + B_1\frac{\partial A_1}{\partial\psi} + \frac{\partial A_1}{\partial\tau} - aB_1^2\big]\cos(\xi + \psi) - \big[(\omega - \Omega)a\frac{\partial B_2}{\partial\psi} \\ + 2\omega A_2 + 2A_1B_1 + aA_1\frac{\partial B_1}{\partial a} + aB_1\frac{\partial B_1}{\partial\psi} + a\frac{\partial B_1}{\partial\tau}\big]\sin(\xi + \psi) + \\ + 2\omega\frac{\partial^2 u_1}{\partial\tau\partial(\xi + \psi)} + 2\sum_{k=1}^{r}\nu_k\frac{\partial^2 u_1}{\partial\tau\partial\theta_k} + 2A_1\sum_{k=1}^{r}\nu_k\frac{\partial^2 u_1}{\partial\theta_k\partial a} + \\ + 2\omega A_1\frac{\partial^2 u_1}{\partial a\partial(\xi + \psi)} + 2B_1\sum_{k=1}^{r}\nu_k\frac{\partial^2 u_1}{\partial\theta_k\partial(\xi + \psi)} + \\ + 2\omega B_1\frac{\partial^2 u_1}{\partial(\xi + \psi)^2} + (\omega - \Omega)\frac{\partial B_1}{\partial\psi}\frac{\partial u_1}{\partial(\xi + \psi)} +$$

$$+ (\omega - \Omega)\frac{\partial A_1}{\partial \psi}\frac{\partial u_1}{\partial a} + \sum_{k=1}^{r} \frac{d\nu_k}{d\tau}\frac{\partial u_1}{\partial \theta_k} + \frac{d\omega}{d\tau}\frac{\partial u_1}{\partial(\xi + \psi)} +$$

$$+ \sum_{k,j=1}^{r} \nu_k \nu_j \frac{\partial^2 u_2}{\partial \theta_k \partial \theta_j} + 2\omega \sum_{k=1}^{r} \nu_k \frac{\partial^2 u_2}{\partial \theta_k \partial(\xi + \psi)} + \omega^2 \frac{\partial^2 u_2}{\partial(\xi + \psi)^2}\Big\} + O(\varepsilon^3).$$

Substituting these expressions into the primary equation (7.1) and equating the coefficients of like powers of $\varepsilon$, we obtain

$$m(\tau)\Big\{\omega^2 \frac{\partial^2 u_1}{\partial(\xi + \psi)^2} + 2\omega \sum_{k=1}^{r} \nu_k \frac{\partial^2 u_1}{\partial \theta_k \partial(\xi + \psi)} + \sum_{k,j=1}^{r} \nu_k \nu_j \frac{\partial^2 u_1}{\partial \theta_k \partial \theta_j} +$$
$$+ \omega^2 u_1\Big\} = m(\tau)\Big[(\omega - \Omega)a\frac{\partial B_1}{\partial \psi} + 2\omega A_1 + \frac{a}{m}\frac{d(m\omega)}{d\tau}\Big] \sin(\xi + \psi) -$$
$$- m(\tau)\Big[(\omega - \Omega)\frac{\partial A_1}{\partial \psi} - 2a\omega B_1\Big] \cos(\xi + \psi) + F_1(\tau, \theta_1, \ldots, \theta_r, x_0, \dot{x}_0), \tag{7.8}$$

$$m(\tau)\Big\{\omega^2 \frac{\partial^2 u_2}{\partial(\xi + \psi)^2} + 2\omega \sum_{k=1}^{r} \nu_k \frac{\partial^2 u_2}{\partial \theta_k \partial(\xi + \psi)} + \sum_{k,j=1}^{r} \nu_k \nu_j \frac{\partial^2 u_2}{\partial \theta_k \partial \theta_j} +$$
$$+ \omega^2 u_2\Big\} = R_{2*}^{(2)0} + m(\tau)\Big[2A_1 B_1 + aB_1\frac{\partial B_1}{\partial \psi} + aA_1\frac{\partial B_1}{\partial a} + a\frac{\partial B_1}{\partial \tau} +$$
$$+ 2\omega A_2 + (\omega - \Omega)a\frac{\partial B_2}{\partial \psi} + \frac{1}{m}\frac{dm}{d\tau}aB_1\Big] \sin(\xi + \psi) - m(\tau)\Big[A_1\frac{\partial A_1}{\partial a} +$$
$$+ B_1\frac{\partial A_1}{\partial \psi} + \frac{\partial A_1}{\partial \tau} - aB_1^2 - 2a\omega B_2 + (\omega - \Omega)\frac{\partial A_2}{\partial \psi} +$$
$$+ \frac{1}{m}\frac{dm}{d\tau}A_1\Big] \cos(\xi + \psi), \tag{7.9}$$

where

$$R_{2*}^{(2)0} = F_2(\tau, \theta, x_0, \dot{x}_0) + u_1\frac{\partial F_1^0}{\partial x_0} + \Big[A_1 \cos(\xi + \psi) - -aB_1 \sin(\xi + \psi) +$$
$$+ \omega\frac{\partial u_1}{\partial(\xi + \psi)} + \sum_{k=1}^{r} \nu_k \frac{\partial u_1}{\partial \theta_k}\Big]\frac{\partial F_1^0}{\partial \dot{x}_0} - m(\tau)\Big[2\omega(\tau)\frac{\partial^2 u_1}{\partial \tau \partial(\xi + \psi)} +$$
$$+ 2\sum_{k=1}^{r} \nu_k \frac{\partial^2 u_1}{\partial \theta_k \partial \tau} + 2A_1 \sum_{k=1}^{r} \nu_k \frac{\partial^2 u_1}{\partial \theta_k \partial a} + 2\omega A_1 \frac{\partial^2 u_1}{\partial(\xi + \psi)\partial a} +$$
$$+ 2B_1 \sum_{k=1}^{r} \nu_k(\tau)\frac{\partial^2 u_1}{\partial \theta_k \partial(\xi + \psi)} + 2\omega B_1 \frac{\partial^2 u_1}{\partial(\xi + \psi)^2} + \sum_{k=1}^{r} \frac{d\nu_k}{d\tau}\frac{\partial u_1}{\partial \theta_k} +$$
$$+ \frac{d\omega}{d\tau}\frac{\partial u_1}{\partial(\xi + \psi)} + \frac{\omega}{m}\frac{dm}{d\tau}\frac{\partial u_1}{\partial(\xi + \psi)} + \frac{1}{m(\tau)}\frac{dm(\tau)}{d\tau}\sum_{k=1}^{r} \nu_k \frac{\partial u_1}{\partial \theta_k} +$$
$$+ ((\omega(\tau) - \Omega(\tau))\frac{\partial B_1}{\partial \psi}\frac{\partial u_1}{\partial(\xi + \psi)} + (\omega - \Omega)\frac{\partial A_1}{\partial \psi}\frac{\partial u_1}{\partial a}\Big]. \tag{7.10}$$

Equation (7.8) allows us to determine $A_1$, $B_1$ and $u_1$, and equation (7.9) gives $A_2$, $B_2$ and $u_2$. We now represent the function $F_1(\tau, \theta_1, \ldots, \theta_r, x_0, \dot{x}_0)$ in a Fourier series:

$$F_1(\tau, \theta, x_0, \dot{x}_0) = \sum_{p,c} F_{1pc} e^{i\left[p(\xi+\psi)+c_1\theta_1+\cdots+c_r\theta_r\right]}, \tag{7.11}$$

where

$$F_{1pc} = \left(\frac{1}{2\pi}\right)^{r+1} \int_0^{2\pi} \cdots \int_0^{2\pi} F_1(\tau, \bar{\theta}, \bar{x}_0, \bar{\dot{x}}_0) \times$$
$$\times e^{-i[p(\bar{\xi}+\bar{\psi})+c_1\bar{\theta}_1+\cdots+c_r\bar{\theta}_r]} d(\bar{\xi}+\bar{\psi}) d\bar{\theta}_1 \ldots d\bar{\theta}_r,$$
$$\bar{x}_0 = a\cos(\bar{\xi}+\bar{\psi}), \quad \bar{\dot{x}}_0 = -a\omega(\tau)\sin(\bar{\xi}+\bar{\psi}). \tag{7.12}$$

The unknown function $u_1$ is also found in a series:

$$u_1(\tau, a, \xi+\psi, \theta) = \sum_{p,c} u_{1pc} e^{i[p(\xi+\psi)+c_1\theta_1+\cdots+c_r\theta_r]}. \tag{7.13}$$

Substituting (7.11) and (7.13) into (7.8), we have

$$m(\tau)\sum_{p,c}\left\{\omega^2(\tau) - \left[p\omega(\tau) + \sum_{k=1}^{r} c_k\nu_k\right]^2\right\} u_{1pc} e^{i[p(\xi+\psi)+c_1\theta_1+\cdots+c_r\theta_r]} =$$
$$= m(\tau)\left[\frac{a}{m(\tau)}\frac{d(m\omega)}{d\tau} + (\omega-\Omega)a\frac{\partial B_1}{\partial\psi} + 2\omega A_1\right]\sin(\xi+\psi) -$$
$$- m(\tau)\left[(\omega-\Omega)\frac{\partial A_1}{\partial\psi} - 2a\omega B_1\right]\cos(\xi+\psi) +$$
$$+ \sum_{p,c} F_{1pc} e^{i[p(\xi+\psi)+c_1\theta_1+\cdots+c_r\theta_r]}. \tag{7.14}$$

We now compare the coefficients of different harmonics in (7.14). It is noted that the fundamental harmonics, $\sin(\xi+\psi)$ and $\cos(\xi+\psi)$ appear in the terms, corresponding to such values $p, c_1, \ldots, c_r$, which satisfy the relation

$$p(\xi+\vartheta) + c_1\theta_1 + \cdots + c_r\theta_r = \pm(\xi+\psi) + \chi. \tag{7.15}$$

On the other hand, from (7.7), we have

$$p^*(\xi+\psi) + c_1^*\theta_1 + \cdots + c_r^*\theta_r = p^*\psi. \tag{7.16}$$

Expressions (7.15) and (7.16) give

$$p = \pm 1 + \sigma p^*, \; c_k = \sigma c_k^*, \; \chi = \sigma p^*\psi, \; k = 1, 2, \ldots, r, \tag{7.17}$$

where $\sigma$ is a proportional coefficient. Then, the values of $p$, $c_1$, ..., $c_r$ satisfying (7.15) have the form of (7.17)

On the basis of the formula (7.4), it is easy to see that, for such values of $p$, $c_1$, ..., $c_r$, one has $V \simeq 0$, where

$$V = -\omega^2(\tau) + \left[p\omega(\tau) + \sum_{k=1}^{r} c_k\nu_k\right]^2,$$

and $V = 0$ when $\omega(\tau) \equiv \Omega(\tau)$. In fact, we have

$$V = \left[(p+1)\omega + \sum_{k=1}^{r} c_k\nu_k\right]\left[(p-1)\omega + \sum_{k=1}^{r} c_k\nu_k\right],$$

or following (7.17)

$$V = \sigma^2\left(p^*\omega + \sum c_k^*\nu_k\right)^2 \simeq 0.$$

From equation (7.14), one has

$$\begin{aligned} & m(\tau)\left[\frac{a}{m(\tau)}\frac{d(m\omega)}{d\tau} + (\omega - \Omega)a\frac{\partial B_1}{\partial\psi} + 2\omega A_1\right]\sin(\xi+\psi) - \\ & - m(\tau)\left[(\omega - \Omega)\frac{\partial A_1}{\partial\psi} - 2a\omega B_1\right]\cos(\xi+\psi) + \\ & + \sum_{\sigma} F_1(1, \pm 1 + \sigma p^*, \sigma c_1^*, \dots, \sigma c_r^*)e^{i[\pm(\xi+\psi)+\sigma p^*\psi]} = 0. \end{aligned} \tag{7.18}$$

Comparing the coefficients of the fundamental harmonics $\cos(\xi+\psi)$ and $\sin(\xi+\psi)$ and taking into acount (7.12), we obtain

$$\begin{aligned} & (\omega - \Omega)\frac{\partial A_1}{\partial\psi} - 2a\omega B_1 = \frac{2}{m(2\pi)^{1+r}}\sum_{\sigma} e^{i\sigma p^*\psi} \times \\ & \times \int_0^{2\pi}\cdots\int_0^{2\pi} F_1(\tau, \overline{\theta}, \overline{x}_0, \dot{\overline{x}}_0)e^{-i\sigma p^*\overline{\psi}_1}\cos(\overline{\xi}+\overline{\psi})d(\overline{\xi}+\overline{\psi})d\overline{\theta}_1\dots d\overline{\theta}_r, \\ & (\omega - \Omega)a\frac{\partial B_1}{\partial\psi} + 2\omega A_1 = -\frac{a}{m}\frac{d(m\omega)}{d\tau} - \frac{2}{m(2\pi)^{r+1}}\sum_{\sigma} e^{i\sigma p^*\psi} \times \\ & \times \int_0^{2\pi}\cdots\int_0^{2\pi} F_1(\tau, \overline{\theta}, \overline{x}_0, \dot{\overline{x}}_0)e^{-i\sigma p^*\overline{\psi}_1}\sin(\overline{\xi}+\overline{\psi})d(\overline{\xi}+\overline{\psi})d\overline{\theta}_1\dots d\overline{\theta}_r, \end{aligned} \tag{7.19}$$

where

$$p^*\overline{\psi}_1 = p^*(\overline{\xi}+\overline{\psi}) + c_1^*\overline{\theta}_1 + \dots + c_r^*\overline{\theta}_r. \tag{7.20}$$

Solving (7.19) gives

$$A_1 = -\frac{a}{2m(\tau)\omega(\tau)}\frac{d[m(\tau)\omega(\tau)]}{d\tau} + \frac{2}{m(\tau)(2\pi)^{1+r}}\sum_\sigma e^{i\sigma p^*\psi}\times$$

$$\times\left\{\frac{[\omega(\tau)-\Omega(\tau)]i\sigma p^*}{4\omega^2(\tau)-\sigma^2p^{*2}[\omega(\tau)-\Omega(\tau)]^2}\int_0^{2\pi}\cdots\int_0^{2\pi} F_1(\tau,\bar\theta,\bar x_0,\bar{\dot x}_0)e^{-i\sigma p^*\bar\psi_1}\times\right.$$

$$\times\cos(\bar\xi+\bar\psi)d(\bar\xi+\bar\psi)d\bar\theta_1\ldots d\bar\theta_r - \frac{2\omega(\tau)}{4\omega^2(\tau)-\sigma^2p^{*2}[\omega(\tau)-\Omega(\tau)]^2}\times$$

$$\left.\times\int_0^{2\pi}\cdots\int_0^{2\pi} F_1(\tau,\bar\theta,\bar x_0,\bar{\dot x}_0)e^{-i\sigma p^*\bar\psi_1}\sin(\bar\xi+\bar\psi)d(\bar\xi+\bar\psi)d\bar\theta_1\ldots d\bar\theta_r\right\},$$

$$B_1 = \frac{-2}{am(\tau)(2\pi)^{r+1}}\sum_\sigma e^{i\sigma p^*\psi}\left\{\frac{(\omega-\Omega)i\sigma p^*}{4\omega^2-\sigma^2p^{*2}(\omega-\Omega)^2}\times\right.$$

$$\times\int_0^{2\pi}\cdots\int_0^{2\pi} F_1(\tau,\bar\theta,\bar x_0,\bar{\dot x}_0)e^{-i\sigma p^*\bar\psi_1}\sin(\bar\xi+\bar\psi)d(\bar\xi+\bar\psi)d\bar\theta_1\ldots d\bar\theta_r+$$

$$+\frac{2\omega}{4\omega^2-\sigma^2p^{*2}(\omega-\Omega)^2}\int_0^{2\pi}\cdots\int_0^{2\pi} F_1(\tau,\bar\theta,\bar x_0,\bar{\dot x}_0)e^{-i\sigma p^*\bar\psi_1}\times$$

$$\left.\times\cos(\bar\xi+\bar\psi)d(\bar\xi+\bar\psi)d\bar\theta_1\ldots d\bar\theta_r\right\}. \tag{7.21}$$

If we consider only the resonance zone, when $\omega(\tau)-\Omega(\tau)$ is a small quantity, we have the following approximate formulae

$$A_1 = \frac{-1}{2\omega(\tau)}\left[\frac{a}{m(\tau)}\frac{d(m\omega)}{d\tau} + \frac{2}{m(2\pi)^{r+1}}\sum_\sigma e^{i\sigma p^*\psi}\times\right.$$

$$\left.\times\int_0^{2\pi}\cdots\int_0^{2\pi} F_1(\tau,\bar\theta,\bar x_0,\bar{\dot x}_0)e^{-i\sigma p^*\bar\psi_1}\sin(\bar\xi+\bar\psi)d(\bar\xi+\bar\psi)d\bar\theta_1\ldots d\bar\theta_r\right],$$

$$B_1 = -\frac{1}{a\omega}\left[\frac{1}{m(2\pi)^{1+r}}\sum_\sigma e^{i\sigma p^*\psi}\times\right. \tag{7.22}$$

$$\left.\times\int_0^{2\pi}\cdots\int_0^{2\pi} F_1(\tau,\bar\theta,\bar x_0,\bar{\dot x}_0)e^{-i\sigma p^*\bar\psi_1}\cos(\bar\xi+\bar\psi)d(\bar\xi+\bar\psi)d\bar\theta_1\ldots d\bar\theta_r\right].$$

The expression for $u_{1pc_1,\dots,c_r}$ is obtained from the equation (7.14) by comparing the coefficients of higher harmonics:

$$u_{1pc} = \frac{1}{m(\tau)} \cdot \frac{F_{1pc}}{\omega^2 - \left[p\omega(\tau) + \sum\limits_{k=1}^{r} c_k \nu_k\right]^2},$$

where $p \neq \pm 1 + \sigma p^*$, $c_k \neq \sigma c_k^*$ so that

$$u_1 = \frac{1}{m(\tau)} \sum_{p,c} \frac{e^{i[p(\xi+\psi)+c_1\theta_1+\dots+c_r\theta_r]}}{\left[\omega^2 - \left(p\omega + \sum\limits_{k=1}^{r} c_k\nu_k\right)^2\right](2\pi)^{1+r}} \times$$

$$\times \int_0^{2\pi} \dots \int_0^{3\pi} F_1(\tau, \bar{\theta}, \bar{x}_0, \bar{\dot{x}}_0) e^{-i[p(\bar{\xi}+\bar{\psi})+c_1\bar{\theta}_1+\dots+c_r\bar{\theta}_r]} d(\bar{\xi}+\bar{\psi}) d\bar{\theta}_1 \dots d\bar{\theta}_r.$$

$$p \neq \pm 1 + \sigma p^*, \quad c_k \neq \sigma c_k^* \quad (k = 1, 2, \dots r) \tag{7.23}$$

Similarly, we obtain the following equations for determination of $A_2$ and $B_2$:

$$(\omega - \Omega)\frac{\partial A_2}{\partial \psi} - 2a\omega B_2 = -\left[A_1 \frac{\partial A_1}{\partial a} + B_1 \frac{\partial A_1}{\partial \psi} + \right.$$
$$\left. \frac{\partial A_1}{\partial \tau} - aB_1^2 + \frac{A_1}{m(\tau)} \frac{dm(\tau)}{d\tau}\right] + \frac{2}{m(2\pi)^{r+1}} \sum_\sigma e^{i\sigma p^*\psi} \times$$
$$\times \int_0^{2\pi} \dots \int_0^{2\pi} R_{2*}^{(2)\circ} e^{-i\sigma p^* \bar{\psi}_1} \cos(\bar{\xi}+\bar{\psi}) d(\bar{\xi}+\bar{\psi}) d\bar{\theta}_1 \dots d\bar{\theta}_r,$$

$$(\omega - \Omega)a\frac{\partial B_2}{\partial \psi} + 2\omega A_2 = -\left[2A_1B_1 + aA_1 \frac{\partial B_1}{\partial a} + aB_1 \frac{\partial B_1}{\partial \psi} + \right.$$
$$\left. a\frac{\partial B_1}{\partial \tau} + \frac{aB_1}{m(\tau)} \frac{dm(\tau)}{d\tau}\right] - \frac{2}{m(2\pi)^{r+1}} \sum_\sigma e^{i\sigma p^*\psi} \times$$
$$\times \int_0^{2\pi} \dots \int_0^{2\pi} R_{2*}^{(2)\circ} e^{-i\sigma p^* \bar{\psi}_1} \sin(\bar{\xi}+\bar{\psi}) d(\bar{\xi}+\bar{\psi}) d\bar{\theta}_1 \dots d\bar{\theta}_r. \tag{7.24}$$

Thus, in the first approximation, we have

$$x = a\cos(\xi + \psi), \tag{7.25}$$

where $a$ and $\psi$ are determined from the differential equations:

$$\frac{da}{dt} = \varepsilon A_1(\tau, a, \psi),$$
$$\frac{d\psi}{dt} = \omega(\tau) - \Omega(\tau) + \varepsilon B_1(\tau, a, \psi), \tag{7.26}$$

with $A_1$ and $B_1$ given by (7.21).

In the second approximation, we have

$$x = a\cos(\xi + \psi) + \varepsilon u_1(\tau, a, \xi + \psi, \theta_1 \dots, \theta_r), \tag{7.27}$$

where $a$ and $\psi$ satisfy the equations

$$\frac{da}{dt} = \varepsilon A_1(\tau, a, \psi) + \varepsilon^2 A_2(\tau, a, \psi),$$
$$\frac{d\psi}{dt} = \omega(\tau) - \Omega(\tau) + \varepsilon B_1(\tau, a, \psi) + \varepsilon^2 B_2(\tau, a, \psi), \tag{7.28}$$

with $u_1$ given by (7.23) and $A_1$, $A_2$, $B_1$ and $B_2$ determined from (7.21) and (7.24).

## 8. Forced Oscillation of Systems Governed by $N$-order Differential Equation

In this paragraph, the following differential equation

$$x^{(N)} + \alpha_1 x^{(N-1)} + \dots + \alpha_{N-1}\dot{x} + \alpha_N x = \varepsilon F(x, \dot{x}, \dots, x^{(N)}, \theta, \varepsilon), \tag{8.1}$$

is considered [37 b, 51]. Here, the function $F$ is periodic relative to $\theta$ with period $2\pi$, $\frac{d\theta}{dt} = \gamma = \text{const}$. The other assumptions for the function $F$ are the same as in the paragraph 8 of Chapter 1. We consider first the case, when the characteristic equation (8.3) of paragraph 8 of Chapter 1 has a pair of simple imaginary roots $\lambda = \pm i\Omega$, and the other roots have negative real parts with sufficiently large values. Moreover, it is supposed that there is a resonance relation

$$\Omega = \frac{p}{q}\gamma + \varepsilon\sigma, \tag{8.2}$$

where $p$ and $q$ are relatively prime.

We shall find the solution of the equation (8.1) in the form of an asymptotic series:

$$x = a\cos\left(\frac{p}{q}\theta + \psi\right) + \varepsilon u_1(a, \Phi, \theta) + \varepsilon^2 u_2(a, \Phi, \theta) + \dots, \tag{8.3}$$

$$\Phi = \frac{p}{q}\theta + \psi, \tag{8.4}$$

where $u_i$ do not contain $\sin\Phi$ and $\cos\Phi$. They are periodic functions in $\Phi$ and $\theta$ with period $2\pi$, limited for finite values of $a$; and $a$, $\psi$ are determined from the following equations

$$\frac{da}{dt} = \varepsilon A_1(a,\psi) + \varepsilon^2 A_2(a,\psi) + \ldots,$$
$$\frac{d\psi}{dt} = \varepsilon B_1(a,\psi) + \varepsilon^2 B_2(a,\psi) + \ldots \tag{8.5}$$

To find the unknown functions $u_i$, $A_i$, $B_i$, we substitute (8.3) and (8.5) into (8.1), noting that

$$\begin{aligned}
x^{(k)} = {} & a\Omega^k \cos\left(\Phi + k\frac{\pi}{2}\right) + \varepsilon\Big[\left(\Omega\frac{\partial}{\partial\Phi} + \gamma\frac{\partial}{\partial\theta}\right)^k u_1 + \\
& + k\Omega^{k-1} A_1 \sin\left(\Phi + k\frac{\pi}{2}\right) + k\Omega^{k-1} a B_1 \cos\left(\Phi + k\frac{\pi}{2}\right) - \\
& - ak\sigma\Omega^{k-1} \cos\left(\Phi + k\frac{\pi}{2}\right)\Big] + \varepsilon^2\Big[\left(\Omega\frac{\partial}{\partial\Phi} + \gamma\frac{\partial}{\partial\theta}\right)^k u_2 + \\
& + k\Omega^{k-1} A_2 \sin\left(\Phi + k\frac{\pi}{2}\right) + k\Omega^{k-1} a B_2 \cos\left(\Phi + k\frac{\pi}{2}\right) + \\
& + k\left(\Omega\frac{\partial}{\partial\Phi} + \gamma\frac{\partial}{\partial\theta}\right)^{k-1}\left(B_1\frac{\partial}{\partial\Phi} + A_1\frac{\partial}{\partial a} - \sigma\frac{\partial}{\partial\Phi}\right)u_1 - \\
& - \frac{k}{2}(k-1)\Omega^{k-2}\Big(A_1\frac{\partial A_1}{\partial a} + B_1\frac{\partial A_1}{\partial\psi} - aB_1^2 - a\sigma^2 + \\
& + 2a\sigma B_1\Big)\cos\left(\Phi + k\frac{\pi}{2}\right) + \frac{k}{2}(k-1)\Omega^{k-2}\Big(aA_1\frac{\partial B_1}{\partial a} + \\
& + aB_1\frac{\partial B_1}{\partial\psi} + 2A_1B_1 - 2\sigma A_1\Big)\sin\left(\Phi + k\frac{\pi}{2}\right)\Big] + O(\varepsilon^3).
\end{aligned} \tag{8.6}$$

Comparing the coefficients of $\varepsilon$ in (8.1), we obtain

$$\begin{aligned}
& -\sigma a\sum_{k=1}^{N} k\alpha_{N-k}\Omega^{k-1}\cos\left(\Phi + k\frac{\pi}{2}\right) + A_1\sum_{k=1}^{N} k\alpha_{N-k}\Omega^{k-1}\sin\left(\Phi + k\frac{\pi}{2}\right) + \\
& + aB_1\sum_{k=1}^{N} k\alpha_{N-k}\Omega^{k-1}\cos\left(\Phi + k\frac{\pi}{2}\right) + \sum_{k=0}^{N}\alpha_{N-k}\left(\Omega\frac{\partial}{\partial\Phi} + \gamma\frac{\partial}{\partial\theta}\right)^k u_1 = F_0,
\end{aligned} \tag{8.7}$$

where

$$\begin{aligned}
F_0 &= F(x_0, \dot{x}_0, \ldots, x_0^{(N)}, \theta, 0), \\
x_0^{(k)} &= a\Omega^k \cos\left(\Phi + k\frac{\pi}{2}\right), \quad k = 0, 1, 2, \ldots, N.
\end{aligned} \tag{8.8}$$

Now, we expand the function $F_0$ in the Fourier series:

$$\begin{aligned}
F_0 &= \frac{1}{4\pi^2}\sum_{m,n} e^{i(m\Phi + n\theta)}\int_0^{2\pi}\int_0^{2\pi}\overline{F}_0 e^{-i(m\overline{\Phi} + n\overline{\theta})}\,d\overline{\Phi}\,d\overline{\theta}, \\
\overline{F}_0 &= F(\overline{x}_0, \dot{\overline{x}}_0, \ldots \dot{\overline{x}}_0^{(N)}, \theta, 0), \quad \overline{x}_0^{(k)} = a\Omega^k \cos\left(\Phi + k\frac{\pi}{2}\right).
\end{aligned} \tag{8.9}$$

In the expression (8.9), the terms $\sin\Phi$ and $\cos\Phi$ correspond to such values $m$, $n$ for which

$$m\Phi + n\theta = \pm\Phi + \mu \quad \text{or } (m \mp 1)\Phi + n\theta = \mu.$$

On the basis of these relations and (8.4), we have

$$m = qr \pm 1, \ n = -pr, \ \mu = rq\psi, \tag{8.10}$$

where $r$ is a proportional coefficient.

Comparing the coefficients of $\sin\Phi$ and $\cos\Phi$ in (8.7), we have:

$$\begin{aligned} S_1A_1 + aS_2B_1 - \sigma a S_2 &= \frac{1}{2\pi^2}\sum_r e^{irq\psi}\int_0^{2\pi}\int_0^{2\pi}\overline{F}_0 e^{-irq\overline{\psi}}\cos\overline{\Phi}d\overline{\Phi}d\overline{\theta}, \\ S_2A_1 - aS_1B_1 + \sigma a S_1 &= \frac{1}{2\pi^2}\sum_r e^{irq\psi}\int_0^{2\pi}\int_0^{2\pi}\overline{F}_0 e^{-irq\overline{\psi}}\sin\overline{\Phi}d\overline{\Phi}d\overline{\theta}, \end{aligned} \tag{8.11}$$

where

$$\overline{\psi} = \overline{\Phi} - \frac{p}{q}\overline{\theta},$$

$$S_1 = \sum_{k=1}^{N} k\alpha_{N-k}\Omega^{k-1}\sin k\frac{\pi}{2},$$

$$S_2 = \sum_{k=1}^{N} k\alpha_{N-k}\Omega^{k-1}\cos k\frac{\pi}{2}\cdot$$

From the equations (8.11), we obtain

$$\begin{aligned} A_1 &= \frac{1}{2\pi^2(S_1^2+S_2^2)}\sum_r e^{irq\psi}\times \\ &\times\int_0^{2\pi}\int_0^{2\pi}\overline{F}_0(S_1\cos\overline{\Phi} + S_2\sin\overline{\Phi})e^{-irq\overline{\psi}}d\overline{\Phi}d\overline{\theta}, \\ B_1 &= \sigma + \frac{1}{2a\pi^2(S_1^2+S_2^2)}\sum_r e^{irq\psi}\times \\ &\times\int_0^{2\pi}\int_0^{2\pi}\overline{F}_0(S_2\cos\overline{\Phi} - S_1\sin\overline{\Phi})e^{-irq\overline{\psi}}d\overline{\Phi}d\overline{\theta}. \end{aligned} \tag{8.12}$$

Now, we find the function $u_1$ in a Fourier series

$$u_1 = \sum_{m,n} u_{1nm}e^{i(m\Phi+n\theta)}. \tag{8.13}$$

In the first approximation, we have:

$$\sum_{k=0}^{N} \alpha_{N-k}\left(\Omega\frac{\partial}{\partial\Phi}+\gamma\frac{\partial}{\partial\theta}\right)^k u_1 =$$
$$=\sum_{m,n}\sum_{k=0}^{N} \alpha_{N-k}(i\Omega)^k\left(m+n\frac{q}{p}\right)^k u_{1nm}e^{i(m\Phi+n\theta)}.$$

By comparing the harmonics different from $\sin\Phi$, $\cos\Phi$ in (8.7), we obtain:

$$u_{1nm}=\frac{\int_0^{2\pi}\int_0^{2\pi}\overline{F}_0 e^{-i(m\overline{\Phi}+n\overline{\theta})}d\overline{\Phi}d\overline{\theta}}{4\pi^2\sum_{k=0}^{N}\alpha_{N-k}(i\Omega)^k\left(m+n\frac{q}{p}\right)^k}. \tag{8.14}$$

$$m\neq qr\pm 1, \quad n\neq -pr.$$

Thus, in the first approximation, we have the solution of the equation (8.1) as

$$x=a\cos\left(\frac{p}{q}\theta+\psi\right), \tag{8.15}$$

where $a$ and $\psi$ are the solution of the equations:

$$\frac{da}{dt}=\varepsilon A_1, \quad \frac{d\psi}{dt}=\varepsilon B_1.$$

The expressions for $A_1$, $B_1$ are of the form (8.12).

The refinement of the first approximation is

$$x=a\cos\left(\frac{p}{q}\theta+\psi\right)+\frac{\varepsilon}{4\pi^2}\sum_{m,n}\frac{e^{i(m\Phi+n\theta)}\int_0^{2\pi}\int_0^{2\pi}\overline{F}_0 e^{-i(m\overline{\Phi}+n\overline{\theta})}d\overline{\Phi}d\overline{\theta}}{\sum_{k=0}^{N}\alpha_{N-k}(i\Omega)^k\left(m+n\frac{p}{q}\right)^k},$$
$$m\neq qr\pm 1, \quad n\neq -pr\,. \tag{8.16}$$

To calculate the second approximation, we compare the coefficients of $\varepsilon^2$ in (8.1):

$$A_2\sum_{k=1}^{N}k\alpha_{N-k}\Omega^{k-1}\sin\left(\Phi+k\frac{\pi}{2}\right)+aB_2\sum_{k=1}^{N}k\alpha_{N-k}\Omega^{k-1}\cos\left(\Phi+k\frac{\pi}{2}\right)+$$
$$+\sum_{k=0}^{N}\alpha_{N-k}\left(\Omega\frac{\partial}{\partial\Phi}-\gamma\frac{\partial}{\partial\theta}\right)^k u_2=G(a,\Phi,\theta), \tag{8.17}$$

$$G(a,\Phi,\theta)=\left(\frac{\partial F}{\partial x}u_1+\frac{\partial F}{\partial \dot{x}}\dot{\tilde{u}}_1+\cdots+\frac{\partial F}{\partial x^{(N)}}\tilde{u}_1^{(N)}+\frac{\partial F}{\partial \varepsilon}\right)_{\varepsilon=0}-$$
$$-\sum_{k=1}^{N}\alpha_{N-k}\left\{k\left(\Omega\frac{\partial}{\partial\Phi}+\gamma\frac{\partial}{\partial\theta}\right)^{k-1}\left(B_1\frac{\partial}{\partial\Phi}+A_1\frac{\partial}{\partial a}-\sigma\frac{\partial}{\partial\Phi}\right)u_1-\right.$$
$$-\frac{1}{2}k(k-1)\Omega^{k-2}\left(A_1\frac{\partial A_1}{\partial a}+B_1\frac{\partial A_1}{\partial\psi}-aB_1^2-a\sigma^2+\right.$$
$$\left.+2a\sigma B_1\right)\cos\left(\Phi+k\frac{\pi}{2}\right)+\frac{1}{2}k(k-1)\Omega^{k-2}\left(aA_1\frac{\partial B_1}{\partial a}+\right.$$
$$\left.\left.+aB_1\frac{\partial B_1}{\partial\psi}+2A_1B_1-2\sigma A_1\right)\sin\left(\Phi+k\frac{\pi}{2}\right)\right\},$$

$$\tilde{u}_1^{(k)}=\left(\Omega\frac{\partial}{\partial\Phi}+\gamma\frac{\partial}{\partial\theta}\right)^k u_1+k\Omega^{k-1}A_1\sin\left(\Phi+k\frac{\pi}{2}\right)+$$
$$+k\Omega^{k-1}aB_1\cos\left(\Phi+k\frac{\pi}{2}\right)-\sigma k\Omega^{k-1}a\cos\left(\Phi+k\frac{\pi}{2}\right). \tag{8.18}$$

The equation (8.17) has the same structure as (8.7). By using the same method as in the first approximation, we find the following formulae:

$$A_2=\frac{1}{2\pi^2(S_1^2+S_2^2)}\sum_r e^{irq\psi}\times$$
$$\times\int_0^{2\pi}\int_0^{2\pi}G(a,\overline{\Phi},\overline{\theta})(S_1\cos\overline{\Phi}+S_2\sin\overline{\Phi})e^{-irq\overline{\psi}}d\overline{\Phi}d\overline{\theta},$$
$$B_2=\frac{1}{2a\pi^2(S_1^2+S_2^2)}\sum_r e^{irq\psi}\times$$
$$\times\int_0^{2\pi}\int_0^{2\pi}G(a,\overline{\Phi},\overline{\theta})(S_2\cos\overline{\Phi}-S_1\sin\overline{\Phi})e^{-irq\overline{\psi}}d\overline{\Phi}d\overline{\theta},$$

$$u_2=\sum_{m,n}\frac{e^{i(m\Phi+n\theta)}\int_0^{2\pi}\int_0^{2\pi}G(a,\overline{\Phi},\overline{\theta})e^{-i(m\overline{\Phi}+n\overline{\theta})}d\overline{\Phi}d\overline{\theta}}{4\pi^2\sum_{k=0}^{N}\alpha_{N-k}(i\Omega)^k\left(m+\frac{n}{p}q\right)^k} \tag{8.19}$$

$(m\neq qr\pm1,\quad n\neq -pr)$.

As a particular case of the equation (8.1), we consider a third order differential equation

$$\dddot{x}+\xi\ddot{x}+\Omega^2\dot{x}+\xi\Omega^2x=\varepsilon\tilde{R}(x,\dot{x},\ddot{x})+\varepsilon P_0\sin\gamma t, \tag{8.20}$$

here $\xi$, $\Omega$, $\gamma$, $P_0$ are constants and $\tilde{R}(x,\dot{x},\ddot{x})$ is the non-linear function of $x,\dot{x},\ddot{x}$. It is supposed that there is a resonance relation

$$\Omega^2=\gamma^2+\varepsilon\sigma, \tag{8.21}$$

The partial periodic solution with period $2\pi/\gamma$ of the equation (8.20) is found under the form

$$x = a\cos(\gamma t + \psi) + \varepsilon u_1(a, \psi, \gamma t) + \varepsilon^2 u_2(a, \psi, \gamma t) + \ldots, \tag{8.22}$$

where $u_i(a, \psi, \gamma t)$ are periodic functions with period $2\pi$ relative to $\psi$ and $\gamma t$ and $a, \psi$ are determined from the equations:

$$\begin{aligned} \frac{da}{dt} &= \varepsilon A_1(a, \psi) + \varepsilon^2 A_2(a, \psi) + \ldots, \\ \frac{d\psi}{dt} &= \varepsilon B_1(a, \psi) + \varepsilon^2 B_2(a, \psi) + \ldots \end{aligned} \tag{8.23}$$

By substituting (8.22), (8.23) into (8.20), we get

$$\begin{aligned} &\varepsilon\big[2\gamma(\gamma a B_1 - \xi A_1)\sin\varphi - 2\gamma(\gamma A_1 + \xi a B_1)\cos\varphi + \\ &+ \frac{\partial^3 u_1}{\partial t^3} + \xi\frac{\partial^2 u_1}{\partial t^2} + \gamma^2\frac{\partial u_1}{\partial t} + \xi\gamma^2 u_1\big] + O(\varepsilon^2) = \\ &= \varepsilon P\sin\gamma t + \varepsilon R(a\cos\varphi, -a\gamma\sin\varphi, -a\gamma^2\cos\varphi) + O(\varepsilon^2), \end{aligned} \tag{8.24}$$

where

$$R(x, \dot{x}, \ddot{x}) = \tilde{R}(x, \dot{x}, \ddot{x}) - \sigma\dot{x} - \xi\sigma x.$$

Comparing the coefficients of $\varepsilon$ in (8.24), we have

$$\begin{aligned} &2\gamma(\gamma a B_1 - \xi A_1)\sin\varphi - 2\gamma(\gamma A_1 + \xi a B_1)\cos\varphi + \\ &+ \frac{\partial^3 u_1}{\partial t^3} + \xi\frac{\partial^2 u_1}{\partial t^2} + \gamma^2\frac{\partial u_1}{\partial t} + \xi\gamma^2 u_1 = P\sin\gamma t + \\ &+ R(a\cos\varphi, -\gamma a\sin\varphi, -\gamma^2 a\cos\varphi), \quad \varphi = \gamma t + \psi. \end{aligned} \tag{8.25}$$

Now, we expand the function $R$ in the Fourier series

$$R(a\cos\varphi, -\gamma a\sin\varphi, -\gamma^2 a\cos\varphi) = \sum_{n=o}^{\infty}(r_{1n}\cos n\varphi + r_{2n}\sin n\varphi). \tag{8.26}$$

Function $u_1$ is also found in a series

$$u_1 = \sum_m (u_{1m}\cos m\varphi + v_{1m}\sin m\varphi) \tag{8.27}$$

with an additional condition: $u_1$ does not contain the terms with vanishing denominator. Substituting (8.26), (8.27) into (8.25) and equating the coefficients of $\sin\varphi$, $\cos\varphi$, we obtain

$$\begin{aligned} 2\gamma(\gamma A_1 + \xi a B_1) &= P\sin\psi - r_{11}, \\ 2\gamma(\gamma a B_1 - \xi A_1) &= P\cos\psi + r_{21}. \end{aligned} \tag{8.28}$$

Comparing the coefficients of the other harmonics in (8.25), we have

$$\begin{aligned}\gamma^2(1-m^2)(\xi u_{1m}+m\gamma v_{1m}) &= r_{1m}, \quad m\neq 1\\ \gamma^2(1-m^2)(\xi v_{1m}-m\gamma u_{1m}) &= r_{2m}.\end{aligned} \tag{8.29}$$

From here, we obtain

$$\begin{aligned}u_{1m} &= \frac{\xi r_{1m}-m\gamma r_{2m}}{\gamma^2(1-m^2)(\xi^2+m^2\gamma^2)},\\ v_{1m} &= \frac{m\gamma r_{1m}+\xi r_{2m}}{\gamma^2(1-m^2)(\xi^2+m^2\gamma^2)}.\end{aligned} \tag{8.30}$$

Equations (8.28) give

$$\begin{aligned}A_1 &= \frac{P(\gamma\sin\psi-\xi\cos\psi)-\gamma r_{11}-\xi r_{21}}{2\gamma(\xi^2+\gamma^2)},\\ B_1 &= \frac{P(\xi\sin\psi+\gamma\cos\psi)-\xi r_{11}+\gamma r_{21}}{2\gamma a(\xi^2+\gamma^2)}.\end{aligned} \tag{8.31}$$

The calculation of higher approximations do not present any difficulty.

Thus, in the first approximation, we have

$$x = a\cos(\gamma t+\psi), \tag{8.32}$$

$$\begin{aligned}\frac{da}{dt} &= \varepsilon\frac{P(\gamma\sin\psi-\xi\cos\psi)-\gamma r_{11}-\xi r_{21}}{2\gamma(\xi^2+\gamma^2)},\\ \frac{d\psi}{dt} &= \varepsilon\frac{P(\xi\sin\psi+\gamma\cos\psi)-\xi r_{11}+\gamma r_{21}}{2\gamma a(\xi^2+\gamma^2)}.\end{aligned} \tag{8.33}$$

The refinement of the first approximation is

$$\begin{aligned}x = {}& a\cos(\gamma t+\psi)+\\ &+\frac{\varepsilon}{\gamma^2}\sum_{\substack{m=0\\ m\neq 1}}^{\infty}\frac{(\xi r_{1m}-m\gamma r_{2m})\cos m\varphi+(m\gamma r_{1m}+\xi r_{2m})\sin m\varphi}{(1-m^2)(\xi^2+m^2\gamma^2)},\end{aligned} \tag{8.34}$$

where $a$ and $\psi$ satisfy the equations (8.33).

The amplitude and phase of the stationary oscillation are determined from the relations

$$\begin{aligned}P(\gamma\sin\psi-\xi\cos\psi) &= \gamma r_{11}+\xi r_{21},\\ P(\xi\sin\psi+\gamma\cos\psi) &= \xi r_{11}-\gamma r_{21}.\end{aligned} \tag{8.35}$$

Eliminating the phase $\psi$ gives

$$P^2 = r_{11}^2+r_{21}^2. \tag{8.36}$$

**Stability.** Let $\delta a$, $\delta\psi$ be small perturbations, and set $a = a_0 + \delta a$, $\psi = \psi_0 + \delta\psi$ where $a_0$, $\psi_0$ are stationary values of $a$, $\psi$ determined from (8.35). Substituting the above expressions into the equations (8.33) and making use of the steady-state equations (8.35), we receive the following variational equations

$$\begin{aligned} \frac{d\delta a}{dt} &= \frac{\varepsilon\left[-(\gamma r'_{11} + \xi r'_{21})\delta a + (\xi r_{11} - \gamma r_{21})\delta\psi\right]}{2\gamma(\xi^2 + \gamma^2)}, \\ \frac{d\delta\psi}{dt} &= -\frac{\varepsilon\left[(\xi r'_{11} + \gamma r'_{21})\delta a + (\gamma r_{11} + \xi r_{21})\delta\psi\right]}{2\gamma(\xi^2 + \gamma^2)a_0}, \end{aligned} \tag{8.37}$$

where prime denotes the derivative with respect to $a_0$.

The characteristic equation of the system (8.37) is

$$\lambda^2 + q\lambda + s = 0, \tag{8.38}$$

where

$$\begin{aligned} q &= \frac{\varepsilon}{2\gamma a_0(\xi^2 + \gamma^2)}\frac{d}{da_0}\left[a_0(\gamma r_{11} + \xi r_{21})\right], \\ s &= \frac{\varepsilon^2}{8\gamma^2 a_0(\xi^2 + \gamma^2)}\frac{d}{da_0}W, \\ W &= r_{11}^2 + r_{21}^2 - P^2. \end{aligned} \tag{8.39}$$

The stability condition is given by the Routh-Hurwitz criterion, that is

$$\begin{aligned} &\frac{d}{da_0}\left[a_0(\gamma r_{11} + \xi r_{21})\right] > 0, \\ &\frac{d}{da_0}W > 0. \end{aligned} \tag{8.40}$$

Since $W = 0$ is the equation of the resonance curve, the second inequality of (8.40) means that the vertical tangents of the resonance curve serve as the boundary between the stable and unstable regions. The first inequality of (8.40) gives some limits on either the amplitude of oscillation or the parameters of the system under consideration.

**Example 24**

Let us consider the Duffing case for $N = 3$

$$\tilde{R}(x, \dot{x}, \ddot{x}) = -\beta x^3.$$

Now, the equation of motion takes the form

$$\dddot{x} + \xi\ddot{x} + \Omega^2\dot{x} + \xi\Omega^2 x + \varepsilon\beta x^3 = \varepsilon P \sin\gamma t.$$

In this case, we have

$$r_{11} = -\xi\sigma a - \frac{3}{4}\beta a^3,$$

$$r_{21} = \sigma\gamma a.$$

In the first approximation, the relation (8.36) gives the following equation of resonance curve

$$\nu^2 = 1 + \frac{\xi}{\Omega^2}\beta^* a^2 \pm \sqrt{\frac{P^*}{a^2} - \frac{\beta^{*2}}{\Omega^2}\cdot a^4}\ , \quad \nu = \frac{\gamma}{\Omega}\ , \tag{8.41}$$

where

$$\beta^* = \frac{3\varepsilon\beta}{4(\xi^2+\Omega^2)}\ , \quad P^* = \frac{\varepsilon^2 P^2}{\Omega^4(\xi^2+\Omega^2)}\ .$$

The first stability condition (8.40) is of the form

$$-3\gamma\beta a_0^3 > 0,$$

i.e. the necessary condition for the stability of stationary oscillation is

$$\beta < 0.$$

Thus, for the Duffing case, only the soft system has the stable periodic oscillations.

The relation (8.41) is plotted in Fig. 52 for $\xi = 10$, $\Omega = 1$, $\beta^* = -10^{-1}$ and $P^* = 10^{-5}$ (curve 1), $P^* = 1.25 \cdot 10^{-6}$ (curve 2). The full lines correspond to the stable state of oscillation where the stability condition (8.40) is valid.

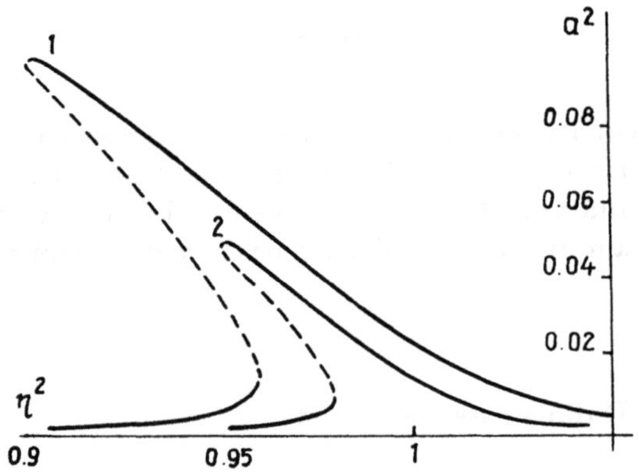


*Fig. 52.* Resonance amplitudes of a third order system.

We consider now the differential equation of the form [37,b]

$$x^{(N)} + \alpha_1 x^{(N-1)} + \cdots + \alpha_{N-1}\dot{x} + \alpha_N x =$$
$$= \varepsilon F(x, \dot{x}, \ldots, x^{(N)}, \theta_1, \ldots, \theta_r, \varepsilon), \tag{8.42}$$

where $\frac{d\theta_j}{dt} = \gamma_j = \text{const}$. It is supposed that the characteristic equation (8.3) of paragraph 8, Chapter 1, has some pairs of simple imaginary roots $\lambda = \pm i\Omega_s$, $s = 1, 2, \ldots, \ell$ and its other roots have negative real parts. Moreover, we suppose that there are the following relations between $\Omega_s$ and $\gamma_j$:

$$q_{1m}\omega_1 + \cdots + q_{\ell m}\omega_\ell + p_{1m}\gamma_1 + \cdots + p_{rm}\gamma_r = 0, \tag{8.43}$$

$(m = 1, 2, \ldots, \rho)$, where

$$\omega_s = \Omega_s - \varepsilon\sigma_s, \quad (s = 1, 2, \ldots, \ell)$$

$\sigma_s$ are detunings, and $q = (q_1, \ldots, q_\ell)$, $p = (p_1, \ldots, p_r)$ are integers.

Now, we find the solution of the equation (8.42) in the form

$$\begin{aligned} x &= \sum_{s=1}^{\ell} a_s \cos\Phi_s + \varepsilon u_1(a, \Phi, \theta) + \varepsilon^2 u_2(a, \Phi, \theta) + \ldots \\ a &= (a_1, \ldots, a_\ell),\ \Phi = (\Phi_1, \ldots, \Phi_\ell),\ \theta = (\theta_1, \ldots, \theta_r), \\ \Phi_s &= \eta_s + \psi_s \qquad (s = 1, 2, \ldots, \ell). \end{aligned} \tag{8.44}$$

Here, the function $\eta_s$ satisfy the following relations

$$\begin{aligned} &\frac{d\eta_s}{dt} = \omega_s, \qquad s = 1, 2, \ldots, \ell \\ &q_{1m}\eta_1 + \cdots + q_{\ell m}\eta_\ell + p_{1m}\theta_1 + \cdots + p_{rm}\theta_r = 0, \\ &\quad m = 1, 2, \ldots, \rho. \end{aligned} \tag{8.45}$$

The functions $u_i(a, \Phi, \theta)$ are periodic with period $2\pi$ relative to $\Phi$ and $\theta$. They are bounded for finite values of $a$ and do not contain $\sin\Phi_s$, $\cos\Phi_s$. The quantities $a$, $\psi$ are the solution of the equations

$$\begin{aligned} \frac{da_s}{dt} &= \varepsilon A_{1s}(a, \psi) + \varepsilon^2 A_{2s}(a, \psi) + \ldots, \\ \frac{d\psi_s}{dt} &= \Omega_s - \omega_s + \varepsilon B_{1s}(a, \psi) + \varepsilon^2 B_{2s}(a, \psi) + \ldots \end{aligned} \tag{8.46}$$

Taking into account that

$$x^{(k)} = \sum_{s=1}^{\ell} a_s\Omega_s^k \cos\left(\Phi_s + k\frac{\pi}{2}\right) + \varepsilon\Big\{k\sum_{s=1}^{\ell} A_{1s}\Omega_s^{k-1} \sin\left(\Phi_s + k\frac{\pi}{2}\right) +$$

$$k\sum_{s=1}^{\ell} a_s B_{1s}\Omega_s^{k-1} \cos\left(\Phi_s + k\frac{\pi}{2}\right) + \Big(\sum_{s=1}^{\ell}\Omega_s\frac{\partial}{\partial\Phi_s} + \sum_{j=1}^{r}\gamma_j\frac{\partial}{\partial\theta_j}\Big)^k u_1\Big\} +$$

$$+\varepsilon^2\Big\{k\sum_{s=1}^{\ell} A_{2s}\Omega_s^{k-1} \sin\left(\Phi_s + k\frac{\pi}{2}\right) + k\sum_{s=1}^{\ell} a_s B_{2s}\Omega_s^{k-1} \cos\left(\Phi_s + k\frac{\pi}{2}\right) +$$

$$+\Big(\sum_{s=1}^{\ell}\Omega_s\frac{\partial}{\partial\Phi_s}+\sum_{j=1}^{r}\gamma_j\frac{\partial}{\partial\theta_j}\Big)^k u_2+$$

$$+k\Big[\sum_{s=1}^{\ell}\Big(A_{1s}\frac{\partial}{\partial a_s}+B_{1s}\frac{\partial}{\partial\Phi_s}\Big)\Big]\Big(\sum_{i=1}^{\ell}\Omega_i\frac{\partial}{\partial\Phi_i}+\sum_{j=1}^{r}\gamma_j\frac{\partial}{\partial\theta_j}\Big)^{k-1}u_1-$$

$$-\frac{1}{2}k(k-1)\sum_{s=1}^{\ell}\Omega_s^{k-2}\Big\{\sum_{i=1}^{\ell}\Big[A_{1i}\frac{\partial A_{1s}}{\partial a_i}+(\sigma_i+B_{1i})\frac{\partial A_{1s}}{\partial\psi_i}\Big]-$$

$$-a_sB_{1s}^2\Big\}\cos\Big(\Phi_s+k\frac{\pi}{2}\Big)+\frac{1}{2}k(k-1)\sum_{s=1}^{\ell}\Omega_s^{k-2}\Big\{a_s\sum_{i=1}^{\ell}\Big[A_{1i}\frac{\partial B_{1s}}{\partial a_i}+$$

$$+(\sigma_i+B_{1i})\frac{\partial B_{1s}}{\partial\psi_i}\Big]+2A_{1s}B_{1s}\Big\}\sin\Big(\Phi_s+k\frac{\pi}{2}\Big)\Big\}+O(\varepsilon^3), \tag{8.47}$$

we can write the left hand side of the equation (8.42) as

$$\sum_{k=0}^{N}\alpha_{N-k}x^{(k)}=\varepsilon\Big[\sum_{k=0}^{N}\alpha_{N-k}\Big(\sum_{s=1}^{\ell}\Omega_s\frac{\partial}{\partial\Phi_s}+\sum_{j=1}^{r}\gamma_j\frac{\partial}{\partial\theta_j}\Big)^k u_1+$$

$$+\sum_{s=1}^{\ell}(L_{1s}A_{1s}+L_{2s}a_sB_{1s})\cos\Phi_s+$$

$$+\sum_{s=1}^{\ell}(L_{2s}A_{1s}-L_{1s}a_sB_{1s})\sin\Phi_s\Big]+O(\varepsilon^2). \tag{8.48}$$

Comparing the coefficients of $\varepsilon$ in (8.42), we get

$$\sum_{k=0}^{N}\alpha_{N-k}\Big(\sum_{s=1}^{\ell}\Omega_s\frac{\partial}{\partial\Phi_s}+\sum_{j=1}^{r}\gamma_j\frac{\partial}{\partial\theta_j}\Big)^k u_1+\sum_{s=1}^{\ell}(L_{1s}A_{1s}+$$

$$+L_{2s}a_sB_{1s})\cos\Phi_s+\sum_{s=1}^{\ell}(L_{2s}A_{1s}-L_{1s}a_sB_{1s})\sin\Phi_s=F_0, \tag{8.49}$$

where

$$F_0=F\Big(\sum_{s=1}^{\ell}a_s\cos\Phi_s,\ldots,\sum_{s=1}^{\ell}a_s\Omega_s^N\cos\Big(\Phi_s+N\frac{\pi}{2}\Big),\theta_1,\ldots,\theta_r,0\Big).$$

Now, we expand the function $F_0$ in a Fourier series:

$$F_0=\sum_{q,p}F_{0qp}e^{i(q_1\Phi_1+\cdots+q_\ell\Phi_\ell+p_1\theta_1+\cdots+p_r\theta_r)}. \tag{8.50}$$

The unknown function $u_1$ is also found in the series

$$u_1 = \sum_{q,p} u_{1qp} e^{i(q_1\Phi_1+\cdots+q_\ell\Phi_\ell+p_1\theta_1+\cdots+p_r\theta_r)}. \tag{8.51}$$

The function $F_0$ contains $\sin\Phi_s$, $\cos\Phi_s$ in the terms which correspond to

$$q_j = q_{js\mu},\ p_v = p_{vs\mu}\ (j = 1, 2, \ldots, \ell,\ v = 1, 2, \ldots, r),$$

for which

$$q_{1s\mu}\Phi_1 + \cdots + q_{es\mu}\Phi_\ell + p_{1s\mu}\theta_1 + \cdots + p_{rs\mu}\theta_r = \pm\Phi_s + \chi_{s\mu}.$$

On the other hand, following the formulae (8.45), we have

$$q_{1m}\Phi_1 + \cdots + q_{\ell m}\Phi_\ell + p_{1m}\theta_1 + \cdots + p_{rm}\theta_r = \\ = q_{1m}\psi_1 + \cdots + q_{\ell m}\psi_\ell, \quad (m = 1, 2, \ldots, \rho).$$

From the last equations, it follows that

$$\begin{aligned} q_j &= q_{js\mu} = \sum_{m=1}^{\rho} r_{ms\mu} q_{jm},\ j \neq s, \\ q_s &= q_{ss\mu} = \pm 1 + \sum_{m=1}^{\rho} r_{ms\mu} q_{sm}, \\ p_v &= p_{vs\mu} = \sum_{m=1}^{\rho} r_{ms\mu} p_{vm}, \end{aligned} \tag{8.52}$$

where $r_{ms\mu}$ are proportional coefficients.

With these values of $q$ and $p$, the coefficients of $u_{1qp}$ in (8.49) are equal to zero. Therefore, equating the coefficients of $\sin\Phi_s$, $\cos\Phi_s$ in (8.49), we obtain:

$$L_{1s}A_{1s} + L_{2s}a_s B_{1s} = \frac{2}{(2\pi)^{\ell+r}} \sum_{\mu} e^{i\sum_{m=1}^{\rho} r_{ms\mu}\psi_m^*} \times \\ \times \int_0^{2\pi} \cdots \int_0^{2\pi} \overline{F}_0 e^{-i\sum_{m=1}^{\rho} r_{ms\mu}\overline{\psi}_m^*} \cos\overline{\Phi}_s d\overline{\Phi}_1 \ldots d\overline{\Phi}_\ell d\overline{\theta}_1 \ldots d\overline{\theta}_r,$$

$$L_{2s}A_{1s} - L_{1s}a_s B_{1s} = \frac{2}{(2\pi)^{\ell+r}} \sum_{\mu} e^{i\sum_{m=1}^{\rho} r_{ms\mu}\psi_m^*} \times \\ \times \int_0^{2\pi} \cdots \int_0^{2\pi} \overline{F}_0 e^{-i\sum_{m=1}^{\rho} r_{ms\mu}\overline{\psi}_m^*} \sin\overline{\Phi}_s d\overline{\Phi}_1 \ldots d\overline{\Phi}_\ell d\overline{\theta}_1 \ldots d\overline{\theta}_r,$$

$$\psi_m^* = q_{1m}\psi_1 + \cdots + q_{\ell m}\psi_\ell,$$
$$\overline{\psi}_m^* = q_{1m}\overline{\Phi}_1 + \cdots + q_{\ell m}\overline{\Phi}_\ell + p_{1m}\overline{\theta}_1 + \cdots + p_{rm}\overline{\theta}_r,$$
$$\overline{F}_0 = F_0\Big|_{\Phi_s=\overline{\Phi}_s,\ \theta_j=\overline{\theta}_j}.$$

From these equations, we find

$$A_{1s} = \frac{2}{(2\pi)^{\ell+r}(L_{1s}^2 + L_{2s}^2)} \sum_\mu e^{i\sum\limits_{m=1}^{\rho} r_{ms\mu}\psi_m^*} \int_0^{2\pi}\cdots\int_0^{2\pi} \overline{F}_0(L_{1s}\cos\overline{\Phi}_s +$$
$$+ L_{2s}\sin\overline{\Phi}_s)e^{-i\sum\limits_{m=1}^{\rho} r_{ms\mu}\overline{\psi}_m^*} d\overline{\Phi}_1\ldots d\overline{\Phi}_\ell d\overline{\theta}_1\ldots d\overline{\theta}_r,$$

$$B_{1s} = \frac{2}{(2\pi)^{\ell+r}(L_{1s}^2 + L_{2s}^2)a_s} \sum_\mu e^{i\sum\limits_{m=1}^{\rho} r_{ms\mu}\psi_m^*} \int_0^{2\pi}\cdots\int_0^{2\pi} \overline{F}_0(L_{2s}\cos\overline{\Phi}_s -$$
$$- L_{1s}\sin\overline{\Phi}_s)e^{-i\sum\limits_{m=1}^{\rho} r_{ms\mu}\overline{\psi}_m^*} d\overline{\Phi}_1\ldots d\overline{\Phi}_\ell d\overline{\theta}_1\ldots d\overline{\theta}_r. \tag{8.53}$$

By comparing the other harmonics in (8.51), we have

$$u_{1qp} = \frac{F_{0qp}}{\sum\limits_{k=0}^{N} \alpha_{N-k} i^k \left(q_1\Omega_1 + \cdots + q_\ell\Omega_\ell + p_1\gamma_1 + \cdots + p_r\gamma_r\right)^k},$$
$$q_j \neq \sum_{m=1}^{\rho} r_{ms\mu}q_{jm} \pm \delta_{js} \quad (j = 1, 2, \ldots, \ell),$$
$$p_v \neq \sum_{m=1}^{\rho} r_{ms\mu}p_{vm} \quad (v = 1, 2, \ldots, r). \tag{8.54}$$

We found the functions $A_{1s}$, $B_{1s}$ and $u_1$, and therefore, the solution of the equation (8.42) in the first approximation:

$$x = \sum_{s=1}^{\ell} a_s\cos\Phi_s,$$
$$\frac{da_s}{dt} = \varepsilon A_{1s}(a, \psi), \tag{8.55}$$
$$\frac{d\Phi_s}{dt} = \Omega_s + \varepsilon B_{1s}(a, \psi),$$

and the refinement of the first approximation:

$$x = \sum_{s=1}^{\ell} a_s \cos \Phi_s + \tag{8.56}$$

$$\varepsilon \sum_{p,q} \frac{F_{0pq} e^{i(q_1\Phi_1+\cdots+q_\ell\Phi_\ell+p_1\theta_1+\cdots+p_r\theta_r)}}{\sum\limits_{k=0}^{N} \alpha_{N-k} i^k (q_1\Omega_1 + \cdots + q_\ell\Omega_\ell + p_1\gamma_1 + \cdots + p_r\gamma_r)^k},$$

where $a_s$, $\psi_s$ are solutions of the equations (8.55).

Similarly, if we compare the coefficients of $\varepsilon^2$ in (8.42) and then we compare the harmonics in the obtained equation, we get:

$$A_{2s} = \frac{2}{(2\pi)^{\ell+r}(L_{1s}^2 + L_{2s}^2)} \sum_{\mu} e^{i\sum\limits_{m=1}^{\rho} r_{ms\mu}\psi_m^*} \int_0^{2\pi} \cdots \int_0^{2\pi} G(a, \overline{\Phi}, \overline{\theta}) \times$$

$$\times \left(L_{1s} \cos \overline{\Phi}_s + L_{2s} \sin \Phi_s\right) e^{-i\sum\limits_{m=1}^{\rho} r_{ms\mu}\overline{\Phi}_m^*} d\overline{\Phi}_1 \ldots d\overline{\Phi}_\ell d\overline{\theta}_1 \ldots d\overline{\theta}_r,$$

$$B_{2s} = \frac{2}{(2\pi)^{\ell+r} a_s (L_{1s}^2 + L_{2s}^2)} \sum_{\mu} e^{i\sum\limits_{m=1}^{\rho} r_{ms\mu}\psi_m^*} \int_0^{2\pi} \cdots \int_0^{2\pi} G(a, \overline{\Phi}, \overline{\theta}) \times$$

$$\times \left(L_{2s} \cos \overline{\Phi}_s - L_{1s} \sin \overline{\Phi}_s\right) e^{-i\sum\limits_{m=1}^{\rho} r_{ms\mu}\overline{\Phi}_m^*} d\overline{\Phi}_1 \ldots d\overline{\Phi}_\ell d\overline{\theta}_1 \ldots d\overline{\theta}_r, \tag{8.57}$$

$$u_{2qp} = \int_0^{2\pi} \cdots \int_0^{2\pi} G(a, \overline{\Phi}, \overline{\theta}) e^{-i(q_1\overline{\Phi}_1+\cdots+q_\ell\overline{\Phi}_\ell+p_1\overline{\theta}_1+\cdots+p_r\overline{\theta}_r)} \times$$

$$\times d\overline{\Phi}_1 \ldots d\overline{\Phi}_\ell d\overline{\theta}_1 \ldots d\overline{\theta}_r \times \frac{1}{(2\pi)^{\ell+r}} \times$$

$$\times \frac{1}{\sum\limits_{k=0}^{N} \alpha_{N-k} i^k \left(q_1\Omega_1 + \cdots + q_\ell\Omega_\ell + p_1\gamma_1 + \cdots + p_r\gamma_r\right)^k}$$

$$\left(q_j \neq \sum_{m=1}^{\rho} r_{ms\mu} q_{jm} \pm \delta_{js}, \quad p_v \neq \sum_{m=1}^{\rho} r_{ms\mu} p_{vm}\right),$$

where

$$G(a,\Phi,\theta)=\left(\frac{\partial F}{\partial x}u_1+\frac{\partial F}{\partial \dot{x}}\dot{\tilde{u}}_1+\cdots+\frac{\partial F}{\partial x^{(N)}}\tilde{u}_1^{(N)}+\frac{\partial F}{\partial \varepsilon}\right)_{\varepsilon=0}-$$
$$-\sum_{k=1}^{N}k\alpha_{N-k}\Big\{\Big[\sum_{s=1}^{\ell}\Big(A_{1s}\frac{\partial}{\partial a_s}+B_{1s}\frac{\partial}{\partial \Phi_s}\Big)\Big]\Big(\sum_{i=1}^{\ell}\Omega_i\frac{\partial}{\partial \Phi_i}+$$
$$+\sum_{j=1}^{r}\gamma_j\frac{\partial}{\partial\theta_j}\Big)^{k-1}u_1-\frac{1}{2}(k-1)\sum_{s=1}^{\ell}\Omega_s^{k-2}\Big\{\sum_{i=1}^{\ell}\Big[A_{1i}\frac{\partial A_{1s}}{\partial a_i}+$$
$$+(\sigma_i+B_{1i})\frac{\partial A_{1s}}{\partial\psi_i}\Big]-a_sB_{1s}^2\Big\}\cos\Big(\Phi_s+k\frac{\pi}{2}\Big)+$$
$$+\frac{1}{2}(k-1)\sum_{s=1}^{\ell}\Omega_s^{k-2}\Big\{a_s\sum_{i=1}^{\ell}\Big[A_{1i}\frac{\partial B_{1s}}{\partial a_i}+(\sigma_i+B_{1i})\frac{\partial B_{1s}}{\partial\psi_i}\Big]+$$
$$+2A_{1s}B_{1s}\Big\}\sin\Big(\Phi_s+k\frac{\pi}{2}\Big)\Big\},$$

$$\tilde{u}_1^{(k)}=\Big(\sum_{s=1}^{\ell}\Omega_s\frac{\partial}{\partial\Phi_s}+\sum_{j=1}^{r}\gamma_j\frac{\partial}{\partial\theta_j}\Big)^k u_1+k\sum_{s=1}^{\ell}A_{1s}\Omega_s^{k-1}\sin\Big(\Phi_s+k\frac{\pi}{2}\Big)+$$
$$+k\sum_{s=1}^{\ell}a_sB_{1s}\Omega_s^{k-1}\cos\Big(\Phi_s+k\frac{\pi}{2}\Big). \tag{8.58}$$

It is not difficult to calculate the higher approximations.

## 9. Single-frequency Oscillations in Nonlinear Systems with Multiple Degrees of Freedom

Consider a system which has $N$ degrees of freedom, and is specified by the differential equations

$$\sum_{s=1}^{N}\{a_{rs}\ddot{q}_s+b_{rs}q_s\}=\varepsilon Q_r^{(1)}(\nu t,q_1,\ldots,q_N,\dot{q}_1,\ldots\dot{q}_N)+$$
$$+\varepsilon^2Q_r^{(2)}(\nu t,q_1,\ldots,q_N,\dot{q}_1,\ldots\dot{q}_N)+\ldots,\quad (r=1,2,\ldots,N), \tag{9.1}$$

where $q_i$ are generalized coordinates, $a_{rs}=a_{sr}$ and $b_{rs}=b_{sr}$ are constants. $Q_r^{(j)}(\nu t,q_1,\ldots,\dot{q}_N)$ are periodic functions with respect to $\nu t$ with period $2\pi$. They can be expanded in the form of finite Fourier sums with the coefficients being polynomials in $q_r$, $\dot{q}_r$ $(r=1,2,\ldots,N)$.

When $\varepsilon=0$, the system of equations (9.1) degenerates to

$$\sum_{s=1}^{N}\{a_{rs}\ddot{q}_s+b_{rs}q_s\}=0. \tag{9.2}$$

These are a set of $N$ linear, second order, ordinary differential equations with constant coefficients. They are called unexcited equations and the corresponding system is called an unexcited system.

Using the principal coordinates $\xi_1, \dots, \xi_N$ connected with the primary coordinates $q_1, \dots, q_N$ by the formulae

$$q_i = \sum_{j=1}^{N} d_i^{(j)} \xi_j, \tag{9.3}$$

where $d_i^{(j)}$ are coefficients of distribution (see Appendix 1), we can write the equations of motion (9.1) in the form

$$\ddot{\xi}_i + \omega_i^2 \xi_i = \frac{\varepsilon}{M_i} \sum_{j=1}^{N} d_j^{(i)} f_j, \quad i = 1, 2, \dots, N, \tag{9.4}$$

where $\omega_1, \omega_2, \dots, \omega_N$ are the natural frequencies of the system under consideration, namely the roots of the characteristic equation

$$D(\omega^2) = 0 \tag{9.5}$$

with

$$D(\omega^2) = \det\| - a_{rs}\omega^2 + b_{rs}\|, \tag{9.6}$$

$$f_j = Q_j^{(1)}(\nu t, q_1, \dots, q_N, \dot{q}_1, \dots, \dot{q}_N) + \\ + \varepsilon Q_j^{(2)}(\nu t, q_1, \dots, q_N, \dot{q}_1, \dots, \dot{q}_N) + \dots,$$

$$\dot{q}_i = \sum_{j=1}^{N} d_i^{(j)} \dot{\xi}_j, \quad M_i = \sum_{j=1}^{N} m_j {d_j^{(i)}}^2, \quad i = 1, 2, \dots, N. \tag{9.7}$$

System (9.4) is a quasi-linear system of $N$ differential equations of second order with respect to $N$ principal coordinates $\xi_1, \dots, \xi_N$. These equations are separable when $\varepsilon = 0$. Equations (9.4) are mathematically exact and are equivalent to the primary equations (9.1). The approximation, which is now to be introduced, consists of considering the single-frequency regime of oscillations.

We suppose that the frequency of external force $\nu$ is in resonance with the natural frequency of the system under consideration, for example, $\omega_1$. Namely, between $\nu$ and $\omega_1$, there exists a resonance relation

$$\omega_1 \simeq \frac{p}{q}\nu, \tag{9.8}$$

where $p$ and $q$ are small relatively prime integers. For the sake of brevity, let us consider the case $p = q = 1$. The other cases will be considered in a similar way.

It is assumed that 1) In the unexcited system (9.2), there exists a harmonic oscillation with frequency $\omega_1$ which depends only on two arbitrary constants, 2) The trivial solution $q_1 = q_2 = \cdots = q_N = 0$ is the only solution corresponding to equilibrium of the unexcited system (9.2), 3) The frequency $\omega_1$ and its overtones $2\omega_1, 3\omega_1, \ldots, k\omega_1$ are not equal to the other natural frequencies $\omega_2, \ldots, \omega_n$. (that is, there are no internal resonances).

We shall find the asymptotic approximate formulae for partial solutions corresponding to the single-frequency oscillations which are close to the principal oscillation of the frequency $\omega_1$.

With the above assumptions, in the first approximation, one can retain in (9.4) only the coordinate $\xi_1$. The other coordinates $\xi_2, \xi_3, \ldots, \xi_N$ are far from the resonance, their oscillations, with the presence of unavoidable damping forces, will be small in comparison with the resonance oscillation of the coordinate $\xi_1$, and can be neglected in the first approximation. So, we have:

$$q_i = d_i^{(1)} \xi_1, \tag{9.9}$$

where $\xi_1$ satisfies the equation

$$\ddot{\xi}_1 + \omega_1^2 \xi_1 = \frac{\varepsilon}{M_1} \sum_{j=1}^{N} d_j^{(1)} f_j, \tag{9.10}$$

$$f_j = Q_j^{(1)}(\nu t, d_1^{(1)}\xi_1, d_2^{(1)}\xi_1, \ldots, d_N^{(1)}\xi_1, \ldots, d_N^{(1)}\dot{\xi}_1) + \varepsilon \ldots \tag{9.11}$$

The problem of finding the solution of equations (9.1) leads to the problem of solving a rather simple equation (9.10), which is investigated in detail in paragraph 3. This is the nature of the approximate method proposed by Bogoliubov for studying the dynamical systems with multiple degrees of freedom.

## 10. Multi-frequency Oscillations in Nonlinear Systems with Multiple Degrees of Freedom [38,c]

In this paragraph, we investigate a system with $N$ degrees of freedom described by the differential equation of general form:

$$\begin{aligned}
&\frac{d}{dt}\Big[\sum_{i=1}^{N} a_{ij}(\tau)\dot{q}_i\Big] + \sum_{i=1}^{N} b_{ij}(\tau) q_i = \\
&= \varepsilon Q_j^{(1)}(\tau, \theta_1, \ldots, \theta_r, q_1, \ldots, q_N, \dot{q}_1, \ldots, \dot{q}_N) + \\
&+ \varepsilon^2 Q_j^{(2)}(\tau, \theta_1, \ldots, \theta_r, q_1, \ldots, q_N, \dot{q}_1, \ldots, \dot{q}_N) + \varepsilon^3 \ldots \\
&j = 1, 2, \ldots, N,
\end{aligned} \tag{10.1}$$

where $\frac{d\theta_r}{dt} = \nu_r(\tau)$. When $\tau = \text{const}$ and $r = 1$, $\theta_1 = \nu t$, we have the system (9.1).

Together with (10.1), we consider the degenerate system

$$\frac{d}{dt}\Big[\sum_{i=1}^{N} a_{ij}(\tau)\dot{q}_i\Big] + \sum_{i=1}^{N} b_{ij}(\tau)q_i = 0, \; j = 1, 2, \ldots, N \tag{10.2}$$

which is obtained from (10.1) by setting $\varepsilon = 0$ and $\tau = \text{const}$. As we know, equations (10.2) have partial solutions

$$q_i^{(j)} = \varphi_i^{(j)} a_j \cos(\omega_j t + \gamma_j), \quad (i, j = 1, 2, \ldots, N) \tag{10.3}$$

which correspond to the principal oscillations. Here, $a_j$ and $\gamma_j$ are arbitrary constants, and $\omega_j$ ($j = 1, 2, \ldots, N$) are natural frequencies determined by the characteristic equation

$$\det \| - a_{ij}\omega^2 + b_{ij} \| = 0. \tag{10.4}$$

Functions $\varphi_i^j$ are principal functions, being nontrivial solutions of the algebraic equations:

$$\sum_{i=1}^{N} \big( - a_{ij}\omega_\ell^2 + b_{ij}\big)\varphi_i^{(\ell)} = 0 \tag{10.5}$$

and possess the orthogonality properties:

$$\sum_{i,j=1}^{N} a_{ij}\varphi_i^{(k)}\varphi_j^{(\ell)} = 0, \quad \sum_{i,j=1}^{N} b_{ij}\varphi_j^{(k)}\varphi_j^{(\ell)} = 0 \tag{10.6}$$

for $k \neq \ell$,

It is supposed that, between the natural frequencies $\omega_1, \ldots, \omega_N$ and the frequencies of external force $\nu_1(\tau), \ldots, \nu_r(\tau)$, there are $n$ resonance relations

$$\begin{aligned} &p_1^{(\alpha)}\omega_1(\tau) + p_2^{(\alpha)}\omega_2(\tau) + \cdots + p_N^{(\alpha)}\omega_N(\tau) + c_1^{(\alpha)}\nu_1(\tau) + \cdots + \\ &+ c_r^{(\alpha)}\nu_1(\tau) = 0, \qquad (\alpha = 1, 2, \ldots, n), \end{aligned} \tag{10.7}$$

where $p_i^{(\alpha)}$, $c_j^{(\alpha)}$ are small integers.

In this case, the solution of equations (10.1) will be found in the form:

$$\begin{aligned} &q_i = \sum_{s=1}^{N} \varphi_i^{(s)} a_s \cos(\xi_s + \psi_s) + \varepsilon u_i^{(1)}(a, \xi + \psi, \theta) + \varepsilon u_i^{(2)} \ldots, \\ &a = (a_1, \ldots, a_N), \quad \xi = (\xi_1, \ldots, \xi_N), \quad \psi = (\psi_1, \ldots, \psi_N), \\ &\theta = (\theta_1, \ldots, \theta_r), \qquad (i = 1, 2, \ldots, N), \end{aligned} \tag{10.8}$$

where $u_i^{(m)}$ are periodic functions of $\xi_1+\psi_1, \ldots, \xi_N+\psi_N, \theta_1, \ldots, \theta_r$ with period $2\pi$, and the amplitude $a_s$ and phase $\psi_s$ are determined from the equations

$$\begin{aligned}\frac{da_s}{dt} &= \varepsilon A_1^{(s)}(\tau, a, \psi) + \varepsilon^2 A_2^{(s)}(\tau, a, \psi) + O(\varepsilon^3),\\ \frac{d\psi_s}{dt} &= \omega_s(\tau) - \Omega_s(\tau) + \varepsilon B_1^{(s)}(\tau, a, \psi) + \varepsilon^2 B_2^{(s)}(\tau, a, \psi) + O(\varepsilon^3),\end{aligned} \tag{10.9}$$

with

$$\frac{d\xi_s}{dt} = \Omega_s, \; p_1^{(\alpha)}\xi_1 + \cdots + p_N^{(\alpha)}\xi_N + c_1^{(\alpha)}\theta_1 + \cdots + c_r^{(\alpha)}\theta_r, \quad s = 1, 2, \ldots, N. \tag{10.10}$$

The problem now leads to the determination of the functions $u_i^{(m)}$, $A_i^{(s)}$, $B_i^{(s)}$ and to integrating the system (10.9). We first calculate the derivatives

$$\begin{aligned}\frac{d^2 a_s}{dt^2} &= \varepsilon \sum_{\ell=1}^{N} (\omega_\ell - \Omega_\ell) \frac{\partial A_1^{(s)}}{\partial \psi_\ell} + \varepsilon^2 \Big\{ \frac{\partial A_1^{(s)}}{\partial \tau} + \\ &+ \sum_{\ell=1}^{N} \Big[ A_1^{(\ell)} \frac{\partial A_1^{(s)}}{\partial a_\ell} + B_1^{(\ell)} \frac{\partial A_1^{(s)}}{\partial \psi_\ell} + (\omega_\ell - \Omega_\ell) \frac{\partial A_2^{(s)}}{\partial \psi_\ell} \Big] \Big\} + O(\varepsilon^3),\\ \frac{d^2 \psi_s}{dt^2} &= \varepsilon \frac{d(\omega_s - \Omega_s)}{d\tau} + \varepsilon \sum_{\ell=1}^{N} (\omega_\ell - \Omega_\ell) \frac{\partial B_1^{(s)}}{\partial \psi_\ell} + \varepsilon^2 \Big\{ \frac{\partial B_1^{(s)}}{\partial \tau} + \\ &+ \sum_{\ell=1}^{N} \Big[ A_1^{(\ell)} \frac{\partial B_1^{(s)}}{\partial a_\ell} + B_1^{(\ell)} \frac{\partial B_1^{(s)}}{\partial \psi_\ell} + (\omega_\ell - \Omega_\ell) \frac{\partial B_2^{(s)}}{\partial \psi_\ell} \Big] \Big\} + O(\varepsilon^3),\end{aligned}$$

$$\begin{aligned}\frac{dq_i}{dt} &= -\sum_{s=1}^{N} \omega_s \varphi_i^{(s)} a_s \sin(\xi_s + \psi_s) + \varepsilon \Big\{ \sum_{k=1}^{r} \nu_k \frac{\partial u_i^{(1)}}{\partial \theta_k} + \\ &+ \sum_{s=1}^{N} A_1^{(s)} \varphi_i^{(s)} \cos(\xi_s + \psi_s) + \sum_{s=1}^{N} \frac{d\varphi_i^{(s)}}{d\tau} a_s \cos(\xi_s + \psi_s) + \\ &+ \sum_{\ell=1}^{N} \omega_\ell \frac{\partial u_i^{(1)}}{\partial(\xi_\ell + \psi_\ell)} - \sum_{s=1}^{\ell} a_s \varphi_i^{(s)} B_1^{(s)} \sin(\xi_s + \psi_s) \Big\} + \\ &+ \varepsilon^2 \Big\{ \sum_{s=1}^{N} A_2^{(s)} \varphi_i^{(s)} \cos(\xi_s + \psi_s) - \sum_{s=1}^{N} B_2^{(s)} \varphi_i^{(s)} a_s \sin(\xi_s + \\ &+ \psi_s) + \frac{\partial u_i^{(1)}}{\partial \tau} + \sum_{\ell=1}^{N} A_1^{(\ell)} \frac{\partial u_i^{(1)}}{\partial a_\ell} + \sum_{\ell=1}^{N} B_1^{(\ell)} \frac{\partial u_i^{(1)}}{\partial(\xi_\ell + \psi_\ell)} + \\ &+ \sum_{\ell=1}^{N} \omega_\ell \frac{\partial u_i^{(2)}}{\partial(\xi_\ell + \psi_\ell)} + \sum_{k=1}^{r} \nu_k \frac{\partial u_i^{(2)}}{\partial \theta_k} \Big\} + O(\varepsilon^3),\end{aligned}$$

$$\frac{d^2q_i}{dt^2} = -\sum_{s=1}^{N} \omega_s^2 \varphi_i^{(s)} a_s \cos(\xi_s + \psi_s) +$$

$$+ \varepsilon\Big\{ -\sum_{s=1}^{N} \frac{d(\omega_s \varphi_i^{(s)})}{\partial \tau} a_s \sin(\xi_s + \psi_s) - \sum_{s=1}^{N} 2\omega_s \varphi_i^{(s)} A_1^{(s)} \sin(\xi_s + \psi_s) -$$

$$- \sum_{s=1}^{N} 2\omega_s \varphi_i^{(s)} B_1^{(s)} a_s \cos(\xi_s + \psi_s) + \sum_{\ell=1}^{N} \sum_{k=1}^{r} 2\nu_k \omega_\ell \frac{\partial^2 u_i^{(1)}}{\partial \theta_k \partial(\xi_\ell + \psi_\ell)} +$$

$$+ \sum_{h,k=1}^{r} \nu_h \nu_k \frac{\partial^2 u_i^{(1)}}{\partial \theta_h \partial \theta_k} + \sum_{s,\ell=1}^{N} (\omega_\ell - \Omega_\ell) \frac{\partial A_1^{(s)}}{\partial \psi_\ell} \varphi_i^{(s)} \cos(\xi_s + \psi_s) -$$

$$- \sum_{s=1}^{N} \omega_s a_s \frac{d\varphi_i^{(s)}}{d\tau} \sin(\xi_s + \psi_s) + \sum_{s,\ell=1}^{N} \omega_j \omega_\ell \frac{\partial^2 u_i^{(1)}}{\partial(\xi_\ell + \psi_\ell)\partial(\xi_s + \psi_s)} -$$

$$- \sum_{\ell,s=1}^{N} (\omega_\ell - \Omega_\ell) a_s \varphi_i^{(s)} \frac{\partial B_1^{(s)}}{\partial \psi_\ell} \sin(\xi_s + \psi_s) \Big\} +$$

$$+ \varepsilon^2 \Big\{ -\sum_{s=1}^{N} 2\omega_s \varphi_i^{(s)} A_2^{(s)} \sin(\xi_s + \psi_s) - \sum_{s=1}^{N} 2\omega_s \varphi_i^{(s)} B_2^{(s)} a_s \cos(\xi_s + \psi_s) +$$

$$+ \sum_{k=1}^{r} \frac{d\nu_k}{dt} \frac{\partial u_i^{(1)}}{\partial \theta_k} + \sum_{k=1}^{r} 2\nu_k \frac{\partial^2 u_i^{(1)}}{\partial \theta_k \partial \tau} + \sum_{\ell=1}^{N} \sum_{k=1}^{r} 2\nu_k A_1^{(\ell)} \frac{\partial^2 u_1^{(1)}}{\partial \theta_k \partial a_\ell} +$$

$$+ \sum_{\ell=1}^{N} \sum_{k=1}^{r} 2\nu_k B_1^{(\ell)} \frac{\partial^2 u_i^{(1)}}{\partial \theta_k \partial(\xi_\ell + \psi_\ell)} + \sum_{s=1}^{N} \varphi_i^{(s)} \frac{\partial A_1^{(s)}}{\partial \tau} \cos(\xi_s + \psi_s) +$$

$$+ \sum_{\ell,s=1}^{N} A_1^{(\ell)} \varphi_i^{(s)} \frac{\partial A_1^{(s)}}{\partial a_\ell} \cos(\xi_s + \psi_s) + \sum_{\ell,s=1}^{N} B_1^{(\ell)} \varphi_i^{(s)} \frac{\partial A_1^{(s)}}{\partial \psi_\ell} \cos(\xi_s + \psi_s) +$$

$$+ \sum_{s=1}^{N} 2A_1^{(s)} \frac{d\varphi_i^{(s)}}{d\tau} \cos(\xi_s + \psi_s) - \sum_{s=1}^{N} 2A_1^{(s)} B_1^{(s)} \varphi_i^{(s)} \sin(\xi_s + \psi_s) +$$

$$+ \sum_{s=1}^{N} \frac{d^2\varphi_i^{(s)}}{d\tau^2} a_s \cos(\xi_s + \psi_s) - \sum_{s=1}^{N} 2B_1^{(s)} \frac{d\varphi_i^{(s)}}{d\tau} a_s \sin(\xi_s + \psi_s) +$$

$$+ \sum_{\ell=1}^{N} \frac{d\omega_\ell}{d\tau} \frac{\partial u_i^{(1)}}{\partial(\xi_\ell + \psi_\ell)} + \sum_{\ell=1}^{N} 2\omega_\ell \frac{\partial^2 u_i^{(1)}}{\partial \tau \partial(\xi_\ell + \psi_\ell)} +$$

$$+ \sum_{\ell,s=1}^{N} 2\omega_\ell A_1^{(j)} \frac{\partial^2 u_i^{(1)}}{\partial a_s \partial(\xi_\ell + \psi_\ell)} + \sum_{\ell,s=1}^{N} 2\omega_\ell B_1^{(s)} \frac{\partial^2 u_i^{(1)}}{\partial(\xi_\ell + \psi_\ell)\partial(\xi_s + \psi_s)} -$$

$$- \sum_{s=1}^{N} a_s \varphi_i^{(s)} \frac{\partial B_1^{(s)}}{\partial \tau} \sin(\xi_s + \psi_s) - \sum_{\ell,s=1}^{N} a_s \varphi_i^{(s)} A_1^{(\ell)} \frac{\partial B_1^{(s)}}{\partial a_\ell} \sin(\xi_s + \psi_s) -$$

$$- \sum_{\ell,s=1}^{N} a_s \varphi_i^{(s)} B_i^{(\ell)} \frac{\partial B_1^{(s)}}{\partial \psi_\ell} \sin(\xi_s + \psi_s) - \sum_{s=1}^{\ell} a_s \varphi_i^{(s)} B_1^{(s)^2} \cos(\xi_s + \psi_s) +$$

$$+ \sum_{\ell,s=1}^{N} (\omega_\ell - \Omega_\ell) \frac{\partial A_2^{(s)}}{\partial \psi_\ell} \varphi_i^{(s)} \cos(\xi_s + \psi_s) +$$

$$+ \sum_{\ell,s=1}^{N} \frac{\partial B_2^{(s)}}{\partial \psi_\ell} \varphi_i^{(s)} a_s (\omega_\ell - \Omega_\ell) \sin(\xi_s + \psi_s) + \sum_{\ell,s=1}^{N} (\omega_s - \Omega_s) \frac{\partial A_1^{(\ell)}}{\partial \psi_s} \frac{\partial u_i^{(1)}}{\partial a_\ell} +$$

$$+ \sum_{\ell,s=1}^{N} (\omega_s - \Omega_s) \frac{\partial B_1^{(\ell)}}{\partial \psi_s} \frac{\partial u_i^{(1)}}{\partial (\xi_\ell + \psi_\ell)} + \sum_{\ell,s=1}^{N} \omega_\ell \omega_s \frac{\partial^2 u_i^{(2)}}{\partial (\xi_\ell + \psi_\ell) \partial (\xi_s + \psi_s)} +$$

$$+ 2 \sum_{k=1}^{r} \sum_{\ell=1}^{N} \omega_\ell \nu_k \frac{\partial^2 u_i^{(2)}}{\partial \theta_k \partial (\xi_\ell + \psi_\ell)} + \sum_{h,k=1}^{r} \nu_k \nu_h \frac{\partial^2 u_i^{(2)}}{\partial \theta_k \partial \theta_h} \Big\} + O(\varepsilon^3).$$

Substituting these values into the left-hand sides of equations (10.1), we obtain:

$$\frac{d}{dt} \Big\{ \sum_{i=1}^{N} a_{ij}(\tau) \dot{q}_i \Big\} + \sum_{i=1}^{N} b_{ij}(\tau) q_i = \varepsilon \sum_{i=1}^{N} \Big\{ a_{ij} \sum_{s=1}^{N} \Big[ - 2 a_s \omega_s \varphi_i^{(s)} B_1^{(s)} +$$

$$+ \sum_{\ell=1}^{N} (\omega_\ell - \Omega_\ell) \varphi_i^{(s)} \frac{\partial A_1^{(s)}}{\partial \psi_\ell} \Big] \cos(\xi_s + \psi_s) - a_{ij} \sum_{s=1}^{N} \Big[ 2 \omega_s \varphi_i^{(s)} A_1^{(s)} +$$

$$+ \sum_{\ell=1}^{N} (\omega_\ell - \Omega_\ell) a_s \varphi_i^{(s)} \frac{\partial B_1^{(s)}}{\partial \psi_\ell} \Big] \sin(\xi_s + \psi_s) - \sum_{s=1}^{N} \Big[ a_{ij} \times$$

$$\times \Big( \frac{d(\omega_s \varphi_i^{(s)})}{d\tau} + \frac{d\varphi_i^{(s)}}{d\tau} \Big) + \omega_s \varphi_i^{(s)} \frac{d a_{ij}}{d\tau} \Big] a_s \sin(\xi_s + \psi_s) +$$

$$+ a_{ij} \Big[ \sum_{h,k=1}^{r} \nu_k \nu_h \frac{\partial u_i^{(1)}}{\partial \theta_k \partial \theta_h} + 2 \sum_{k=1}^{r} \sum_{\ell=1}^{N} \nu_k \omega_\ell \frac{\partial^2 u_i^{(1)}}{\partial \theta_k \partial (\xi_\ell + \psi_\ell)} +$$

$$+ \sum_{s,\ell=1}^{N} \omega_s \omega_\ell \frac{\partial^2 u_i^{(1)}}{\partial (\xi_s + \psi_s) \partial (\xi_\ell + \psi_\ell)} \Big] + b_{ij} u_i^{(1)} \Big\} +$$

$$+ \varepsilon^2 \sum_{i=1}^{N} \Big\{ a_{ij} \sum_{s=1}^{N} \Big[ - 2 \omega_s a_s \varphi_i^{(s)} B_2^{(s)} +$$

$$+ \sum_{\ell=1}^{N} (\omega_\ell - \Omega_\ell) \varphi_i^{(s)} \frac{\partial A_2^{(s)}}{\partial \psi_\ell} \Big] \cos(\xi_s + \psi_s) -$$

$$- a_{ij} \sum_{s=1}^{N} \Big[ 2 \omega_s \varphi_i^{(s)} A_2^{(s)} + \sum_{\ell=1}^{N} (\omega_\ell - \Omega_\ell) a_s \varphi_i^{(s)} \frac{\partial B_2^{(s)}}{\partial \psi_\ell} \Big] \sin(\xi_s + \psi_s) +$$

$$+a_{ij}\sum_{s=1}^{N}\Big[\varphi_i^{(s)}\frac{\partial A_1^{(s)}}{\partial\tau}-a_s\varphi_i^{(s)}B_1^{(s)^2}+\sum_{\ell=1}^{N}A_1^{(\ell)}\varphi_i^{(s)}\frac{\partial A_1^{(s)}}{\partial a_\ell}+$$

$$+\sum_{\ell=1}^{N}B_1^{(\ell)}\varphi_i^{(s)}\frac{\partial A_1^{(s)}}{\partial\psi_\ell}\Big]\cos(\xi_s+\psi_s)-a_{ij}\sum_{s=1}^{N}\Big[\sum_{\ell=1}^{N}a_s\varphi_i^{(s)}B_1^{(\ell)}\frac{\partial B_1^{(s)}}{\partial\psi_\ell}+$$

$$+\sum_{\ell=1}^{N}a_s\varphi_i^{(s)}A_1^{(\ell)}\frac{\partial B_1^{(s)}}{\partial a_\ell}+2A_1^{(s)}B_1^{(s)}\varphi_i^{(s)}+a_s\varphi_i^{(s)}\frac{\partial B_1^{(s)}}{\partial\tau}\Big]\sin(\xi_s+\psi_s)+$$

$$+\sum_{s=1}^{N}\Big[a_{ij}a_s\frac{d^2\varphi_i^{(s)}}{d\tau^2}+2a_{ij}A_1^{(s)}\frac{d\varphi_i^{(s)}}{d\tau}+a_s\frac{da_{ij}}{d\tau}\frac{d\varphi_i^{(s)}}{d\tau}+$$

$$+\frac{da_{ij}}{d\tau}A_1^{(s)}\varphi_i^{(s)}\Big]\cos(\xi_s+\psi_s)-\sum_{s=1}^{N}\Big[2a_{ij}B_1^{(s)}\frac{d\varphi_i^{(s)}}{d\tau}+$$

$$+B_1^{(s)}\varphi_i^{(s)}\frac{da_{ij}}{d\tau}\Big]a_s\sin(\xi_s+\psi_s)\Big\}+O(\varepsilon^3). \tag{10.11}$$

Replacing $q_i$ and $\dot{q}_i$ in the right-hand sides of (10.1) by their values found above, we get

$$\varepsilon Q_j(\tau,\theta,q,\dot{q},\varepsilon)=\varepsilon Q_j^{(1)o}+\varepsilon^2\Big\{Q_j^{(2)o}+\sum_{i=1}^{N}\Big[u_i^{(1)}\frac{\partial Q_j^{(1)o}}{\partial q_i^o}+$$

$$+\frac{\partial Q_j^{(1)o}}{\partial\dot{q}_i^o}\Big(\sum_{k=1}^{r}\nu_k\frac{\partial u_i^{(1)}}{\partial\theta_k}+\sum_{s=1}^{N}A_1^{(s)}\varphi_i^{(s)}\cos(\xi_s+\psi_s)+$$

$$+\sum_{s=1}^{N}\frac{d\varphi_i^{(s)}}{d\tau}a_s\cos(\xi_s+\psi_s)-\sum_{s=1}^{N}a_s\varphi_i^{(s)}B_1^{(s)}\sin(\xi_s+\psi_s)+$$

$$+\sum_{\ell=1}^{N}\omega_\ell\frac{\partial u_i^{(1)}}{\partial(\xi_\ell+\psi_\ell)}\Big)\Big]\Big\}+O(\varepsilon^3), \tag{10.12}$$

where

$$Q_j^{(1)o}=Q_j^{(1)}(\tau,\theta,q^o,\dot{q}^o),\quad Q_j^{(2)o}=Q_j^{(2)}(\tau,\theta,q^o,\dot{q}^o),$$

$$q_i^o=\sum_{j=1}^{N}\varphi_i^{(j)}a_j\cos(\xi_j+\psi_j),\quad (i,j=1,2,\ldots,N). \tag{10.13}$$

Comparing the coefficients of the like powers of $\varepsilon$ in (10.11) and (10.12), we obtain

$$\sum_{i=1}^{N}\Big\{a_{ij}\Big[\sum_{s=1}^{N}\sum_{\ell=1}^{N}\omega_s\omega_\ell\frac{\partial^2 u_i^{(1)}}{\partial(\xi_s+\psi_s)\partial(\xi_\ell+\psi_\ell)}+$$
$$+2\sum_{\ell=1}^{N}\sum_{k=1}^{r}\nu_k\omega_\ell\frac{\partial^2 u_i^{(1)}}{\partial\theta_k\partial(\xi_\ell+\psi_\ell)}+\sum_{h,k=1}^{N}\nu_h\nu_k\frac{\partial^2 u_i^{(1)}}{\partial\theta_k\partial\theta_h}\Big]+$$
$$+b_{ij}u_i^{(1)}\Big\}=Q_j^{(1)o}(\tau,\theta,q^o,\dot q^o)-\sum_{i=1}^{N}\Big\{a_{ij}\sum_{s=1}^{N}\Big[\sum_{\ell=1}^{N}(\omega_\ell-\Omega_\ell)\times$$
$$\times\varphi_i^{(s)}\frac{\partial A_1^{(s)}}{\partial\psi_\ell}-2a_s\omega_s\varphi_i^{(s)}B_1^{(s)}\Big]\cos(\xi_s+\psi_s)-$$
$$-a_{ij}\sum_{s=1}^{N}\Big[\sum_{\ell=1}^{N}(\omega_\ell-\Omega_\ell)a_s\varphi_i^{(s)}\frac{\partial B_1^{(s)}}{\partial\psi_\ell}+2\omega_s\varphi_i^{(s)}A_1^{(s)}\Big]\times$$
$$\times\sin(\xi_s+\psi_s)-\sum_{s=1}^{N}\Big[a_{ij}\Big(\frac{d(\omega_s\varphi_i^{(s)})}{d\tau}+\omega_s\frac{d\varphi_i^{(s)}}{d\tau}\Big)+$$
$$+\omega_s\varphi_i^{(s)}\frac{da_{ij}}{d\tau}\Big]a_s\sin(\xi_s+\psi_s)\Big\},\quad (j=1,2,\ldots,N),\tag{10.14}$$

$$\sum_{i=1}^{N}\Big\{a_{ij}\Big[\sum_{s=1}^{N}\sum_{\ell=1}^{N}\omega_s\omega_\ell\frac{\partial^2 u_i^{(2)}}{\partial(\xi_s+\psi_s)\partial(\xi_\ell+\psi_\ell)}+$$
$$+2\sum_{\ell=1}^{N}\sum_{k=1}^{r}\nu_k\omega_\ell\frac{\partial^2 u_i^{(2)}}{\partial\theta_k\partial(\xi_\ell+\psi_\ell)}+\sum_{h,k=1}^{r}\nu_k\nu_h\frac{\partial^2 u_i^{(2)}}{\partial\theta_h\partial\theta_k}\Big]+b_{ij}u_i^{(2)}\Big\}=$$
$$=R_{j*2}^{(2)o}-\sum_{i=1}^{N}\Big\{a_{ij}\sum_{s=1}^{N}\Big[\sum_{\ell=1}^{N}(\omega_\ell-\Omega_\ell)\varphi_i^{(s)}\frac{\partial A_2^{(s)}}{\partial\psi_\ell}-$$
$$-2a_s\omega_s\varphi_i^{(s)}B_2^{(s)}\Big]\cos(\xi_s+\psi_s)-a_{ij}\sum_{s=1}^{N}\Big[\sum_{\ell=1}^{N}(\omega_\ell-\Omega_\ell)a_s\varphi_i^{(s)}\frac{\partial B_2^{(s)}}{\partial\psi_\ell}+$$
$$+2\omega_s\varphi_i^{(s)}A_2^{(s)}\Big]\sin(\xi_s+\psi_s)+a_{ij}\sum_{s=1}^{N}\Big[\varphi_i^{(s)}\frac{\partial A_1^{(s)}}{\partial\tau}+\sum_{\ell=1}^{N}A_1^{(\ell)}\varphi_i^{(s)}\frac{\partial A_1^{(s)}}{\partial a_\ell}+$$
$$+\sum_{\ell=1}^{N}B_1^{(\ell)}\varphi_i^{(s)}\frac{\partial A_1^{(s)}}{\partial\psi_\ell}-a_s\varphi_1^{(s)}B_1^{(s)^2}\Big]\cos(\xi_s+\psi_s)-$$
$$-a_{ij}\sum_{s=1}^{N}\Big[2A_1^{(s)}B_1^{(s)}\varphi_i^{(s)}+\sum_{\ell=1}^{N}a_s\varphi_i^{(s)}A_1^{(\ell)}\frac{\partial B_1^{(s)}}{\partial a_\ell}+$$
$$+\sum_{\ell=1}^{N}a_s\varphi_i^{(s)}B_1^{(\ell)}\frac{\partial B_1^{(s)}}{\partial\psi_\ell}+a_s\varphi_i^{(s)}\frac{\partial B_1^{(s)}}{\partial\tau}\Big]\sin(\xi_s+\psi_s)+\sum_{s=1}^{N}\Big[a_{ij}a_s\frac{d^2\varphi_i^{(s)}}{d\tau^2}+$$

$$+ 2a_{ij}A_1^{(s)}\frac{d\varphi_i^{(s)}}{d\tau} + a_s\frac{da_{ij}}{d\tau}\frac{d\varphi_i^{(s)}}{d\tau} + \frac{da_{ij}}{d\tau}A_1^{(s)}\varphi_i^{(s)}\Big]\cos(\xi_s+\psi_s)-$$

$$-\sum_{s=1}^{N}\Big[2a_{ij}B_1^{(s)}\frac{d\varphi_i^{(s)}}{d\tau} + \frac{da_{ij}}{d\tau}B_1^{(s)}\varphi_i^{(s)}\Big]a_s\sin(\xi_s+\psi_s)\Big\}, \qquad (10.15)$$

where

$$R_{j*2}^{(2)o} = Q_j^{(2)o} + \sum_{i=1}^{N}\Big[u_i^{(1)}\frac{\partial Q_j^{(1)o}}{\partial q_j^o} + \frac{\partial Q_j^{(1)o}}{\partial \dot{q}_j^o}\Big(\sum_{k=1}^{r}\nu_k\frac{\partial u_i^{(1)}}{\partial\theta_k}+$$

$$+\sum_{s=1}^{N}A_1^{(s)}\varphi_i^{(s)}\cos(\xi_s+\psi_s) + \sum_{s=1}^{N}\frac{d\varphi_i^{(s)}}{d\tau}a_s\cos(\xi_s+\psi_s)-$$

$$-\sum_{s=1}^{N}a_s\varphi_i^{(s)}B_1^{(s)}\sin(\xi_s+\psi_s) + \sum_{\ell=1}^{N}\omega_\ell\frac{\partial u_i^{(1)}}{\partial(\xi_\ell+\psi_\ell)}\Big)\Big]-$$

$$-\sum_{i=1}^{N}a_{ij}\Big\{2\sum_{\ell=1}^{N}\omega_\ell\frac{\partial^2 u_i^{(1)}}{\partial(\xi_\ell+\psi_\ell)\partial\tau} + 2\sum_{k=1}^{r}\nu_k\frac{\partial^2 u_i^{(1)}}{\partial\tau\partial\theta_k}+$$

$$+2\sum_{\ell=1}^{N}\sum_{k=1}^{r}\nu_k A_1^{(\ell)}\frac{\partial^2 u_i^{(1)}}{\partial\theta_k\partial a_\ell} + 2\sum_{\ell=1}^{N}\sum_{k=1}^{r}\nu_k B_1^{(\ell)}\frac{\partial^2 u_i^{(1)}}{\partial\theta_k\partial(\xi_\ell+\psi_\ell)}+$$

$$+2\sum_{s=1}^{N}\sum_{\ell=1}^{N}\omega_\ell B_1^{(s)}\frac{\partial^2 u_i^{(1)}}{\partial(\xi_\ell+\psi_\ell)\partial(\xi_s+\psi_s)} + 2\sum_{s,\ell=1}^{N}\omega_\ell A_1(s)\frac{\partial^2 u_i^{(1)}}{\partial a_s\partial(\xi_\ell+\psi_\ell)}$$

$$+\sum_{s,\ell=1}^{N}(\omega_s-\Omega_s)\frac{\partial B_1^{(\ell)}}{\partial\psi_s}\frac{\partial u_i^{(1)}}{\partial(\xi_\ell+\psi_\ell)} + \sum_{s,\ell=1}^{N}(\omega_s-\Omega_s)\frac{\partial A_1^{(\ell)}}{\partial\psi_s}\frac{\partial u_i^{(1)}}{\partial a_\ell}+$$

$$+\sum_{k=1}^{r}\frac{d\nu_k}{d\tau}\frac{\partial u_i^{(1)}}{\partial\theta_k} + \sum_{\ell=1}^{N}\frac{d\omega_\ell}{d\tau}\frac{\partial u_i^{(1)}}{\partial(\xi_\ell+\psi_\ell)}\Big\}-$$

$$-\sum_{i=1}^{N}\frac{da_{ij}}{d\tau}\Big[\sum_{k=1}^{r}\nu_k\frac{\partial u_i^{(1)}}{\partial\theta_k} + \sum_{\ell=1}^{N}\omega_\ell\frac{\partial u_i^{(1)}}{\partial(\xi_\ell+\psi_\ell)}\Big],$$

$$\cdots\cdots\cdots\cdots\cdots\cdots \qquad (10.16)$$

For obtaining $A_1^{(s)}$, $B_1^{(s)}$ and $u_i^{(1)}$, we multiply the j-th equation of (10.14) by $\varphi_j^{(v)}$ then sum on $j$ from 1 to $N$. This leads to

$$M_v\Big[\sum_{\ell=1}^{N}(\omega_\ell-\Omega_\ell)\frac{\partial A_1^{(v)}}{\partial\psi_\ell} - 2a_v\omega_v B_1^{(v)}\Big]\cos(\xi_v+\psi_v)-$$

$$-a_v M_v\frac{d\omega_v}{d\tau}\sin(\xi_v+\psi_v) - M_v\Big[\sum_{\ell=1}^{N}(\omega_\ell-\Omega_\ell)a_v\frac{\partial B_1^{(v)}}{\partial\psi_\ell}+$$

$$+ 2\omega_v A_1^{(v)}\Big] \sin(\xi_v - \psi_v) - \omega_v a_v \frac{dM_v}{d\tau} \sin(\xi_v + \psi_v) -$$

$$- \sum_{\substack{s=1 \\ s \neq v}}^{N} \omega_s a_s \Big[ \sum_{i,j=1}^{N} a_{ij} \Big( \varphi_j^{(v)} \frac{d\varphi_j^{(s)}}{d\tau} - \varphi_i^{(s)} \frac{d\varphi_j^{(v)}}{d\tau} \Big) \Big] \sin(\xi_s + \psi_s) +$$

$$+ \sum_{i,j=1}^{N} a_{ij} \varphi_j^{(v)} \Big( \sum_{s,\ell=1}^{N} \omega_s \omega_\ell \frac{\partial^2 u_i^{(1)}}{\partial(\xi_s + \psi_s) \partial(\xi_\ell + \psi_\ell)} +$$

$$+ 2 \sum_{\ell=1}^{N} \sum_{k=1}^{r} \nu_k \omega_\ell \frac{\partial^2 u_i^{(1)}}{\partial \theta_k \partial(\xi_\ell + \psi_\ell)} + \sum_{h,k=1}^{r} \nu_h \nu_k \frac{\partial^2 u_i^{(1)}}{\partial \theta_h \partial \theta_k} +$$

$$+ \omega_v^2 u_i^{(1)} \Big) = \sum_{j=1}^{N} Q_j^{(1)o} \varphi_j^{(v)}, \quad M_v = \sum_{ij} a_{ij} \varphi_i^{(v)} \varphi_j^{(v)}. \tag{10.17}$$

Here, the following formula is used:

$$\sum_{s,i,j=1}^{N} \Big[ a_{ij} \varphi_j^{(v)} \Big( \frac{d(\omega_s \varphi_j^{(s)})}{d\tau} + \omega_s \frac{d\varphi_i^{(s)}}{d\tau} \Big) + \omega_s \varphi_i^{(s)} \varphi_j^{(v)} \frac{da_{ij}}{d\tau} \Big] a_s \times$$

$$\times \sin(\xi_s + \psi_s) = \sum_{s,i,j=1}^{N} \Big[ a_{ij} \varphi_i^{(s)} \varphi_j^{(v)} \Big] \frac{d\omega_s}{d\tau} a_s \sin(\xi_s + \psi_s) +$$

$$+ \sum_{s,i,j=1}^{N} \omega_s \Big( 2a_{ij} \varphi_j^{(v)} \frac{d\varphi_j^{(s)}}{d\tau} + \varphi_i^{(s)} \varphi_j^{(v)} \frac{da_{ij}}{\partial \tau} \Big) a_s \sin(\xi_s + \psi_s) =$$

$$= M_v a_v \frac{d\omega_v}{d\tau} \sin(\xi_v + \psi_v) + a_v \omega_v \frac{dM_v}{d\tau} \sin(\xi_v + \psi_v) +$$

$$+ \sum_{\substack{s=1 \\ s \neq v}}^{N} a_s \omega_s \Big[ \sum_{i,j=1}^{N} a_{ij} \Big( \varphi_j^{(v)} \frac{d\varphi_i^{(s)}}{\partial \tau} - \frac{d\varphi_j^{(v)}}{\partial \tau} \varphi_i^{(s)} \Big) \Big] \sin(\xi_s + \psi_s) \tag{10.18}$$

because

$$\sum_{i,j=1}^{N} \Big[ \varphi_i^{(s)} \varphi_j^{(v)} \frac{da_{ij}}{d\tau} + a_{ij} \varphi_j^{(v)} \frac{d\varphi_i^{(s)}}{d\tau} + a_{ij} \varphi_i^{(s)} \frac{d\varphi_j^{(v)}}{d\tau} \Big] = \begin{cases} 0 & \text{if} \quad s \neq v, \\ \dfrac{dM_v}{d\tau} & \text{if} \quad s = v, \end{cases}$$

$$\sum_{i,j=1}^{N} a_{ij} \Big( \varphi_j^{(v)} \frac{d\varphi_i^{(s)}}{d\tau} - \varphi_i^{(s)} \frac{d\varphi_j^{(v)}}{d\tau} \Big) =$$

$$= \sum_{i,j=1}^{N} a_{ij} \Big( \varphi_i^{(v)} \frac{d\varphi_j^{(s)}}{d\tau} - \varphi_i^{(s)} \frac{d\varphi_j^{(v)}}{d\tau} \Big) = 0 \quad \text{if} \quad s = v.$$

We now expand functions (9.13) in a Fourier series:

$$Q_j^{(1)o}(\tau,\theta,q^o,\dot q^o)=\sum_{p,c}Q_{jpc}^{(1)o}\,e^{i[p_1(\xi_1+\psi_1)+\cdots+p_N(\xi_n+\psi_N)+c_1\theta_1+\cdots+c_r\theta_r]},\tag{10.19}$$

where

$$\begin{aligned}Q_{jpc}^{(1)o}&=\frac{1}{(2\pi)^{N+r}}\int_0^{2\pi}\cdots\int_0^{2\pi}\overline{Q}_j^{(1)o}e^{-i[p_1(\overline{\xi}_1+\overline{\psi}_1+\cdots+c_r\overline{\theta}_r]}\times\\&\times d(\overline{\xi}_1+\overline{\psi}_1)\dots d(\overline{\xi}_N+\overline{\psi}_N)d\overline{\theta}_1\dots d\overline{\theta}_r,\\\overline{Q}_j^{(1)o}&=Q_j^{(1)o}(\tau,a,\overline{\xi}+\overline{\psi},\overline{\theta}).\end{aligned}\tag{10.20}$$

The unknown functions $u_i^{(1)}$ are also found in the form of a series:

$$u_i^{(1)}=\sum_{p,c}u_{ipc}e^{i[p_1(\xi_1+\psi_1)+\cdots+p_N(\xi_N+\psi_N)+c_1\theta_1+\cdots+c_r\theta_r]}.\tag{10.21}$$

Substituting $Q_j^{(1)o}$ and $u_i^{(1)}$ from (10.19) and (10.21) into (10.17), we have

$$\begin{aligned}&M_v\Big[\sum_{\ell=1}^{N}(\omega_\ell-\Omega_\ell)\frac{\partial A_1^{(v)}}{\partial\psi_\ell}-2a_v\omega_vB_1^{(v)}\Big]\cos(\xi_v+\psi_v)-\\&-\Big[M_v\Big(\sum_{\ell=1}^{N}(\omega_\ell-\Omega_\ell)a_v\frac{\partial B_1^{(v)}}{\partial\psi_\ell}+2\omega_vA_1^{(v)}\Big)+a_v\frac{d(\omega_vM_v)}{d\tau}\Big]\sin(\xi_v+\psi_v)-\\&-\sum_{\substack{s=1\\s\neq v}}^{N}\omega_sa_s\Big[\sum_{j,\ell=1}^{N}a_{\ell j}\Big(\varphi_j^{(v)}\frac{d\varphi_\ell^{(s)}}{d\tau}-\varphi_\ell^{(s)}\frac{d\varphi_j^{(v)}}{d\tau}\Big)\Big]\sin(\xi_s+\psi_s)+\\&+\sum_{p,c}\Big[\omega_v^2-\Big(\sum_{s=1}^{N}p_s\omega_s+\sum_{k=1}^{r}c_k\nu_k\Big)^2\Big]\Big\{\sum_{i,j=1}^{N}a_{ij}\varphi_j^{(v)}\times\\&\times u_{ipc}^{(1)}e^{i[p_1(\xi_1+\psi_1)+\cdots+p_N(\xi_N+\psi_N)+c_1\theta_1+\cdots+c_r\theta_r]}\Big\}=\\&=\sum_{p,c}e^{i[p_1(\xi_1+\psi_1)+\cdots+c_r\theta_r]}\int_0^{2\pi}\cdots\int_0^{2\pi}\sum_{j=1}^{N}\varphi_j^{(v)}\overline{Q}_j^{(1)o}\times\\&\times e^{-i[p_1(\overline{\xi}_1+\overline{\psi}_1)+\cdots+c_r\overline{\theta}_r]}d(\overline{\xi}_1+\overline{\psi}_1)\dots d(\overline{\xi}_N+\overline{\psi}_N)\dots d\overline{\theta}_r.\end{aligned}\tag{10.22}$$

In the right-hand sides of (10.22), the coefficients of the fundamental harmonics $\sin(\xi_v+\psi_v)$ and $\cos(\xi_v+\psi_v)$ correspond to the values $p_{1v}^{(\beta)},\dots,p_{Nv}^{(\beta)}, c_{1v}^{(\beta)},\dots, c_{rv}^{(\beta)}$ of $p_1,\dots,p_N, c_1,\dots,c_r$, which satisfy the relations:

$$\begin{aligned}&p_{1v}^{(\beta)}(\xi_1+\psi_1)+\cdots+p_{Nv}^{(\beta)}(\xi_N+\psi_N)+\\&+c_{v1}^{(\beta)}\theta_1+\cdots+c_{rv}^{(\beta)}\theta_r=\pm(\xi_v+\psi_v)+\chi_v^{(\beta)}.\end{aligned}\tag{10.23}$$

On the other hand, from (10.10) it follows

$$p_1^{(\alpha)}(\xi_1+\psi_1)+\cdots+p_N^{(\alpha)}(\xi_N+\psi_N)+c_1^{(\alpha)}\theta_1+\cdots+c_r^{(\alpha)}\theta_r= \\ =p_1^{(\alpha)}\psi_1+\cdots+p_N^{(\alpha)}\psi_N, \quad (\alpha=1,2,\ldots,n). \tag{10.24}$$

Comparing (10.23) with (10.24), we have

$$p_{1v}^{(\beta)}=\sum_{\alpha=1}^{n}\sigma_{\alpha v}^{(\beta)}p_1^{(\alpha)},\ldots,p_{vv}^{(\beta)}\mp 1=\sum_{\alpha=1}^{n}\sigma_{\alpha v}^{(\beta)}p_v^{(\alpha)},\ldots,$$
$$p_{Nv}^{(\beta)}=\sum_{\alpha=1}^{n}\sigma_{\alpha v}^{(\beta)}p_N^{(\alpha)},\ c_{kv}^{(\beta)}=\sum_{\alpha=1}^{n}\sigma_{\alpha v}^{(\beta)}c_k^{(\alpha)},\quad (k=1,2,\ldots,r),$$
$$\chi_v^{(\beta)}=\sum_{\alpha=1}^{n}\left(p_1^{(\alpha)}\psi_1+\cdots+p_N^{(\alpha)}\psi_N\right)\sigma_{\alpha v}^{(\beta)}=\sum_{\alpha=1}^{n}\sigma_{\alpha v}^{(\beta)}\psi^{(\alpha)},$$
$$\psi^{(\alpha)}=p_1^{(\alpha)}\psi_1+\cdots+p_N^{(\alpha)}\psi_N, \tag{10.25}$$

where $\sigma_{\alpha v}^{(\beta)}$ are the coefficients of proportionality.

It is noted that

$$V=-\omega_v^2+\left(\sum_{s=1}^{N}p_s\omega_s+\sum_{k=1}^{r}c_k\nu_k\right)^2\simeq 0. \tag{10.26}$$

In fact, we have

$$V=\left[p_{1v}^{(\beta)}\omega_1+\cdots+(p_{vv}^{(\beta)}+1)\omega_v+\cdots+p_{Nv}^{(\beta)}\omega_N+c_{1v}^{(\beta)}\nu_1+\right. \\ \left.+\cdots+c_{rv}^{(\beta)}\nu_r\right]\left[p_{1v}^{(\beta)}\omega_1+\cdots+(p_{vv}^{(\beta)}-1)\omega_v+\cdots+c_{rv}^{(\beta)}\nu_r\right],$$

or following (10.25):

$$V=\left\{\sum_{\alpha=1}^{n}\sigma_{\alpha v}^{(\beta)}\left[p_1^{(\alpha)}\omega_1+\cdots+p_N^{(\alpha)}\omega_N+c_1^{(\alpha)}\nu_1+\cdots+c_r^{(\alpha)}\nu_r\right\}^2\simeq 0.$$

Comparing the coefficients of the fundamental harmonics $\sin(\xi_v+\psi_v)$ and $\cos(\xi_v+\psi_v)$ in (10.22), we obtain

$$\sum_{\ell=1}^{N}(\omega_\ell-\Omega_\ell)\frac{\partial A_1^{(v)}}{\partial\psi_\ell}-2a_v\omega_vB_1^{(v)}=$$

$$=\frac{2}{M_v}\sum_\beta\frac{e^{i\sum_{\alpha=1}^{n}\sigma_{\alpha v}^{(\beta)}\Psi^{(\alpha)}}}{(2\pi)^{N+r}}\int_0^{2\pi}\cdots\int_0^{2\pi}\sum_{j=1}^{N}\varphi_j^{(v)}\overline{Q}_j^{(1)o}\times$$

$$\times e^{-i\sum_{\alpha=1}^{n}\sigma_{\alpha v}^{(\beta)}\overline{\Psi}_1(\alpha)}\cos(\overline{\xi}_v+\overline{\psi}_v)d(\overline{\xi}_1+\overline{\psi}_1)\ldots d(\overline{\xi}_N+\overline{\psi}_N)d\overline{\theta}_1\ldots d\overline{\theta}_r,$$

$$\sum_{\ell=1}^{N}(\omega_\ell-\Omega_\ell)a_v\frac{\partial B_1^{(v)}}{\partial\psi_\ell}+2\omega_vA_1^{(v)}=-\frac{a_v}{M_v}\frac{d(\omega_vM_v)}{d\tau}- \tag{10.27}$$

$$-\frac{2}{M_v}\sum_\beta\frac{e^{i\sum_{\alpha=1}^{n}\sigma_{\alpha v}^{(\beta)}\Psi^{(\alpha)}}}{(2\pi)^{N+r}}\int_0^{2\pi}\cdots\int_0^{2\pi}\sum_{j=1}^{N}\varphi_j^{(v)}\overline{Q}_j^{(1)o}\times$$

$$\times e^{-i\sum_{\alpha=1}^{n}\sigma_{\alpha v}^{(\beta)}\overline{\Psi}_1^{(\alpha)}}\sin(\overline{\xi}_v+\overline{\psi}_v)d(\overline{\xi}_1+\overline{\psi}_1)\ldots d(\overline{\xi}_N+\overline{\psi}_N)d\overline{\theta}_1\ldots d\overline{\theta}_r,$$

where

$$\Psi^{(\alpha)}=p_1^{(\alpha)}\psi_1+\cdots+p_N^{(\alpha)}\psi_N,$$
$$\overline{\Psi}_1^{(\alpha)}=p_1^{(\alpha)}(\overline{\xi}_1+\overline{\psi}_1)+\cdots+p_N^{(\alpha)}(\overline{\xi}_N+\overline{\psi}_N)+c_1^{(\alpha)}\overline{\theta}_1+\cdots+c_r^{(\alpha)}\overline{\theta}_r.$$

Comparing the other harmonics in (10.22) gives

$$\sum_{p,c}{}'\Big[\omega_v^2-\Big(\sum_{s=1}^{N}p_s\omega_s+\sum_{k=1}^{r}c_k\nu_k\Big)^2\Big]\Big\{\sum_{i,j=1}^{N}a_{ij}\varphi_j^{(v)}\times$$
$$\times u_{ipc}^{(1)}e^{i[p_1(\xi_1+\psi_1)+\cdots+p_N(\xi_N+\psi_N)+c_1\theta_1+\cdots+c_r\theta_r]}\Big\}=$$

$$+\sum_{p,c}{}'\frac{1}{(2\pi)^{N+r}}e^{i[p_1(\xi_1+\psi_1)+\cdots+c_r\theta_r]}\int_0^{2\pi}\cdots\int_0^{2\pi}\sum_{j=1}^{N}\varphi_j^{(v)}\times$$
$$\times\overline{Q}_j^{(1)o}e^{-i[p_1(\overline{\xi}_1+\overline{\psi}_1)+\cdots+c_r\overline{\theta}_r]}d(\overline{\xi}_1+\overline{\psi}_1)d\overline{\theta}_1\ldots d\overline{\theta}_r+$$
$$+\sum_{\ell,j,s=1}^{N}{}'\omega_sa_aa_{\ell j}\Big(\varphi_j^{(v)}\frac{d\varphi_\ell^{(s)}}{d\tau}-\varphi_\ell^{(s)}\frac{d\varphi_j^{(v)}}{d\tau}\Big)\sin(\xi_s+\psi_s), \tag{10.28}$$

where primed summation symbols denote the fact that in the sum there are no fundamental harmonics $\sin(\xi_v+\psi_v)$ and $\cos(\xi_v+\psi_v)$. From these equations, after simple but rather long calculations, one finds

$$u_i^{(1)} = \sum_{p,c}\sum_{v=1}^{N} \varphi_i^{(v)} \frac{e^{i[p_1(\xi_1+\psi_1)+\cdots+p_N(\xi_N+\psi_N)+c_1\theta_1+\cdots+c_r\theta_r]}}{\left[\omega_v^2 - \left(\sum_{\ell=1}^{N} p_\ell\omega_\ell + \sum_{k=1}^{r} c_k\nu_k\right)^2\right] M_v (2\pi)^{N+r}} \times$$
$$\times \int_0^{2\pi}\cdots\int_0^{2\pi} \sum_{j=1}^{N} \varphi_j^{(v)} \overline{Q}_j^{(1)o} e^{-i[p_1(\overline{\xi}_1+\overline{\psi}_1)+\cdots+c_r\overline{\theta}_r]} d(\overline{\xi}_1+\overline{\psi}_1)\times$$
$$\times \ldots d(\overline{\xi}_N+\overline{\psi}_N) d\overline{\theta}_1 \ldots d\overline{\theta}_r + \sum_{v,s,\ell,j=1}^{N} \frac{1}{M_v(\omega_v^2-\omega_s^2)} \times$$
$$\times \varphi_i^{(v)} a_{\ell j}\omega_s a_s \left(\varphi_j^{(v)} \frac{d\varphi_\ell^{(s)}}{d\tau} - \varphi_\ell^{(s)} \frac{d\varphi_j^{(v)}}{d\tau}\right) \sin(\xi_s+\psi_s), \tag{10.29}$$

where

$$p_\ell = p_{\ell v}^{(\beta)} \neq \sum_{\alpha=1}^{n} \sigma_{\alpha v}^{(\beta)} p_\ell^{(\alpha)} \quad (\ell \neq v),$$
$$p_v = p_{vv}^{(\beta)} \neq \pm 1 + \sum_{\alpha=1}^{n} \sigma_{\alpha v}^{(\beta)} p_v^{(\alpha)},$$
$$c_k = c_{kv}^{(\beta)} \neq \sum_{\alpha=1}^{n} \sigma_{\alpha v}^{(\beta)} c_k^{(\alpha)}.$$

From the first equation of (10.27), one has

$$a_v B_1^{(v)} = \frac{1}{2\omega_v} \sum_{s=1}^{N} (\omega_s - \Omega_s) \frac{\partial A_1^{(v)}}{\partial \psi_s} - \frac{1}{\omega_v M_v} \sum_\beta K_\beta e^{i\sum_{\alpha=1}^{n} \sigma_{\alpha v}^{(\beta)} \psi^{(\alpha)}}, \tag{10.30}$$

where

$$K_\beta = \frac{1}{(2\pi)^{N+r}} \int_0^{2\pi}\cdots\int_0^{2\pi} \sum_{j=1}^{N} \varphi_j^{(v)} \overline{Q}_j^{(1)o} e^{-i\sum_{\alpha=1}^{n} \sigma_{\alpha v}^{(\beta)} \overline{\Psi}_1^{(\alpha)}} \times$$
$$\times \cos(\overline{\xi}_v + \overline{\psi}_v) d(\overline{\xi}_1+\overline{\psi}_1) \ldots d(\overline{\xi}_N+\overline{\psi}_N) d\overline{\theta}_1 \ldots d\overline{\theta}_r. \tag{10.31}$$

Hence,

$$\sum_{\ell=1}^{N} (\omega_\ell - \Omega_\ell) a_v \frac{\partial B_1^{(v)}}{\partial \psi_\ell} = \frac{1}{2\omega_v} \sum_{s,\ell=1}^{N} (\omega_\ell - \Omega_\ell)(\omega_s - \Omega_s) \frac{\partial^2 A_1^{(v)}}{\partial \psi_\ell \partial \psi_s} -$$
$$- \frac{i}{\omega_v M_v} \sum_\beta \sum_{\ell=1}^{N} \sum_{y=1}^{n} (\omega_\ell - \Omega_\ell) \sigma_{yv}^{(\beta)} p_\ell^{(y)} K_\beta e^{i\sum_{\alpha=1}^{N} \sigma_{\alpha v}^{(\beta)} \Psi^{(\alpha)}}$$

Substituting this expression into the second equation of (10.27), we eliminate the unknown function $B_1^{(v)}$

$$\sum_{s,\ell=1}^{N}(\omega_\ell-\Omega_\ell)(\omega_s-\Omega_s)\frac{\partial^2 A_1^{(v)}}{\partial\psi_\ell\partial\psi_s}+4\omega_v^2A_1^{(v)}=$$

$$=-\frac{2a_v\omega_v}{M_v}\frac{d(\omega_vM_v)}{d\tau}-\frac{4\omega_v}{M_v}\sum_\beta S_\beta e^{i\sum\limits_{\alpha=1}^{N}\sigma_{\alpha v}^{(\beta)}\Psi^{(\alpha)}}+$$

$$+\frac{2i}{M_v}\sum_\beta\sum_{\ell=1}^{N}\sum_{y=1}^{n}(\omega_\ell-\Omega_\ell)\sigma_{yv}^{(\beta)}p_\ell^{(y)}K_\beta e^{i\sum\limits_{\alpha=1}^{n}\sigma_{\alpha v}^{(\beta)}\Psi^{(\alpha)}}, \tag{10.32}$$

where

$$S_\beta=\frac{1}{(2\pi)^{N+r}}\int_0^{2\pi}\cdots\int_0^{2\pi}\sum_{j=1}^{N}\varphi_j^{(v)}\overline{Q}_j^{(1)o}e^{-i\sum\limits_{\alpha=1}^{n}\sigma_{\alpha v}^{(\beta)}\overline{\Psi}_1^{(\alpha)}}\times$$
$$\times\sin(\overline{\xi}_v+\overline{\psi}_v)d(\overline{\xi}_1+\overline{\psi}_1)\dots d(\overline{\xi}_N+\overline{\psi}_N)d\overline{\theta}_1\dots d\overline{\theta}_r. \tag{10.33}$$

We shall find $A_1^{(v)}$ in the form:

$$A_1^{(v)}=-\frac{a_v}{2\omega_vM_v}\frac{d(\omega_vM_v)}{d\tau}+\sum_\beta A_{1\beta}^{(v)}e^{i\sum\sigma_{\alpha v}^{(\beta)}\Psi^{(\alpha)}}, \tag{10.34}$$

where the coefficients $A_{1\beta}^{(v)}$ are to be determined. Substituting (10.34) into (10.32) and taking into account

$$\sum_{s,\ell=1}^{N}(\omega_\ell-\Omega_\ell)(\omega_s-\Omega_s)\frac{\partial^2 A_1^{(v)}}{\partial\psi_\ell\partial\psi_s}=-\sum_\beta\Big[\sum_{s=1}^{N}\sum_{y=1}^{n}(\omega_s-\Omega_s)\sigma_{yv}^{(\beta)}p_s^{(y)}\Big]\times$$
$$\times\Big[\sum_{\ell=1}^{N}\sum_{\rho=1}^{n}(\omega_\ell-\Omega_\ell)\sigma_{\rho v}^{(\beta)}p_\ell^{(\rho)}A_{1\beta}^{(v)}e^{i\sum\limits_{\alpha=1}^{n}\sigma_{\alpha v}^{(\beta)}\Psi^{(\alpha)}}\Big]=$$
$$=-\sum_\beta\Big[\sum_{s=1}^{N}\sum_{y=1}^{n}(\omega_s-\Omega_s)\sigma_{yv}^{(\beta)}p_s^{(y)}\Big]^2A_{1\beta}^{(v)}e^{i\sum\limits_{\alpha=1}^{n}\sigma_{\alpha v}^{(\beta)}\Psi^{(\alpha)}},$$

then comparing the coefficients of $\exp\Big(i\sum\limits_{\alpha=1}^{n}\sigma_{\alpha v}^{(\beta)}\Psi^{(\alpha)}\Big)$, we obtain:

$$\left\{4\omega_v^2 - \left[\sum_{s=1}^{N}\sum_{y=1}^{n}\sigma_{yv}^{(\beta)}p_s^{(y)}(\omega_s - \Omega_s)\right]^2\right\}A_{1\beta}^{(v)} =$$

$$= -\frac{4\omega_v}{M_v}S_\beta + \frac{2i}{M_v}\left[\sum_{\ell=1}^{N}\sum_{y=1}^{n}(\omega_\ell - \Omega_\ell)\sigma_{yv}^{(\beta)}p_\ell^{(y)}\right]K_\beta.$$

Finally, we have

$$A_1^{(v)} = -\frac{a_v}{2\omega_v M_v}\frac{d(\omega_v M_v)}{d\tau} -$$

$$-\frac{4\omega_v}{M_v}\sum_\beta \frac{e^{i\sum\limits_{\alpha=1}^{n}\sigma_{\alpha v}^{(\beta)}\Psi^{(\alpha)}}}{(2\pi)^{N+r}\left\{4\omega_v^2 - \left[\sum\limits_{s=1}^{N}\sum\limits_{y=1}^{n}\sigma_{yv}^{(\beta)}p_s^{(y)}(\omega_s - \Omega_s)\right]^2\right\}} \times$$

$$\times \int_0^{2\pi}\cdots\int_0^{2\pi}\sum_{j=1}^{N}\varphi_j^{(v)}\overline{Q}_j^{(1)o}e^{-i\sum\limits_{\alpha=1}^{n}\sigma_{\alpha v}^{(\beta)}\overline{\Psi}_1^{(\alpha)}}\sin(\overline{\xi}_v + \overline{\psi}_v)d(\overline{\xi}_1 + \overline{\psi}_1)d\overline{\theta}_1\ldots d\overline{\theta}_r$$

$$+\frac{2i}{M_v}\sum_\beta \frac{\sum\limits_{\ell=1}^{N}\sum\limits_{y=1}^{n}(\omega_\ell - \Omega_\ell)\sigma_{yv}^{(\beta)}p_\ell^{(y)}e^{i\sum\limits_{\alpha=1}^{n}\sigma_{\alpha v}^{(\beta)}\Psi^{(\alpha)}}}{(2\pi)^{N+r}\left\{4\omega_v^2 - \left[\sum\limits_{s=1}^{N}\sum\limits_{y=1}^{n}\sigma_{yv}^{(\beta)}p_s^{(y)}(\omega_s - \Omega_s)\right]^2\right\}} \times$$

$$\times \int_0^{2\pi}\cdots\int_0^{2\pi}\sum_{j=1}^{N}\varphi_j^{(v)}\overline{Q}_j^{(1)o}e^{-i\sum\limits_{\alpha=1}^{n}\sigma_{\alpha v}^{(\beta)}\overline{\Psi}_1^{(\alpha)}}\cos(\overline{\xi}_v + \overline{\psi}_v)d(\overline{\xi}_1 + \overline{\psi}_1)d\overline{\theta}_1\ldots d\overline{\theta}_r. \tag{10.35}$$

Similarly, we obtain for $B_1^{(v)}$:

$$B_1^{(v)} = -\frac{4\omega_v}{a_v M_v}\sum_\beta \frac{e^{i\sum\limits_{\alpha=1}^{n}\sigma_{\alpha v}^{(\beta)}\Psi^{(\alpha)}}}{(2\pi)^{N+r}\left\{4\omega_v^2 - \left[\sum\limits_{s=1}^{N}\sum\limits_{y=1}^{n}\sigma_{yv}^{(\beta)}p_s^{(y)}(\omega_s - \Omega_s)\right]^2\right\}} \times$$

$$\times \int_0^{2\pi}\cdots\int_0^{2\pi}\sum_{j=1}^{N}\varphi_j^{(v)}\overline{Q}_j^{(1)o}e^{-i\sum\limits_{\alpha=1}^{n}\sigma_{\alpha v}^{(\beta)}\overline{\Psi}_1^{(\alpha)}}\cos(\overline{\xi}_v + \overline{\psi}_v)d(\overline{\xi}_1 + \overline{\psi}_1)d\overline{\theta}_1\ldots d\overline{\theta}_r -$$

$$
-\frac{2i}{a_v M_v}\sum_{\beta}\frac{\sum_{\ell=1}^{N}\sum_{y=1}^{n}(\omega_\ell-\Omega_\ell)\sigma_{yv}^{(\beta)}p_\ell^{(y)}e^{i\sum_{\alpha=1}^{n}\sigma_{\alpha v}^{(\beta)}\Psi^{(\alpha)}}}{(2\pi)^{N+r}\left\{4\omega_v^2-\left[\sum_{s=1}^{N}\sum_{y=1}^{n}\sigma_{yv}^{(\beta)}p_s^{(y)}(\omega_s-\Omega_s)\right]^2\right\}}\times
$$

$$
\times\int_0^{2\pi}\cdots\int_0^{2\pi}\sum_{j=1}^{N}\varphi_j^{(v)}\overline{Q}_j^{(1)o}e^{-i\sum_{\alpha=1}^{n}\sigma_{\alpha v}^{(\beta)}\overline{\Psi}_1^{(\alpha)}}\sin(\overline{\xi}_v+\overline{\psi}_v)d(\overline{\xi}_1+\overline{\psi}_1)d\overline{\theta}_1\ldots d\overline{\theta}_r. \tag{10.36}
$$

Thus, we have in the first approximation

$$
q_i=\sum_{s=1}^{N}\varphi_i^{(s)}a_s\cos(\xi_s+\psi_s),\quad (i=1,2,\ldots,N) \tag{10.37}
$$

with $a_s$ and $\psi_s$ determined by

$$
\begin{aligned}
&\frac{da_s}{dt}=\varepsilon A_1^{(s)}(\tau,a,\psi),\\
&\frac{d\psi_s}{dt}=\omega_s(\tau)-\Omega_s(\tau)+\varepsilon B_1^{(s)}(\tau,a,\psi),\\
&\frac{d\xi_s}{dt}=\Omega_s,\\
&p_1^{(\alpha)}\xi_1+\cdots+p_N^{(\alpha)}\xi_N+c_1^{(\alpha)}\theta_1+\cdots+c_r^{(\alpha)}\theta_r\equiv 0,
\end{aligned} \tag{10.38}
$$

where, $A_1^{(s)}$ and $B_1^{(s)}$ are of the form (10.35) and (10.36).

Continuing this process, it is not difficult to calculate the higher approximations. But the calculations are rather long and they are not given here.

CHAPTER 4

# PARAMETRICALLY - EXCITED OSCILLATIONS

Parametrically-excited oscillations are those which depend for their excitation upon the time-dependence of the parameters of the oscillatory system. The governing differential equations of this system have time-dependent coefficients, being generally periodic.

The term parametrically-excited oscillation or parametric oscillation is generally used in connection with the phenomenon that if a parameter of an oscillatory system is made to vary at double the natural frequency $\omega$, the system begins to oscillate with the natural frequency $\omega$.

A distinguishing feature of parametrically-excited oscillations is the fact that excitation cannot become effective if the oscillator remains in its position of equilibrium. As we have seen in Chapter 3, this property is not true for forced oscillations, which can build up from the position of rest since the exciting force for the forced oscillations is effective when the oscillator is at rest.

The approximate method using in this chapter is the Krylov - Bogoliubov method presented in Chapter 3 for forced systems. We do not reproduce it here, but merely mention the conclusions.

## 1. Some Examples of Parametrically-excited Oscillators

### The pendulum with periodic motion of suspension point

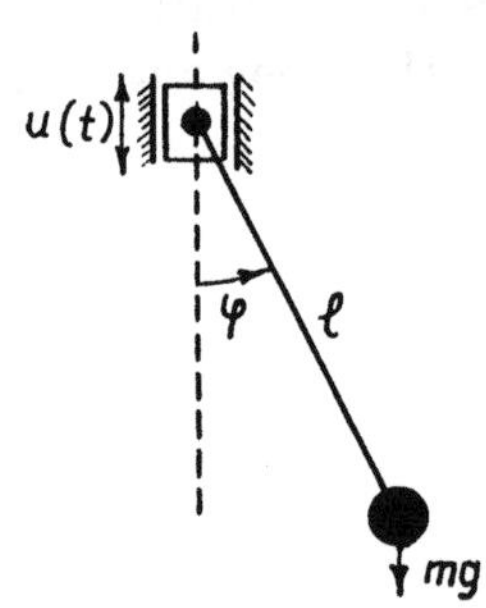


*Fig. 53.* Pendulum with oscillating suspension point.

Let us consider a pendulum which is free to turn about a horizontal axis and whose motion may be described by the angle $\varphi$ (Fig. 53). Further, let the suspension point move up and down in a vertical direction according to a definite time-law, its motion being given by $u = u(t)$. If now the pendulum equation is to be set up for a system which moves with the suspension point, then the differential equation of relative motion of the pendulum is

$$m\ell^2\ddot{\varphi} = (-mg - m\ddot{u})\ell \sin\varphi,$$

or

$$\ddot{\varphi} + \frac{1}{\ell}(g + \ddot{u})\sin\varphi = 0. \tag{1.1}$$

If now $u(t)$ is a periodic function of time, the coefficient appearing in this equation as a factor of $\sin\varphi$ must also be a periodic function of time. Then, equation (1.1) is a nonlinear equation with periodic coefficients.

In the neighbourhood of $\varphi = 0$, equation (1.1) reduces to

$$\ddot{\varphi} + \frac{1}{\ell}(g + \ddot{u})\varphi = 0. \tag{1.2}$$

Next, let us assume that the motion of the suspension point is harmonic and of the form

$$u = u_0 \cos\nu t. \tag{1.3}$$

Then equation (1.2) yields

$$\ddot{\varphi} + \frac{1}{\ell}(g - u_0\nu^2 \cos\nu t)\varphi = 0. \tag{1.4}$$

This equation is linear, and the coefficient is a harmonic function of time. Such an equation is known as a Mathieu equation, and it is encountered in various forms in many problems in mathematical physics.

We note that equation (1.4) admits the equilibrium position $\varphi = 0$, which is stable for a simple pendulum ($u_0 = 0$). For certain values of the parameters $g/\ell$ and $u_0\nu^2/\ell$, this equilibrium position can be rendered unstable by the moving suspension point. In the next paragraph, we will study the stability properties of the system for small $u_0\nu^2/\ell$ compared with $g/\ell$.

By introducing the notation

$$x = \varphi, \quad \frac{g}{\ell} = \omega^2, \quad \frac{u_0\nu^2}{g} = \varepsilon\alpha,$$

we have:

$$\ddot{x} + \omega^2(1 - \varepsilon\alpha\cos\nu t)x = 0, \;\; \varepsilon \ll 1. \tag{1.5}$$

For $\nu = 2$, the notation

$$\mu = \omega^2, \quad \gamma = \varepsilon\omega^2\alpha = \frac{u_0\nu^2}{\ell}$$

allows us to reduce the equation to the standard form of Mathieu equation

$$\ddot{x} + (\mu - \gamma\cos 2t)x = 0. \tag{1.6}$$

**Transverse oscillation of a beam subjected to a periodically longitudinal load**

Consider a pinned beam of length $\ell$, cross-section $A$, stiffness $EI$ and density $\gamma$ which is subjected to a pulsating compressive force $F = F(t)$ (Fig. 54). The behaviour of the oscillations of the beam is described by the differential equation

$$EI\frac{\partial^4 y}{\partial z^4} + \frac{\gamma A}{g}\frac{\partial^2 y}{\partial t^2} + F(t)\frac{\partial^2 y}{\partial z^2} = 0, \tag{1.7}$$

where $g$ is the gravitational acceleration. The boundary conditions are

$$\begin{aligned} y\Big|_{z=0} &= 0, & \frac{\partial^2 y}{\partial z^2}\Big|_{z=0} &= 0, \\ y\Big|_{z=\ell} &= 0, & \frac{\partial^2 y}{\partial z^2}\Big|_{z=\ell} &= 0. \end{aligned} \tag{1.8}$$

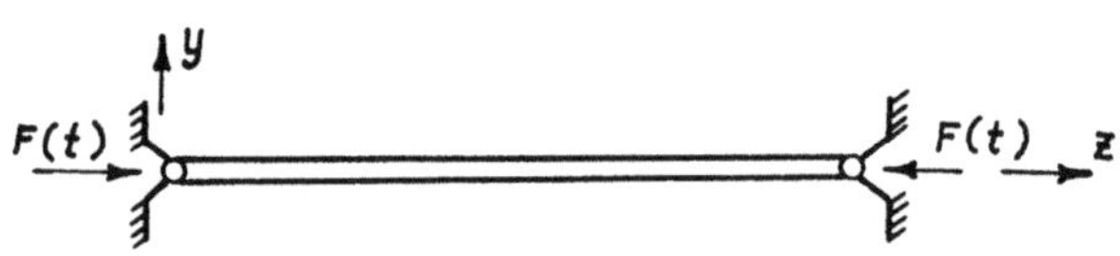


*Fig. 54.* Beam subjected to a pulsating compression force.

Equation (1.7) with the conditions (1.8) can be solved by the expression

$$y = x \sin \pi\frac{z}{\ell},$$

where $x = x(t)$ is a function of time. By substitution into (1.7), we have

$$\ddot{x} + \omega^2\left[1 - \frac{\ell^2}{\pi^2 EI}F(t)\right]x = 0, \tag{1.9}$$

where,

$$\omega^2 = \frac{g\pi^4 EI}{\gamma A \ell^4}.$$

Equation (1.9), again, is an equation with periodically varying coefficients.

Another example of the system with parametrically-excited oscillation is shown in Fig. 55. This system consists of a rod of length $2\ell$ balanced in the middle by a pinned support (Fig. 55). The coefficients of stiffness of the elastic supports are $C_0$. The rod is subjected to a horizontal periodic force $P(t)$. In the position of equilibrium, the rod is horizontal.

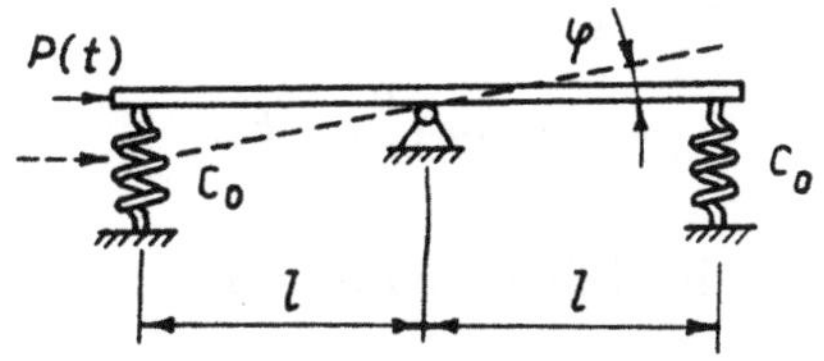


*Fig. 55.* A parametrically - excited oscillating system.

For small deviations of the rod from the equilibrium position, the rod is affected by the moment of the restoring force $-C_0\ell^2\varphi$ and by the moment of the longitudinal force $P(t)\ell\varphi$. Hence, the differential equation for the motion of the rod is

$$\ddot{\varphi} + \frac{C_0\ell - P(t)}{I}\ell\varphi = 0, \tag{1.10}$$

where $I$ is the moment of inertia of the rod with respect to the rotating axis. Equation (1.10) is an equation with periodically varying parameter.

### Motion in the neighbourhood of stationary oscillations

Consider a nonlinear system governed by the differential equation

$$\ddot{y} + f(y) = 0. \tag{1.11}$$

Let $y = \varphi(t)$ be its stationary periodic solution. In order to discuss the stability of this solution, one sets

$$y = \varphi(t) + x(t), \tag{1.12}$$

where $x(t)$ is a small *disturbance*. After substitution of (1.12) into equation (1.11) and neglection of all higher terms of $x(t)$ beyond the two in the Taylor series, we obtain

$$\ddot{x} + \left(\frac{\partial f}{\partial y}\right)_{y=\varphi(t)} x = 0. \tag{1.13}$$

Since $\varphi(t)$ is periodic, the factor of $x$ in the equation (1.13) must also be a periodic function of time.

## 2. Behaviour of Oscillators Governed by a Mathieu Equation

The asymptotic methods of Krylov - Bogoliubov, presented in paragraph 3 of Chapter 3, are used for obtaining approximate solution for the Mathieu equation (1.5):

$$\ddot{x} + \omega^2(1 - \varepsilon\alpha\cos\nu t)x = 0 \tag{2.1}$$

in the principal parametric resonance case when $\omega \simeq \frac{\nu}{2}$. Using the results from paragraph 3, Chapter 3, for $p = 1$, $q = 2$, we have in the first approximation

$$x = a\cos\left(\frac{\nu}{2}t + \theta\right), \tag{2.2}$$

where $a$ and $\theta$ are given by the differential equation

$$\begin{aligned} \frac{da}{dt} &= -\varepsilon\frac{a\alpha\omega^2}{2\nu}\sin 2\theta, \\ \frac{d\theta}{dt} &= \omega - \frac{\nu}{2} - \varepsilon\frac{\alpha\omega^2}{2\nu}\cos 2\theta. \end{aligned} \tag{2.3}$$

In order to solve these equations, we introduce the new variables $u$ and $v$ according to the formulae

$$u = a\cos\theta, \qquad v = a\sin\theta. \tag{2.4}$$

We have

$$\begin{aligned} \frac{du}{dt} &= \frac{da}{dt}\cos\theta - a\cdot\frac{d\theta}{dt}\sin\theta = \\ &= \left[-\frac{\varepsilon\alpha\omega^2}{2\nu} - \left(\omega - \frac{\nu}{2}\right)\right]v, \\ \frac{dv}{dt} &= \left[-\frac{\varepsilon\alpha\omega^2}{2\nu} + \left(\omega - \frac{\nu}{2}\right)\right]u. \end{aligned} \tag{2.5}$$

The nature of the solution of the system (2.5), being a linear system with constant coefficients, reduces to the investigation of the roots of the characteristic equation (Chapter 2)

$$\lambda^2 - \frac{\varepsilon^2\alpha^2\omega^4}{4\nu^2} + \left(\omega - \frac{\nu}{2}\right)^2 = 0. \tag{2.6}$$

The roots of this equation are denoted by $\pm\lambda$

$$\lambda = \sqrt{\frac{\varepsilon^2\alpha^2\omega^4}{4\nu^2} - \left(\omega - \frac{\nu}{2}\right)^2}. \tag{2.7}$$

Thus, the general solution of (2.5) is of the form

$$\begin{aligned} u &= C_1e^{\lambda t} + C_2e^{-\lambda t}, \\ v &= \frac{1}{\lambda}C_1\left(-\frac{\varepsilon\alpha\omega^2}{2\nu} + \omega - \frac{\nu}{2}\right)e^{\lambda t} + \frac{1}{\lambda}C_2\left(\frac{\varepsilon\alpha\omega^2}{2\nu} - \omega + \frac{\nu}{2}\right)e^{-\lambda t}, \end{aligned} \tag{2.8}$$

where $C_1$ and $C_2$ are constants of integration.

The amplitude $a$ and the phase $\theta$ are given by the relations

$$a^2 = u^2 + v^2, \quad \theta = \arctan\frac{v}{u}\cdot \tag{2.9}$$

According to the formulae (2.7) - (2.9), it is obvious that if $\lambda$ is imaginary, $a$ is a bounded function of $t$. If $\lambda$ is real, $a$ increases exponentially and one has *parametric resonance.*

The condition for $\lambda$ to be real is

$$\frac{\varepsilon\alpha\omega^2}{2\nu} > \left|\omega - \frac{\nu}{2}\right| \quad \text{or} \quad 1 - \frac{\alpha}{2} < \left(\frac{2\omega}{\nu}\right)^2 < 1 + \frac{\alpha}{2} \tag{2.10}$$

or

$$\frac{\varepsilon\alpha\omega}{4} > \left|\omega - \frac{\nu}{2}\right|$$

since, $\nu = 2\omega + O(\varepsilon)$. Thus, if $\nu$ is in the interval

$$2\omega\left(1 - \frac{\varepsilon\alpha}{4}\right) < \nu < 2\omega\left(1 + \frac{\varepsilon\alpha}{4}\right) \tag{2.11}$$

the amplitude increases exponentially.

Following the formulae (3.7), (3.23), (3.27) and (3.28) of Chapter 3, we have in the second approximation

$$x = a\cos\left(\frac{\nu}{2}t + \theta\right) - \frac{\varepsilon a\alpha\omega}{8\left(\omega + \frac{\nu}{2}\right)}\cos\left(\frac{3}{2}\nu t + \theta\right), \tag{2.12}$$

where $a$ and $\theta$ are determined from the differential equation

$$\begin{aligned} \frac{da}{dt} &= -\frac{\varepsilon a\alpha\omega^2}{2\nu}\sin 2\theta, \\ \frac{d\theta}{dt} &= \omega - \frac{\nu}{2} + \frac{\varepsilon^2\alpha^2(\omega+\nu)\omega}{32\left(\omega + \frac{\nu}{2}\right)} - \frac{\varepsilon\alpha\omega^2}{2\nu}\cos 2\theta. \end{aligned} \tag{2.13}$$

The same change of variables, used in (2.4), reduces this system to

$$\begin{aligned} \frac{du}{dt} &= \left[-\frac{\varepsilon\alpha\omega^2}{2\nu} - \omega + \frac{\nu}{2} - \frac{\varepsilon^2\alpha^2(\omega+\nu)\omega}{32\left(\omega + \frac{\nu}{2}\right)}\right]v, \\ \frac{dv}{dt} &= \left[-\frac{\varepsilon\alpha\omega^2}{2\nu} + \omega - \frac{\nu}{2} + \frac{\varepsilon^2\alpha^2(\omega+\nu)\omega}{32\left(\omega + \frac{\nu}{2}\right)}\right]u. \end{aligned} \tag{2.14}$$

The roots of the characteristic equation of the system (2.14) are $\pm\lambda$,

$$\lambda = \sqrt{\frac{\varepsilon^2\alpha^2\omega^4}{4\nu^2} - \left[\omega - \frac{\nu}{2} + \frac{\varepsilon^2\alpha^2(\omega+\nu)\omega}{32\left(\omega + \frac{\nu}{2}\right)}\right]^2}\,. \tag{2.15}$$

The zone of instability is now

$$2\omega\left(1-\frac{\varepsilon\alpha}{4}-\frac{\varepsilon^2\alpha^2}{64}\right)<\nu<2\omega\left(1+\frac{\varepsilon\alpha}{4}-\frac{\varepsilon^2\alpha^2}{64}\right). \tag{2.16}$$

We find the relation between $\omega$ and $h$, with which the solution of the equation (2.1) is periodic with period $4\pi/\nu$. For this, it is necessary that $\lambda$ equal zero in (2.7) and (2.15) ($a$ = const).

For the first approximation, this gives

$$\frac{2\omega}{\nu}=1\pm\frac{\varepsilon\alpha}{4} \tag{2.17}$$

and, for the second approximation

$$\frac{2\omega}{\nu}=1\pm\frac{\varepsilon\alpha}{4}+\frac{5\varepsilon^2\alpha^2}{64}. \tag{2.18}$$

With the same degree of accuracy, the first approximation is

$$\frac{4\omega^2}{\nu^2}=1\pm\frac{\varepsilon\alpha}{2} \tag{2.19}$$

and, the second approximation is

$$\frac{4\omega^2}{\nu^2}=1\pm\frac{\varepsilon\alpha}{2}+\frac{7}{32}\varepsilon^2\alpha^2. \tag{2.20}$$

These relations may be regarded as the curves in the $\left(\frac{4\omega^2}{\nu^2},\alpha\right)$ plane. On these curves, the solution of (2.1) is periodic i.e. relations (2.19) and (2.20) are the equations of boundary curves of the zones of instability of solutions of equation (2.1) (Fig. 56):

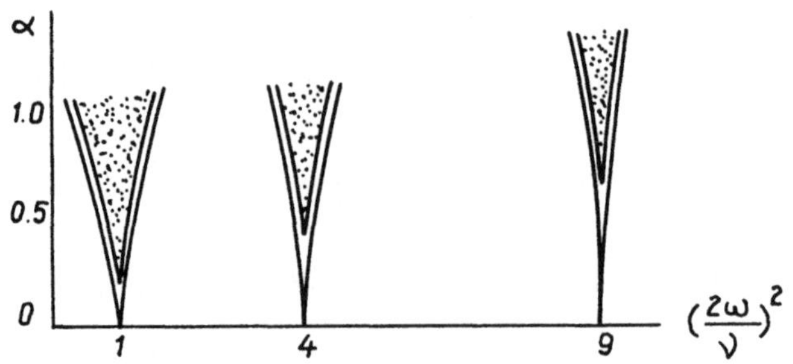


*Fig. 56.* Instability zones.

From the formulae (2.2), (2.12), (2.19) and (2.20), we have the following periodic solutions of (2.1) with period $\frac{4\pi}{\nu}$.

For the first approximation

$$x = a_0 \cos\left(\frac{\nu}{2}t + \theta_0\right), \tag{2.21}$$

and for the second approximation

$$x = a_0 \cos\left(\frac{\nu}{2}t + \theta_0\right) - \frac{\varepsilon\alpha a_0}{16}\cos\left(\frac{3\nu}{2}t + \theta_0\right), \tag{2.22}$$

where $a_0$ and $\theta_0$ are constants.

We now study the solutions of equation (2.1) and find the boundary of instability in the case

$$\omega \simeq \frac{\nu}{2}\cdot p, \qquad p = 2, 3. \tag{2.23}$$

*For the case* $\omega \simeq \nu$, it is necessary to let $p = 2$, $q = 2$ in the formulae of paragraph 3, Chapter 3. Then in the second approximation, we obtain

$$x = a\cos(\nu t + \theta) - \frac{\varepsilon\alpha a\omega^2}{2\nu(\nu + 2\omega)}\cos(2\nu t + \theta) - \frac{\varepsilon\alpha\omega^2 a}{2\nu(\nu - 2\omega)}\cos\theta, \tag{2.24}$$

where $a$ and $\theta$ are determined from the differential equation

$$\begin{aligned} \frac{da}{dt} &= -\frac{\varepsilon^2\alpha^2 a\omega^4}{8\nu^2(2\omega - \nu)}\sin 2\theta, \\ \frac{d\theta}{dt} &= \omega - \nu - \frac{\varepsilon^2\alpha^2\omega^4}{8\nu^2(2\omega - \nu)}\cos 2\theta. \end{aligned} \tag{2.25}$$

As in the previous case for $\omega = \frac{\nu}{2}$, from (2.25), we find the condition for the roots of the characteristic equation to be real:

$$\left|\omega - \nu - \frac{\varepsilon^2\alpha^2\omega^4}{4\nu(4\omega^2 - \nu^2)}\right| < \frac{\varepsilon^2\alpha^2\omega^4}{8\nu^2|2\omega - \nu|}\,. \tag{2.26}$$

Hence, the zone of instability is determined with second order accuracy by the inequality

$$4 + \frac{2\varepsilon^2\alpha^2\omega^4}{\nu^2(4\omega^2 - \nu^2)} - \frac{\varepsilon^2\alpha^2\omega^4}{\nu^3(2\omega - \nu)} < \left(\frac{2\omega}{\nu}\right)^2 < 4 + \frac{2\varepsilon^2\alpha^2\omega^4}{\nu^2(4\omega^2 - \nu^2)} + \frac{\varepsilon^2\alpha^2\omega^4}{\nu^3(2\omega - \nu)} \tag{2.27}$$

or

$$4 + \frac{2}{3}\varepsilon^2\alpha^2 - \varepsilon^2\alpha^2 < \left(\frac{2\omega}{\nu}\right)^2 < 4 + \frac{2}{3}\varepsilon^2\alpha^2 + \varepsilon^2\alpha^2 \tag{2.28}$$

since, $\omega = \nu + O(\varepsilon)$.

*For the case* $\omega \simeq \frac{3}{2}\nu$, it is necessary to calculate the third approximation, because in the second approximation, one obtains only the correction to the natural frequency $\omega$. After some long calculations, we get

$$x = a\cos\left(\frac{3}{2}\nu t + \theta\right) - \frac{\varepsilon\omega^2\alpha a}{2}\left\{\frac{\cos\left(\frac{5}{2}\nu t + \theta\right)}{\nu(2\omega+\nu)} - \frac{\cos\left(\frac{1}{2}\nu t + \theta\right)}{\nu(2\omega-\nu)}\right\} +$$
$$+ \frac{\varepsilon^2\alpha^2\omega^4 a}{16\nu^2}\left\{\frac{\cos\left(\frac{7}{2}\nu t + \theta\right)}{(2\omega+\nu)(\nu+\omega)} + \frac{\cos\left(\frac{1}{2}\nu t - \theta\right)}{(2\omega-\nu)(\omega-\nu)}\right\}, \tag{2.29}$$

where $a$ and $\theta$ are determined from the system of equations of the third approximation:

$$\frac{da}{dt} = -\frac{\varepsilon^3\alpha^3\omega^6 a}{3\cdot 2^5\cdot\nu^3(2\omega-\nu)(\omega-\nu)}\sin 2\theta, \tag{2.30}$$
$$\frac{d\theta}{dt} = \omega - \frac{3}{2}\nu - \frac{\varepsilon^2\alpha^2\omega^4}{3\cdot 2\nu(4\omega^2-\nu^2)} - \frac{\varepsilon^3\alpha^3\omega^6}{3\cdot 2^5\cdot\nu^3(2\omega-\nu)(\omega-\nu)}\cos 2\theta.$$

The structure of these equations shows that in concrete cases even the equations of third approximation are very simple.

From the equations (2.30), we find for the zone of instability the inequality

$$9 + \frac{2\varepsilon^2\alpha^2\omega^4}{\nu^2(4\omega^2-\nu^2)} - \frac{\varepsilon^3\alpha^3\omega^6}{2^3\nu^4(2\omega-\nu)(\omega-\nu)} < \left(\frac{2\omega}{\nu}\right)^2 <$$
$$< 9 + \frac{2\varepsilon^2\alpha^2\omega^4}{\nu^2(4\omega^2-\nu^2)} + \frac{\varepsilon^3\alpha^3\omega^6}{2^3\nu^4(2\omega-\nu)(\omega-\nu)}, \tag{2.31}$$

or with the same accuracy

$$9 + \frac{81\varepsilon^2\alpha^2}{64} - \frac{3^6\cdot\varepsilon^3\alpha^3}{2^9} < \left(\frac{2\omega}{\nu}\right)^2 < 9 + \frac{81\varepsilon^2\alpha^2}{64} - \frac{3^6\varepsilon^3\alpha^3}{2^9} \tag{2.32}$$

since, $\omega = \frac{3}{2}\nu + O(\varepsilon)$.

The analysis of the expressions (2.10) (for $p = 1$), (2.28) (for $p = 2$) and (2.32) (for $p = 3$) shows that the width of the zone of instability decreases with its degree $p$ as $(\varepsilon\alpha)^p$. Fig. 56 shows the first three zones of instability, following the inequalities (2.10), (2.28) and (2.32). Thus, the zones of instability of the Mathieu equation (2.1) are determined entirely by the two quantities $\alpha$ and $\omega$, and not by the initial conditions.

It is noted that, for the differential equation of the form

$$\ddot{x} + \varepsilon h\dot{x} + \omega^2(1 - \varepsilon\alpha\cos\nu t)x = 0, \tag{2.33}$$

which is different from (2.1) by the damping term $\varepsilon h\dot{x}$, the zones of instability become narrower (see stippled areas in Fig. 56). It is not difficult to verify that in this case, instead of the inequalities (2.10), (2.28) and (2.32), we have:

$$1-\varepsilon\sqrt{\frac{\alpha^2}{4}-\frac{4h^2}{\nu^2}}<\left(\frac{2\omega}{\nu}\right)^2<1+\varepsilon\sqrt{\frac{\alpha^2}{4}-\frac{4h^2}{\nu^2}}, \qquad (2.34)$$

$$4+\frac{2}{3}\varepsilon^2\alpha^2-\varepsilon\sqrt{\varepsilon^2\alpha^4-64\frac{h^2}{\nu^2}}<\left(\frac{2\omega}{\nu}\right)^2<4+\frac{2}{3}\varepsilon^2\alpha^2+\varepsilon\sqrt{\varepsilon^2\alpha^4-64\frac{h^2}{\nu^2}}, \qquad (2.35)$$

$$9+\frac{81}{64}\varepsilon^2\alpha^2-\varepsilon\sqrt{\frac{3^{12}}{2^{18}}\varepsilon^4\alpha^6-4\cdot 3^4\frac{h^2}{\nu^2}}<\left(\frac{2\omega}{\nu}\right)^2<$$
$$<9+\frac{81}{64}\varepsilon^2\alpha^2+\varepsilon\sqrt{\frac{3^{12}}{2^{18}}\varepsilon^4\alpha^6-4\cdot 3^4\frac{h^2}{\nu^2}}\,. \qquad (2.36)$$

For these inequalities to be valid, the following are also required:

for the first zone $\quad \alpha > 4\frac{h}{\nu}\,,$ (2.37)

for the second zone $\quad \alpha > 2\sqrt{2\frac{h}{\nu}}\,,$ (2.38)

for the third zone $\quad \alpha > \frac{8}{3}\sqrt[6]{\frac{4h^2}{9\nu^2}}\,.$ (2.39)

As is easy to see, with the appearance of the damping force, the parametric resonances $\omega \simeq \nu$ and $\omega \simeq \frac{3}{2}\nu$ are less marked and require larger values of modulation $\alpha\omega^2$ than in the principal resonance case $\omega \simeq \nu/2$.

Hence, the most interesting resonance case is the principal one: $\omega \simeq \nu/2$.

In Figure 57, the stability diagram for the standard Mathieu equation (1.6) is

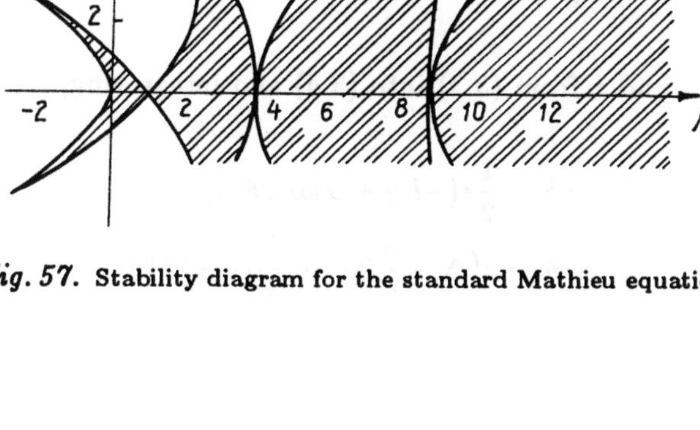


*Fig. 57.* Stability diagram for the standard Mathieu equation.

plotted with respect to the coordinates $\mu$ and $\gamma$. The shaded areas correspond to the stability regions. For each pair of values $\mu$ and $\gamma$ in the shaded areas, there corresponds an almost-periodic solution of (1.6), which is bounded for all values of $t$. Outside of these regions, there exist unbounded solutions of the equation (1.6).

## 3. Oscillators Governed by a Nonlinear Mathieu Equation

In this paragraph, we consider the oscillation of the systems described by the differential equation

$$\ddot{x} + \varepsilon h\dot{x} + (\omega^2 + \varepsilon\alpha\cos\gamma t)x + \varepsilon\beta x^3 = 0, \tag{3.1}$$

where $h, \omega, \alpha, \gamma, \beta$ are constants, and $\varepsilon$ is a small positive parameter. This equation is obtained from the standard Mathieu equation (2.1) by adding the damping term $h\dot{x}$ and the cubic term $\beta x^3$. We shall consider here the principal parametric resonance case when there is a resonance relation between the natural frequency $\omega$ and exciting frequency $\gamma$:

$$\omega^2 - \frac{1}{4}\gamma^2 = \varepsilon\Delta, \tag{3.2}$$

where $\Delta$ is a detuning parameter.

In the first approximation, we look for a solution of (3.1) of the form:

$$x = a\cos\varphi, \quad \dot{x} = -\frac{1}{2}\gamma a\sin\varphi, \tag{3.3}$$

where $\varphi = \frac{\gamma}{2}t + \theta$, $a$ and $\theta$ given by differential equation of the type (3.27) or (4.6) of Chapter 3:

$$\begin{aligned} \gamma\dot{a} &= \frac{\varepsilon}{\pi}\int_0^{2\pi}\{h\dot{x} + \Delta x + \alpha x\cos\gamma t + \beta x^3\}\sin\varphi d\varphi, \\ \gamma a\dot{\theta} &= \frac{\varepsilon}{\pi}\int_0^{2\pi}\{h\dot{x} + \Delta x + \alpha x\cos\gamma t + \beta x^3\}\cos\varphi d\varphi. \end{aligned} \tag{3.4}$$

Substituting $x$ and $\dot{x}$ from (3.3) into (3.4) and integrating, we obtain:

$$\begin{aligned} \gamma\dot{a} &= \frac{1}{2}\varepsilon(-h\gamma + \alpha\sin 2\theta)a, \\ \gamma a\dot{\theta} &= \varepsilon\big(\Delta + \frac{3}{4}\beta a^2 + \frac{1}{2}\alpha\cos 2\theta\big)a. \end{aligned} \tag{3.5}$$

The stationary state exists if (3.5) has a singular point $a = a_0 = \text{const}$ and $\theta = \theta_0 = \text{const}$ determined by the equations:

$$\begin{aligned} \left(-h\gamma + \alpha \sin 2\theta_0\right) a_0 &= 0, \\ \left(\Delta + \frac{3}{4}\beta a_0^2 + \frac{1}{2}\alpha \cos 2\theta_0\right) a_0 &= 0. \end{aligned} \tag{3.6}$$

There is a trivial solution

$$a_0 = 0, \quad \theta_0 \quad \text{is arbitrary.} \tag{3.7}$$

The nontrivial solutions $a_0 \neq 0$ are given by

$$\begin{aligned} h\gamma &= \alpha \sin 2\theta_0, \\ -2\left(\Delta + \frac{3}{4}\beta a_0^2\right) &= \alpha \cos 2\theta_0, \end{aligned} \tag{3.8}$$

whence, upon the elimination of $\theta_0$, one has

$$W(a_0^2, \gamma^2) = 0, \tag{3.9}$$

where

$$W(a_0^2, \gamma^2) = h^2\gamma^2 + 4\left(\Delta + \frac{3}{4}\beta a_0^2\right)^2 - \alpha^2. \tag{3.10}$$

From equation (3.9), we obtain

$$a_0^2 = \frac{2}{3\beta}\left[\frac{2}{\varepsilon}\left(\frac{\gamma^2}{4} - \omega^2\right) \pm \sqrt{\alpha^2 - h^2\gamma^2}\right]. \tag{3.11}$$

The phase angle $\theta_0$ is determined by

$$\begin{aligned} \sin 2\theta_0 &= \frac{h}{\alpha}\gamma, \\ \cos 2\theta_0 &= \mp\frac{1}{\alpha}\sqrt{\alpha^2 - h^2\gamma^2}, \end{aligned} \tag{3.12}$$

where in the formulae (3.11) and (3.12), one should consistently use either upper or lower signs.

Introducing the notation

$$B = \frac{\varepsilon h}{\omega}, \quad C = \frac{\varepsilon \alpha}{\omega^2}, \quad \eta = \frac{\gamma}{\omega} \tag{3.13}$$

and

$$\rho^2 = \frac{3}{2}\varepsilon\beta\frac{a_0^2}{\omega^2} \quad \text{for} \quad \beta > 0, \tag{3.14}$$

$$\bar{\rho}^2 = -\frac{3}{2}\varepsilon\beta\frac{a_0^2}{\omega^2} \quad \text{for} \quad \beta < 0, \tag{3.15}$$

we have

$$\rho^2 = 2\left(\frac{1}{4}\eta^2 - 1\right) \pm \sqrt{C^2 - B^2\eta^2}, \tag{3.16}$$

$$\bar{\rho}^2 = 2\left(1 - \frac{1}{4}\eta^2\right) \mp \sqrt{C^2 - B^2\eta^2}. \tag{3.17}$$

In both cases

$$\sin 2\theta_0 = \frac{B}{C}\eta, \tag{3.18}$$

$$\cos 2\theta_0 = \mp\frac{1}{C}\sqrt{C^2 - B^2\eta^2}, \tag{3.19}$$

where for equations (3.16), (3.17) and (3.19), one should consistently use either upper or lower signs.

Figure 58 and Figure 59 show the dependence of $\rho^2$ and $\bar{\rho}^2$ on $\eta^2$ for the case $C^2 = 0.1$ and for different values of $B^2$.

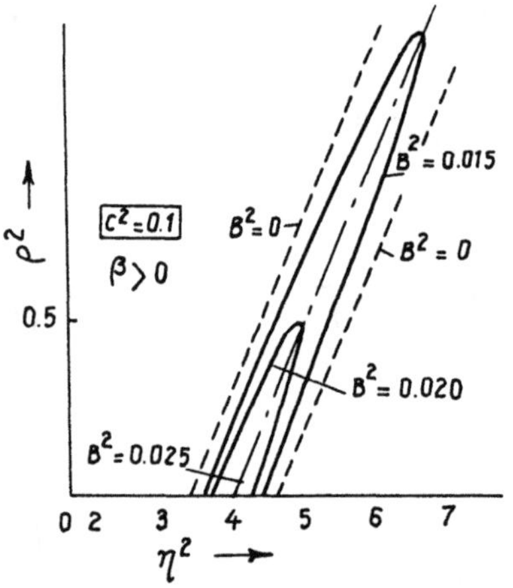


*Fig. 58.* Amplitude curves of parametric oscillation when $\beta > 0$ with various damping coefficients.

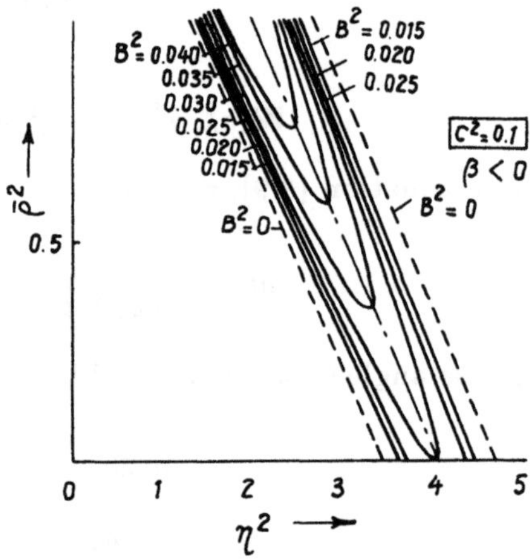


*Fig. 59.* Amplitude curves of parametric oscillation when $\beta < 0$ with various damping coefficients.

For $B = 0$, we have two parallel straight lines, and for nonzero values of $B$, we have parabolas growing between these straight lines. The straight line given by the equation

$$\rho^2 = 2\left(\frac{1}{4}\eta^2 - 1\right) \tag{3.20}$$

is plotted in Fig. 58 and Fig. 59 by the alternating dotted-dashed line, it cuts all parabolas at the points with coordinates $(\eta^2, \rho^2)$, where

$$\eta^2 = \frac{C^2}{B^2}, \quad \rho^2 = \frac{C^2 - 4B^2}{2B^2}. \tag{3.21}$$

At these points, the parabolas have vertical tangents.

From the equation $\frac{d\rho^2}{d\eta^2} = 0$, one obtains the maximum amplitude given by

$$\rho_{\max}^2 = \frac{C^2 - 4B^2 + B^4}{2B^2} . \tag{3.22}$$

Beginning from the value $B^2 = B_{\max}^2$, the parabolas run below the axis $\eta^2$ (Fig. 58). Hence, the positive values $\rho^2$ become impossible. The value $B_{\max}^2$ is determined from the equation $\rho_{\max}^2 = 0$, or

$$B_{\max}^4 - 4B_{\max}^2 + C^2 = 0. \tag{3.23}$$

Of the two roots of this equation, only the following solution is valid

$$B_{\max}^2 = 2 - \sqrt{4 - C^2}. \tag{3.24}$$

To verify the stability of the stationary solution, we use the criterion (4.17), Chapter 3. Because function $W$ (3.10) is positive outside and negative inside the resonance curve, the stable branch of the resonance curve is the upper branch, which corresponds to the upper sign before the radical in (3.16), namely

$$\rho^2 = 2\left(\frac{1}{4}\eta^2 - 1\right) + \sqrt{C^2 - B^2\eta^2}. \tag{3.25}$$

Thus, among the two forms of oscillations corresponding to definite values of $\gamma$, the form with large amplitude is stable, and the form with small amplitude is unstable.

In order to study the stability of the trivial solution $a_0 = 0$ (3.7), it is convenient to use the cartesian coordinates $(u, v)$ instead of the polar ones (3.3) as follows

$$u = a\cos\theta, \qquad v = a\sin\theta. \tag{3.26}$$

Taking into account (3.5), we find the differential equation for $u$ and $v$:

$$\begin{aligned} \frac{du}{dt} &= \frac{da}{dt}\cos\theta - a\frac{d\theta}{dt}\sin\theta = \\ &= \frac{\varepsilon}{2\gamma}\Big[-h\gamma u + \big(\alpha - 2\Delta - \frac{3}{2}\beta a^2\big)v\Big], \\ \frac{dv}{dt} &= \frac{\varepsilon}{2\gamma}\Big[\big(\alpha + 2\Delta + \frac{3}{2}\beta a^2\big)u - h\gamma v\Big] . \end{aligned} \tag{3.27}$$

In view of (3.26), the problem of stability of the trivial solution $a_0 = 0$ leads to the study of the singular point $u = v = 0$ of the system (3.27). The linearized equations of (3.27) are

$$\begin{aligned}\frac{du}{dt} &= \frac{\varepsilon}{2\gamma}\big[-h\gamma u+(\alpha-2\Delta)v\big],\\ \frac{dv}{dt} &= \frac{\varepsilon}{2\gamma}\big[(\alpha+2\Delta)u-h\gamma v\big].\end{aligned} \tag{3.28}$$

Their characteristic equation is

$$\gamma^2\lambda^2+\frac{1}{2}\varepsilon h\gamma^2\lambda+\Big[\frac{\varepsilon^2h^2\gamma^2}{4}+\big(\omega^2-\frac{\gamma^2}{4}\big)^2-\frac{\varepsilon^2\alpha^2}{4}\Big]=0. \tag{3.29}$$

The singular point $u=v=0$ is stable if

$$\varepsilon^2(h^2\gamma^2-\alpha^2)+4\big(\omega^2-\frac{\gamma^2}{4}\big)^2>0. \tag{3.30}$$

Observing the equation (3.9), one can see that the condition (3.30) requires the free term of the quadratic equation in $a_0^2$ (3.9) to be positive. This means that the trivial solution $a_0=0$ is stable if equation (3.9) has either two positive roots, two negative roots or two conjugate complex roots. Hence, for the case represented in Fig. 58 for $\beta>0$, the zero solution $a_0=0$ is stable if the value $\eta^2$ does not lie in the interval $\eta_1^2\le\eta^2\le\eta_2^2$ of the axis $\eta^2$, from which the parabola is growing up.

In Fig. 60, the stable branches are shown by heavy lines. With increasing $\eta$, the state of equilibrium is still stable until the value $\eta_1$ is reached. From this moment, the parametric oscillations appear and grow with increasing amplitude, following the stable branch NQ. For $\eta=\dfrac{C}{B}$ (point Q) the amplitude suddenly drops to zero. As $\eta$ continues to increase, the state of equilibrium remains stable. If, however, $\eta$ decreases (say, from $\eta=\dfrac{C}{B}$) the state of equilibrium is still stable until the value $\eta=\eta_2$ is reached. At this point, there will be an upward jump to point P on the stable branch NQ. As $\eta$ continues to decrease, the amplitude decreases and vanishes at $\eta=\eta_1$ (point N).

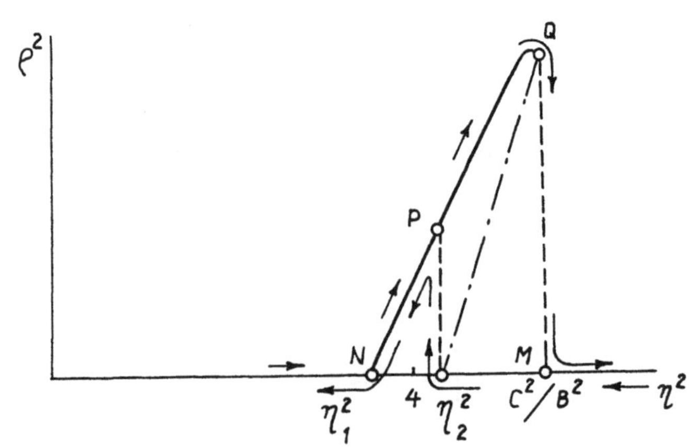


*Fig. 60.* Stability examination for the case $\beta>0$ with slowly varying frequency of parametric excitation.

## 4. Some Generalized Mathieu Equations

In the previous paragraph, we studied a simple generalized Mathieu equation, where the damping term and the cubic free term are taken into consideration (see (3.1)). The next step for generalization of the Mathieu equation is to account for the nonlinearity of modulation i.e. the nonlinearity of the coefficient of the function $\cos\gamma t$ in (3.1). In this paragraph, more complicated generalized Mathieu equations will be investigated.

1. First, let us consider the differential equation [37,h]

$$\ddot{x} + \omega^2 x + \varepsilon(cx + dx^3)\cos\gamma t + \varepsilon\alpha x^3 + \varepsilon R(x, \dot{x}) = 0, \tag{4.1}$$

where $\omega, c, d, \gamma$ and $\alpha$ are constants, $R(x, \dot{x})$ is a function characterizing the nonlinear friction force, and $\varepsilon$ is a small positive parameter. All Figures of this paragraph are plotted for the case $c > 0$ and $\alpha < 0$.

It is assumed that there is a resonance relation of the type

$$\omega^2 = \frac{\gamma^2}{4} + \varepsilon\Delta, \tag{4.2}$$

where $\Delta$ is a detuning parameter. Introducing the new variables $a$ and $\psi$ by means of the formulae

$$\begin{aligned} x &= a\sin\varphi, \\ \dot{x} &= \frac{\gamma}{2} a\cos\varphi, \end{aligned} \tag{4.3}$$

and imposing the condition

$$\dot{a}\sin\varphi + a\dot{\psi}\cos\varphi = 0,$$

where, $\psi = \varphi - \frac{\gamma}{2}t$, we find that equation (4.1) becomes

$$\begin{aligned} \frac{\gamma}{2}\dot{a} &= -\varepsilon\left[\Delta x + \alpha x^3 + (cx + dx^3)\cos\gamma t + R(x, \dot{x})\right]\cos\varphi, \\ \frac{\gamma}{2}a\dot{\psi} &= \varepsilon\left[\Delta x + \alpha x^3 + (cx + dx^3)\cos\gamma t + R(x, \dot{x})\right]\sin\varphi. \end{aligned} \tag{4.4}$$

We have made no approximations, and equations (4.4) are exact. Since $a$ and $\psi$ do not vary rapidly, the technique of slowly varying parameter can be used to obtain approximate solutions to equations (4.4). The process leads to the first approximation by replacing the right-hand sides by their averaged terms, and neglecting the rapidly changing terms represented by the trigonometric functions. Thus, we have in the first approximation

$$\begin{aligned} \frac{\gamma}{2}\dot{a} &= -\varepsilon\Big[\Big(\frac{c}{4}a + \frac{d}{8}a^3\Big)\sin 2\psi + S(a,\gamma)\Big], \\ \frac{\gamma}{2}a\dot{\psi} &= \varepsilon\Big[\frac{\Delta}{2}a + \frac{3}{8}\alpha a^3 - \Big(\frac{c}{4}a + \frac{d}{4}a^3\Big)\cos 2\psi + B(a,\gamma)\Big], \end{aligned} \tag{4.5}$$

where

$$S(a,\gamma)=\frac{1}{2\pi}\int_0^{2\pi}\cos\varphi\cdot R(a\sin\varphi,\frac{\gamma}{2}a\cos\varphi)d\varphi,$$
$$B(a,\gamma)=\frac{1}{2\pi}\int_0^{2\pi}\sin\varphi\cdot R(a\sin\varphi,\frac{\gamma}{2}a\cos\varphi)d\varphi. \tag{4.6}$$

The steady-state response of the system can be obtained by setting $\dot a$ and $\dot\psi$ to zero, that is

$$\left(\frac{c}{4}a_0+\frac{d}{8}a_0^3\right)\sin 2\psi_0=-S(a_0,\gamma),$$
$$\left(\frac{c}{4}a_0+\frac{d}{4}a_0^3\right)\cos 2\psi_0=\frac{\Delta}{2}a_0+\frac{3}{8}\alpha a_0^3+B(a_0,\gamma), \tag{4.7}$$

where $a_0$ and $\psi_0$ denote the steady-state values of $a$ and $\psi$ respectively.

Eliminating $\psi_0$ from (4.7), we obtain the equation describing the steady-state resonance curve as

$$W(a_0,\gamma)=0, \tag{4.8}$$

where

$$W(a,\gamma)=\frac{16S^2(a_0,\gamma)}{a_0^2(c+\frac{d}{2}a_0^2)^2}+\frac{16\left(\frac{\Delta}{2}a_0+\frac{3}{8}\alpha a_0^3+B\right)^2}{a_0^2(c+da_0^2)^2}-1, \tag{4.9}$$

with $c+da_0^2\neq 0$, $2c+da_0^2\neq 0$.

We now consider the stability of the stationary oscillations. Let $\delta a$ and $\delta\psi$ be small perturbations, representing the deviation from the steady-state solution, and set $a=a_0+\delta a$, $\psi=\psi_0+\delta\psi$. If we substitute these expressions into equations (4.5) and use the steady-state equations (4.7), we have the following variational equations

$$\frac{\gamma}{2}\frac{d}{dt}\delta a=-\varepsilon\Big\{\left[S'+\left(\frac{c}{4}+\frac{3}{8}da_0^2\right)\sin 2\psi_0\right]\delta a+$$
$$+\left(\frac{c}{2}+\frac{d}{4}a_0^2\right)a_0\cos 2\psi_0\delta\psi\Big\},$$
$$\frac{\gamma}{2}a_0\frac{d}{dt}\delta\psi=\varepsilon\Big\{\left[\frac{\Delta}{2}+\frac{9}{8}\alpha a_0^2+B'-\left(\frac{c}{4}+\frac{3}{4}da_0^2\right)\cos 2\psi_0\right]\delta a+$$
$$+\left(\frac{c}{2}+\frac{d}{2}a_0^2\right)a_0\sin 2\psi_0\delta\psi\Big\}, \tag{4.10}$$

where prime denotes derivative with respect to $a_0$.

The stability of stationary solutions depends on the roots of the characteristic equation of the system (4.10), which may be written in the form

$$\frac{\gamma^2}{4}a_0\lambda^2+\frac{\varepsilon}{2}\gamma\lambda\frac{\partial}{\partial a_0}(a_0S)+\frac{1}{32}\omega^4a_0^2(C+Da_0^2)(2C+Da_0^2)\frac{\partial W}{\partial a_0}=0, \tag{4.11}$$

where

$$C = \varepsilon\frac{c}{\omega^2}\ ;\ D = \varepsilon\frac{d}{\omega^2}\ ;\ \beta = \frac{3\varepsilon\alpha}{4\omega^2}\ ;\ \eta = \frac{\gamma}{2\omega}\ . \tag{4.12}$$

If all values of $\lambda$ have negative real parts, then any perturbation from the steady-state motion tends to die out, hence the steady-state solution is stable. If one $\lambda$ has a positive real part, the perturbations will grow and the steady-state solution is unstable.

The roots of equation (4.11) will have only negative real parts when

$$2a_0\frac{\partial}{\partial a_0^2}\left(a_0 S(a_0,\gamma)\right) > 0, \tag{4.13}$$

$$2a_0(C + Da_0^2)(2C + Da_0^2)\frac{\partial W}{\partial a_0^2} > 0, \tag{4.14}$$

with $c + da_0^2 \neq 0$, $2c + da_0^2 \neq 0$.

It can be shown that the equation of vertical tangents is given by

$$\frac{\partial W}{\partial a_0^2} = 0. \tag{4.15}$$

Hence, in the $(a_0^2, \gamma^2)$ plane, the locus of points of the resonance curve at which the tangents are vertical serves as the boundaries, separating the stable and unstable branches of the resonance curve.

Using formulae (3.26) and the variational equations of the type (3.28), we may easily verify that the zero solution $a = 0$ is unstable in the interval

$$1 - \frac{1}{2}C < \eta^2 < 1 + \frac{1}{2}C \tag{4.16}$$

For $R(x, \dot{x}) \equiv 0$, the resonance curve consists of two intersecting straight lines with the coordinates of the point of intersection $I$ (Fig. 61):

$$a_0^2 = -\frac{C}{D}\ ,\quad \eta^2 = 1 - \beta\frac{C}{D}\ , \tag{4.17}$$

and a segment PQ : $a_0^2 = -2C/D$

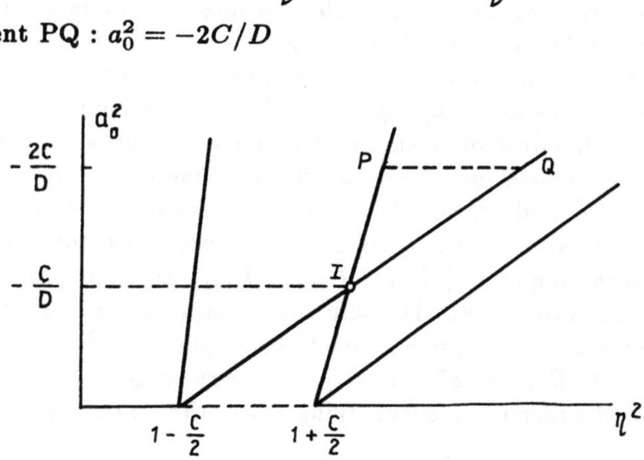


Fig. 61. Resonance curve for the case $R \equiv 0$.

**For linear friction**

$$R(x, \dot{x}) = h\dot{x}. \tag{4.18}$$

In this case, the equations of the first approximation (4.5) become

$$\begin{aligned} \gamma\dot{a} &= -\varepsilon\Big[\frac{h}{2}\gamma a + \big(\frac{c}{2}a + \frac{d}{4}a^3\big)\sin 2\psi\Big], \\ \gamma a\dot{\psi} &= \varepsilon\Big[\Delta a + \frac{3}{4}\alpha a^3 - \big(\frac{c}{2}a + \frac{d}{2}a^3\big)\cos 2\psi\Big]. \end{aligned} \tag{4.19}$$

The equation for the resonance curves (4.8) is now

$$\frac{4(1-\eta^2+\beta a_0^2)^2}{(C+Da_0^2)^2} + \frac{H^2\eta^2}{(2C+Da_0^2)^2} - 1 = 0, \tag{4.20}$$

where

$$H = \frac{4\varepsilon h}{\omega}. \tag{4.21}$$

By solving the equation (4.20), we obtain an approximate formula

$$\begin{aligned} \eta^2 = 1 + \beta a_0^2 - \frac{H^2(C+Da_0^2)^2}{8(2C+Da_0^2)^2} \pm \\ \pm \frac{1}{2}(C+Da_0^2)\sqrt{1 - \frac{(1+\beta a_0^2)H^2}{(2C+Da_0^2)^2}}. \end{aligned} \tag{4.22}$$

If $d > 0$, we have the resonance curves shown in Fig. 62, where the straight lines "1" correspond to $H = 0$. The resonance curves have parabolic form. With increasing friction ($H_{(2)} < H_{(3)} < H_{(4)}$) the resonance curve moves up. The stippled areas show the regions of unstable stationary oscillations, in which the stability conditions (4.13) and (4.14) are violated.

If $d < 0$, we have resonance curves represented in Fig. 63, where the straight lines "1", including the segment PQ : $a_0^2 = -2C/D$, correspond to the value $H = 0$. With sufficiently small values of $H$ the resonance curve consists of three branches (see lines "2"), one of which lies above the straight line $a_0^2 = -2C/D$, the second - between $a_0^2 = -C/D$ and $a_0^2 = -2C/D$, and the third below the straight line $a_0^2 = -C/D$. The last two branches cross at the intersection point I of the straight lines "1". With increasing of $H$ ($H_{(2)} < H_{(3)} < H_{(4)}$), the first branch moves up, while the second one moves down. For sufficiently large values of $H$, there are two branches of the resonance curve, one above the straight line $a_0^2 = -2C/D$, and the other below $a_0^2 = -C/D$ (see "3", Fig. 63). For very large values of $H$, there is only one branch of the resonance curve, lying above the straight line $a_0^2 = -2C/D$ (see "4", Fig. 63).

If $d = 0$, the resonance curves have the form shown in Fig. 64, where the two parallel straight lines correspond to the zero value $H$. With increasing $H$

$(H_{(2)} < H_{(3)} < H_{(4)})$, the maximum of the amplitudes of stationary oscillations decreases.

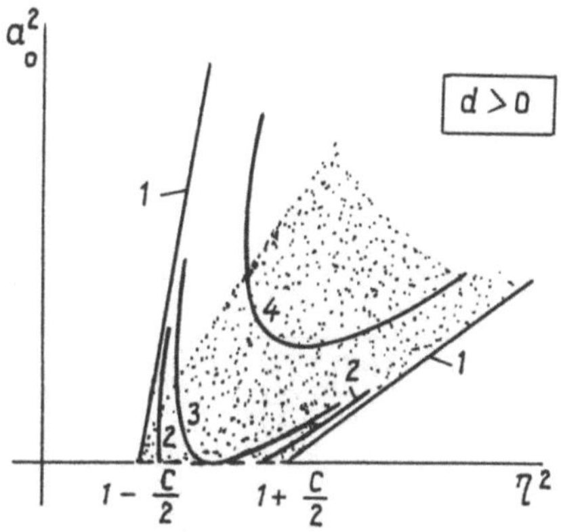


*Fig. 62.* Resonance curves for the case $d > 0$ with various damping coefficients.

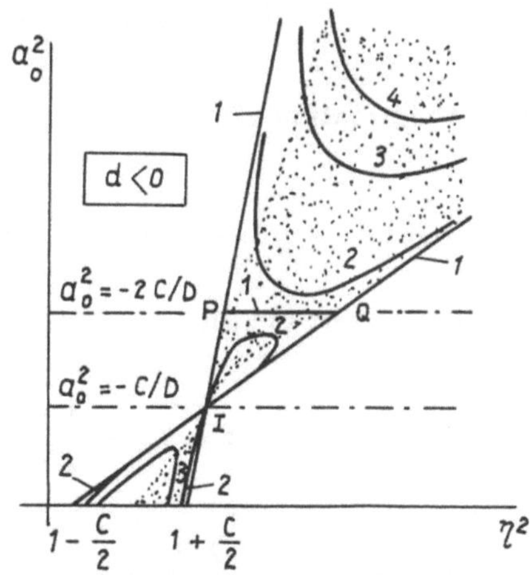


*Fig. 63.* Resonance curves for the case $d < 0$ with various damping coefficients.

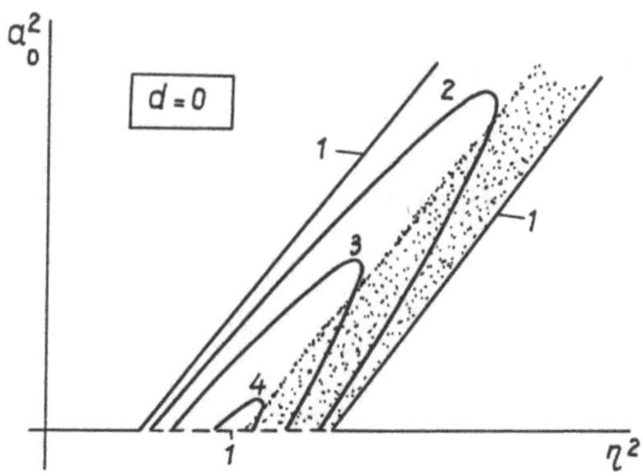


*Fig. 64.* Resonance curves for the case $d = 0$ with various damping coefficients.

**For the case of dry friction**

$$R(x, \dot{x}) = h_0 \operatorname{sign} \dot{x}, \tag{4.23}$$

where $h_0$ is a positive constant, one has

$$S(a, \gamma) = \begin{cases} \frac{2}{\pi} h_0 & \text{if} \quad a \neq 0, \\ 0 & \text{if} \quad a = 0, \end{cases} \tag{4.24}$$
$$B(a, \gamma) = 0.$$

Hence, the expression (4.9) becomes

$$W = \frac{16H^2}{a_0^2(2C + Da_0^2)^2} + \frac{4(1 - \eta^2 + \beta a_0^2)^2}{(C + Da_0^2)^2} - 1, \tag{4.25}$$

where the notation of (4.12) is used, and

$$H_0 = \frac{4\varepsilon h_0}{\pi\omega^2}. \tag{4.26}$$

From (4.25), one obtains

$$\eta^2 = 1 + \beta a_0^2 \pm \frac{1}{2}(C + Da_0^2)\sqrt{1 - \frac{16H_0^2}{a_0^2(2C + Da_0^2)^2}}. \tag{4.27}$$

For $d < 0$ and $D > -\dfrac{2C^3}{27H_0^2}$ and with small values of $H_0$, there are three branches of the resonance curve; the two lower branches cross at point I (4.17). With the growth of dry friction ($H_0$), the upper branch moves up and the two lower branches become narrower. For $d < 0$ and $D < -\dfrac{2C^3}{27H_0^2}$, the resonance curve has only one parabolic branch, lying above the straight line $a_0^2 = -\dfrac{2C}{D}$.

When $d > 0$, the resonance curve is similar to that of Fig. 62. The case $d = 0$ is illustrated in Fig. 65, where the straight lines "1" correspond to $H_0 = 0$. With increasing $H_0$ ($H_{(2)} < H_{(3)} < H_{(4)}$), the resonance curve moves up.

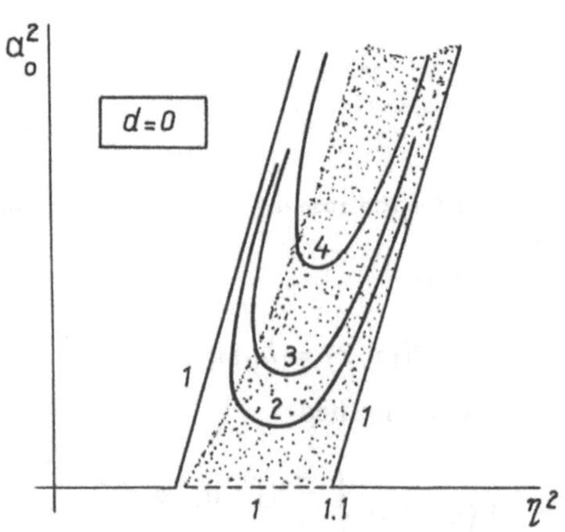


*Fig. 65.* Resonance curves for the case of dry frictions.

**Turbulent friction**

Assuming a friction force in the form

$$R(x, \dot{x}) = h_2 \dot{x}^2 \operatorname{sign} \dot{x}, \quad h_2 > 0, \tag{4.28}$$

we can verify without difficulty that $B(a, \gamma) = 0$,

$$S(a, \gamma) = \begin{cases} \dfrac{1}{3\pi} h_2 \gamma^2 a^2 & \text{for} \quad a \neq 0, \\ 0 & \text{for} \quad a = 0. \end{cases} \tag{4.29}$$

Hence, the formula (4.8) becomes

$$\frac{H_2^2 \eta^4 a_0^2}{(2C + Da_0^2)^2} + \frac{(1 - \eta^2 + \beta a_0^2)^2}{(C + Da_0^2)^2} - \frac{1}{4} = 0, \tag{4.30}$$

where

$$H_2 = \frac{16}{3\pi} \varepsilon h_2. \tag{4.31}$$

The relation (4.30) can be somewhat simplified if we limit ourselves to the study of the behaviour of the oscillations with frequency $\frac{\gamma}{2}$, close to $\omega$. Then, $1 - \eta^2$ will be a small quantity of $\varepsilon$-order. Substituting $\eta^4$ in the first term of (4.30) by unity, we neglect the small quantities of order $\varepsilon^3$ in comparison with those of order $\varepsilon^2$. If this is done, one has

$$\eta^2 = 1 + \beta a_0^2 \pm \frac{1}{2}(C + Da_0^2) \sqrt{1 - \frac{4H_2^2 a_0^2}{(2C + Da_0^2)^2}}. \tag{4.32}$$

We always suppose $c > 0$, $\beta > 0$. For $d < 0$ and small values of $H_2$, the resonance curve consists of three branches. The first branch lies above the straight line $a^2 = -\frac{2C}{D}$. The second lies between the straight lines $a^2 = -\frac{C}{D}$ and $a^2 = -\frac{2C}{D}$. The third lies below $a^2 = -\frac{C}{D}$. The two lower branches intersect at the point I (4.17). With increasing $H_2$, the second branch becomes lower and the first moves up. For sufficiently large values of $H_2$, the resonance curve consists of only two branches, one of which is above the straight line $a^2 = -2\frac{C}{D}$, and the other is below $a^2 = -\frac{C}{D}$.

For $d > 0$, if $\frac{H_2^2}{2C} > D$, the resonance curve consists of two parabolic branches (curves 3, Fig. 66). With increasing $H_2$, these branches diverge: the upper branch moves up and the lower branch moves down. If $\frac{H_2^2}{2C} < D$, the resonance curves have the form shown by "2" in Fig. 66, where the heavy and dashed branches correspond to the stability and instability of oscillations, respectively.

The case $d = 0$ is represented in Fig. 67. With the growth of $H_2$, the maximum amplitude decreases $(H_{(2)} < H_{(3)} < H_{(4)})$.

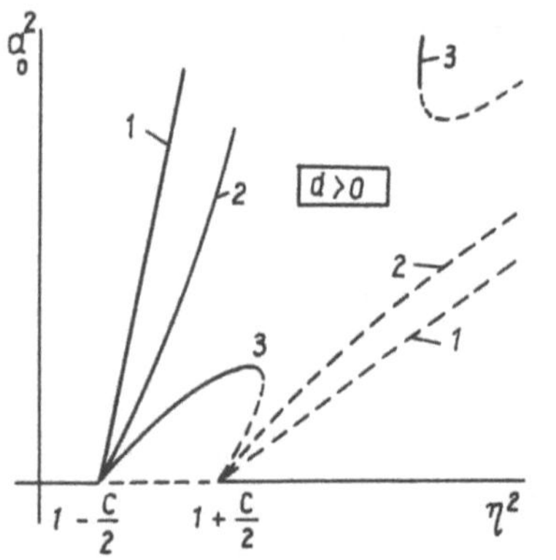


*Fig. 66.* Resonance curves for the case of turbulent friction and $d > 0$.

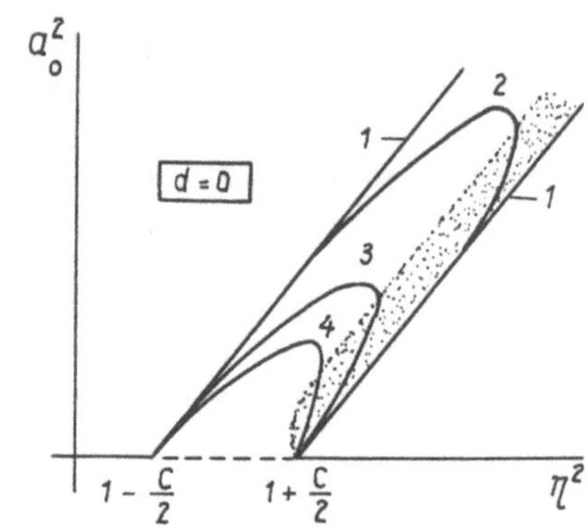


*Fig. 67.* Resonance curves for the case of turbulent friction and $d = 0$.

**Combination friction**

Let $R(x, \dot{x})$ be

$$R(x, \dot{x}) = (h_0 + h_2\dot{x}^2)\,\mathrm{sign}\,\dot{x}, \quad h_0 > 0, \; h_2 > 0. \tag{4.33}$$

It is found that the expressions (4.6) now take the form

$$S(a, \gamma) = \begin{cases} \dfrac{2}{\pi}h_0 + \dfrac{1}{3\pi}h_2\gamma^2 a^2 & \text{for} \quad a \neq 0, \\ \dfrac{1}{3\pi}h_2\gamma^2 a^2 & \text{for} \quad a = 0, \end{cases} \tag{4.34}$$

and the averaged equations (4.7) become for $a \neq 0$:

$$\begin{aligned} \frac{\gamma}{2}\frac{da}{dt} &= -\varepsilon\Big[\frac{2}{\pi}h_0 + \frac{1}{3\pi}h_2\gamma^2 a^2 + \Big(\frac{ca}{4} + \frac{da^3}{8}\Big)\sin 2\psi\Big], \\ \frac{\gamma}{2}\frac{a\,d\psi}{dt} &= \varepsilon\Big[\frac{\Delta}{2}a + \frac{3}{8}\alpha a^3 - \Big(\frac{ca}{4} + \frac{da^3}{4}\Big)\cos 2\psi\Big]. \end{aligned} \tag{4.35}$$

The stationary values of amplitudes of oscillation and the exciting frequency are related by the equation

$$\frac{4(2H_0 + H_2\eta^2 a_0^2)^2}{a_0^2(2C + Da_0^2)^2} + \frac{4(1 - \eta^2 + \beta a_0^2)^2}{(C + Da_0^2)^2} - 1 = 0. \tag{4.36}$$

From here, we obtain

$$\eta^2 = 1 + \beta a_0^2 \pm \frac{1}{2}(C + Da_0^2)\sqrt{1 - \frac{4(2H_0 + H_2 a_0^2)^2}{a_0^2(2C + Da_0^2)^2}} \,. \tag{4.37}$$

Depending on the disposition of the curves

$$y = 4(2H_0 + H_2 A)^2, \quad z = A(2C + DA)^2, \tag{4.38}$$

in the $(A, y)$ - plane, the resonance curves have different forms. Some possible resonance curves are represented in Fig. 68 (for $d < 0$) and Fig. 69 (for $d = 0$); the straight lines "1" correspond to $h_0 = h_2 = 0$. In Fig. 68, the index numbers of the curves show some different possibilities. In Fig. 69, with increasing $H_2$ $(H_2^{(2)} < H_2^{(3)} < H_2^{(4)} < H_2^{(5)})$, the maximum of the amplitudes decreases. In this Figure, the heavy and dashed branches correspond to the stability and instability of stationary solutions, respectively.

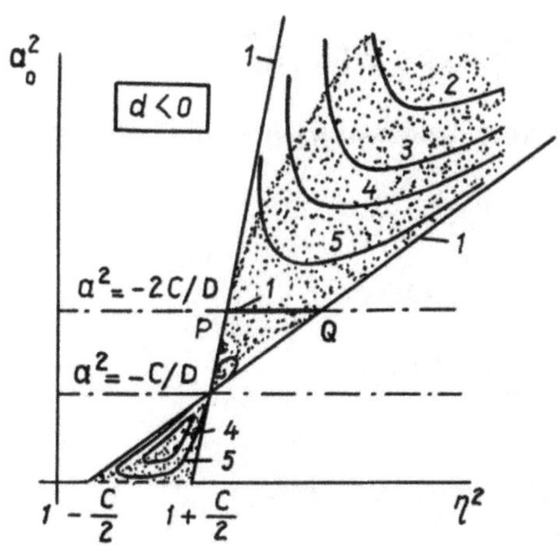


*Fig. 68.* Resonance curves for the case of combination friction and $d < 0$.

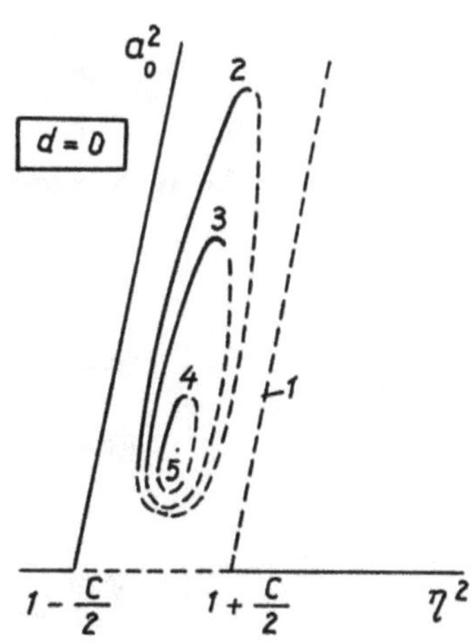


*Fig. 69.* Resonance curves for the case of combination friction and $d = 0$.

We are now concerned with the study of the parametric oscillations of a nonlinear system of the type

$$\ddot{x} + \omega^2 x + \varepsilon(cx + dx^3 + gx^5)\cos\gamma t + \varepsilon(\alpha x^3 + \delta x^5) + \varepsilon R(x, \dot{x}) = 0 \tag{4.39}$$

under the resonance condition (4.2), in which $\omega$, $c$, $d$, $g$, $\gamma$, $\delta$, $\alpha$ are constants, $\varepsilon$ is a small positive parameter, and $R(x, \dot{x})$ is the function characterizing the friction.

We shall find the solution of equation (4.39) in the form

$$x = a\sin\theta, \quad \dot{x} = \frac{\gamma}{2}a\cos\theta, \tag{4.40}$$

where $a$ and $\theta$ are new variables. Upon substitution of the expressions (4.40) into equation (4.39), we obtain the following equations for new variables:

$$\begin{aligned} \frac{\gamma}{2}\dot{a} = & -\varepsilon\cos\theta\big[\Delta a\sin\theta + (cx + dx^3 + gx^5)\cos\gamma t + \\ & + \alpha x^3 + \delta x^5 + R(x,\dot{x})\big], \\ \frac{\gamma}{2}a\dot{\psi} = & \ \varepsilon\sin\theta\big[\Delta a\sin\theta + (cx + dx^3 + gx^5)\cos\gamma t + \\ & + \alpha x^3 + \delta x^5 + R(x,\dot{x})\big], \end{aligned} \tag{4.41}$$

where $\psi = \theta - \frac{\gamma}{2}t$.

The equations (4.41) are in the standard form for the averaging method. Following this method, in the first approximation, we have

$$\begin{aligned} \frac{\gamma}{2}\dot{a} = & -\varepsilon\Big[\Big(\frac{ca}{4} + \frac{da^3}{8} + \frac{5}{64}ga^5\Big)\sin 2\psi + S(a,\gamma)\Big], \\ \frac{\gamma}{2}a\dot{\psi} = & \ \varepsilon\Big[\frac{\Delta}{2}a + \frac{3}{8}\alpha a^3 + \frac{5}{16}\delta a^5 - \\ & - \Big(\frac{ca}{4} + \frac{da^3}{4} + \frac{15}{64}ga^5\Big)\cos 2\psi + B(a,\gamma)\Big], \end{aligned} \tag{4.42}$$

obtained from (4.41) by averaging their right-hand sides over time and considering

$$\begin{aligned} &\langle\sin^4\theta\rangle = \frac{3}{8}, \qquad \langle\sin^6\theta\rangle = \frac{5}{16}, \\ &\langle\cos\gamma t\cdot\sin^5\theta\cdot\cos\theta\rangle = \frac{5}{64}\sin 2\psi, \\ &\langle\cos\gamma t\cdot\sin^6\theta\rangle = -\frac{15}{64}\cos 2\psi, \ldots \end{aligned}$$

$$\begin{aligned} S(a,\gamma) &= \frac{1}{2\pi}\int_0^{2\pi}\cos\theta\cdot R\Big(a\sin\theta, \frac{\gamma}{2}a\cos\theta\Big)d\theta, \\ B(a,\gamma) &= \frac{1}{2\pi}\int_0^{2\pi}\sin\theta\cdot R\Big(a\sin\theta, \frac{\gamma}{2}a\cos\theta\Big)d\theta. \end{aligned} \tag{4.43}$$

The steady-state response can be obtained by setting $\dot{a}$ and $\dot{\psi}$ equal to zero in equations (4.42). Thus, we have

$$\begin{aligned} &\Big(\frac{c}{4}a_0 + \frac{d}{8}a_0^3 + \frac{5}{64}ga_0^5\Big)\sin 2\psi_0 = -S(a_0,\gamma), \\ &\Big(\frac{c}{4}a_0 + \frac{d}{4}a_0^3 + \frac{15}{64}ga_0^5\Big)\cos 2\psi_0 = \frac{\Delta}{2}a_0 + \frac{3}{8}\alpha a_0^3 + \frac{5}{16}\delta a_0^5 + B(a_0,\gamma). \end{aligned} \tag{4.44}$$

Eliminating the phase $\psi_0$ gives

$$M = 0, \tag{4.45}$$

where,

$$M = \frac{16S^2}{a_0^2\big(c + \frac{d}{2}a_0^2 + \frac{5}{16}ga_0^4\big)^2} + \frac{16\big(\frac{\Delta}{2}a_0 + \frac{3}{8}\alpha a_0^3 + \frac{5}{16}\delta a_0^5 + B\big)^2}{a_0^2\big(c + da_0^2 + \frac{15}{16}ga_0^4\big)^2} - 1. \tag{4.46}$$

To study the stability of stationary oscillations, the variational equations based on the stationary solution (4.44) will be considered. By setting $a = a_0 + \delta a$, $\psi = \psi_0 + \delta\psi$ in (4.42), where $\delta a$ and $\delta\psi$ are small perturbations, and using the steady-state equations (4.44), we get

$$\begin{aligned}\frac{\gamma}{2}\frac{d}{dt}\delta a = & -\varepsilon\Big\{\Big[\Big(\frac{c}{4} + \frac{3}{8}da_0^2 + \frac{25}{64}ga_0^4\Big)\sin 2\psi_0 + S'(a_0,\gamma)\Big]\delta a + \\ & + \Big(\frac{c}{2}a_0 + \frac{d}{4}a_0^3 + \frac{5}{32}ga_0^5\Big)\cos 2\psi_0\delta\psi\Big\},\end{aligned}$$

$$\begin{aligned}\frac{\gamma}{2}a_0\frac{d}{dt}\delta\psi = \varepsilon\Big\{\Big[\frac{\Delta}{2} + \frac{9}{8}\alpha a_0^2 + \frac{25}{16}\delta a_0^4 - \Big(\frac{c}{4} + \frac{3}{4}da_0^2 + \frac{75}{64}ga_0^4\Big)\cos 2\psi_0 + B'\Big]\delta a + \\ + \Big(\frac{c}{2}a_0 + \frac{d}{2}a_0^3 + \frac{15}{32}ga_0^5\Big)\sin 2\psi_0\cdot\delta\psi\Big\}.\end{aligned}$$

The characteristic equation of this system is of the form

$$\begin{aligned}&\frac{\gamma^2}{4}a_0\lambda^2 + \frac{\varepsilon}{2}\gamma\frac{\partial}{\partial a_0}(a_0S)\lambda + \\ &+ \frac{\varepsilon^2}{16}a_0^2\Big(c + \frac{d}{2}a_0^2 + \frac{5}{16}ga_0^4\Big)\Big(c + da_0^2 + \frac{15}{16}ga_0^4\Big)\frac{\partial M}{\partial a_0} = 0.\end{aligned} \tag{4.47}$$

The stability condition is given by the Routh-Hurwitz criterion, that is

$$\frac{\partial}{\partial a_0}(a_0S) > 0, \tag{4.48}$$

$$E\cdot\frac{\partial M}{\partial a} > 0, \tag{4.49}$$

where

$$E = (2C + Da_0^2 + 2Ga_0^4)(C + Da_0^2 + 3Ga_0^4), \tag{4.50}$$

$$C = \frac{\varepsilon c}{\omega^2}, \quad D = \frac{\varepsilon d}{\omega^2}, \quad G = \frac{5\varepsilon g}{16\omega^2}.$$

The first stability condition, $\frac{\partial}{\partial a_0}(a_0 S) > 0$, limits the friction force and, in general, it is always satisfied. The second stability condition, $E \cdot \frac{\partial M}{\partial a_0} > 0$, is determined as follows. The resonance curve ($M = 0$) divides the plane $(a_0^2, \eta^2)$ into distinct regions, in each of which the expression $M$ (4.46) has a definite sign (+ or -). If on moving up along a straight line parallel to the axis $a_0^2$, we pass from a region $M < 0$ to a region $M > 0$, then at the point of intersection between the straight line and the resonance curve, the derivative $\frac{\partial M}{\partial a_0}$ is positive. So, that point corresponds to a stable state of oscillation if $E > 0$, and to an unstable one if $E < 0$. On the contrary, if we pass from a region $M > 0$ to a region $M < 0$, then the point of intersection corresponds to the stable state of oscillation if $E < 0$ and to an unstable one if $E > 0$. In the figures presented below the branches of resonance curves corresponding to the stable state of oscillation will be shown by heavy lines.

It is worth mentioning that the zero solution $a = 0$ of equations (4.42) (if it exists) is unstable in the interval determined by (4.16).

Let us consider the case

$$R(x, \dot{x}) = h_0 \operatorname{sign} \dot{x}, \tag{4.51}$$

where $h_0$ is a positive constant. In this case, we have $B(a, \gamma) = 0$ and

$$S(a, \gamma) = \begin{cases} \frac{2}{\pi} h_0 & \text{if} \quad a \neq 0, \\ 0 & \text{if} \quad a = 0, \end{cases} \tag{4.52}$$

and therefore equations (4.42) become

for $a \neq 0$:

$$\begin{aligned} \frac{\gamma}{2}\dot{a} &= -\varepsilon\Big[\frac{2}{\pi}h_0 + \Big(\frac{c}{4}a + \frac{d}{8}a^3 + \frac{5}{64}ga^5\Big)\sin 2\psi\Big], \\ \frac{\gamma}{2}a\dot{\psi} &= \varepsilon\Big[\frac{\Delta}{2}a + \frac{3}{8}\alpha a^3 + \frac{5}{16}\delta a^5 - \Big(\frac{c}{4}a + \frac{d}{4}a^3 + \frac{15}{64}ga^5\Big)\cos 2\psi\Big], \end{aligned} \tag{4.53}$$

and for $a = 0$:

$$\begin{aligned} \frac{\gamma}{2}\dot{a} &= -\varepsilon\Big(\frac{c}{4}a + \frac{d}{8}a^3 + \frac{5}{64}ga^5\Big)\sin 2\psi, \\ \frac{\gamma}{2}a\dot{\psi} &= \varepsilon\Big[\frac{\Delta}{2}a + \frac{3}{8}\alpha a^3 + \frac{5}{16}\delta a^5 - \Big(\frac{c}{4}a + \frac{d}{4}a^3 + \frac{15}{64}ga^5\Big)\cos 2\psi\Big]. \end{aligned} \tag{4.54}$$

Introducing the notation

$$H_0 = \frac{4\varepsilon h_0}{\pi\omega^2}, \quad \eta = \frac{\gamma}{2\omega}, \quad \beta = \frac{3\varepsilon\alpha}{4\omega^2}, \quad \delta_* = \frac{5\varepsilon\delta}{8\omega^2}, \tag{4.55}$$

we find that the equation (1.6) takes the form

$$\frac{16H_0^2}{a_0^2\left(2C+Da_0^2+2Ga_0^4\right)^2}+\frac{4(1-\eta^2+\beta a_0^2+\delta_* a_0^4)^2}{\left(C+Da_0^2+3Ga_0^4\right)^2}=1. \tag{4.56}$$

From here, we obtain

$$\eta^2=1+\beta a_0^2+\delta_* a^4\pm\frac{1}{2}\left(C+Da_0^2+3Ga_0^4\right)\sqrt{1-\frac{16H_0^2}{a_0^2\left(2C+Da_0^2+2Ga_0^4\right)^2}}\cdot \tag{4.57}$$

The form of the resonance curve depends on the roots of the following equation

$$C+Da_0^2+3GA_0^4=0, \tag{4.58}$$

and on the disposition of the curves

$$z=\left(2C+DA+2GA^2\right)^2, \tag{4.59}$$

and

$$z=\frac{16H_0^2}{A}, \quad A=a_0^2. \tag{4.60}$$

The relation (4.57) is plotted in Fig. 70 for the case $D^2-12GC<0$, $d<0$, $\alpha>0$, $c>0$. When $H_0=0$, we have two separate parabolas (curves "0"). With the growth of the friction force $(H_0)$, the resonance curves move up $\left(H_{0(1)}<H_{0(2)}<H_{0(3)}\right)$.

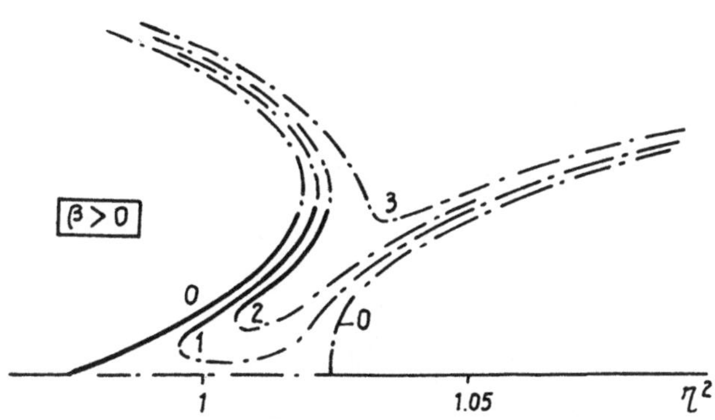


*Fig. 70.* Resonance curves for the case of dry friction.

**Turbulent friction**

$$R(x,\dot{x})=h_2\dot{x}^2\operatorname{sign}\dot{x}.$$

It is not difficult to show that

$$S(a,\gamma) = \langle \cos\theta \cdot R(a\sin\theta, \frac{\gamma}{2}a\cos\theta)\rangle = \begin{cases} \frac{1}{3\pi}h_2\gamma^2a^2 & \text{for} \quad a \neq 0, \\ 0 & \text{for} \quad a = 0. \end{cases} \tag{4.61}$$

Therefore, we have the following equations of the first approximation

$$\begin{aligned} \frac{\gamma}{2}\dot{a} &= -\varepsilon\Big[\frac{1}{3\pi}h_2\gamma^2a^2 + \big(\frac{c}{4}a + \frac{d}{8}a^3 + \frac{5}{64}ga^5\big)\sin 2\psi\Big], \\ \frac{\gamma}{2}a\dot{\psi} &= \varepsilon\Big[\frac{\Delta}{2}a + \frac{3}{8}\alpha a^3 + \frac{5}{16}\delta a^5 - \big(\frac{c}{4}a + \frac{d}{4}a^3 + \frac{15}{64}ga^5\big)\cos 2\psi\Big]. \end{aligned} \tag{4.62}$$

The relation (4.45) now takes the form

$$\frac{H_2^2a_0^2\eta^4}{\big(2C + Da_0^2 + 2Ga_0^4\big)^2} + \frac{\big(1 - \eta^2 + \beta a_0^2 + \delta_* a_0^4\big)^2}{\big(C + Da_0^2 + 3Ga_0^4\big)^2} - \frac{1}{4} = 0,$$

where $H_2 = \dfrac{16}{3\pi}\varepsilon h_2$. From here, we obtain approximately:

$$\eta^2 = 1 + \beta a_0^2 + \delta_* a_0^4 \pm \frac{1}{2}\big(C + Da^2 + 3Ga^4\big)\sqrt{1 - \frac{4H_2^2a^2}{\big(2C + Da_0^2 + 2Ga_0^4\big)^2}}. \tag{4.63}$$

This relation is plotted in Fig. 71 for the case $D^2 - 12GC < 0$, where the curve with index "0" corresponds to the value $H_2 = 0$. For small values of $H_2$, the resonance curve consists of two branches denoted by "1". For sufficiently large values of $H_2$, there is only one branch, namely "2".

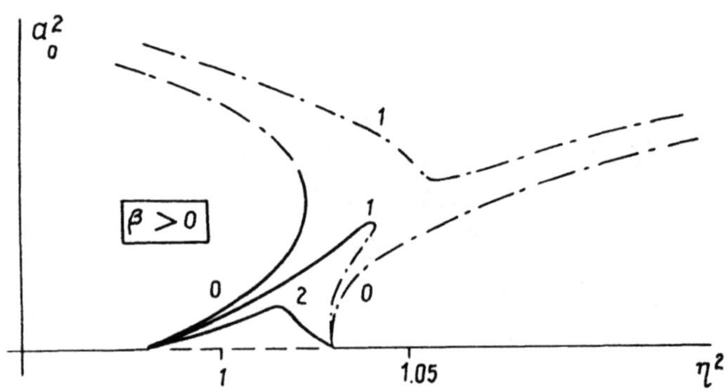


*Fig. 71.* Resonance curves for the case of turbulent friction.

**Linear friction**

$$R(x, \dot{x}) = h\dot{x}. \tag{4.64}$$

Some resonance curves for $H < 2C$, where $H = \dfrac{4\varepsilon h}{\omega}$, $c > 0$, $d < 0$, are represented in Fig. 72. With increasing $H$, the amplitude of oscillation decreases $H_{(3)} > H_{(2)} > H_{(1)}$.

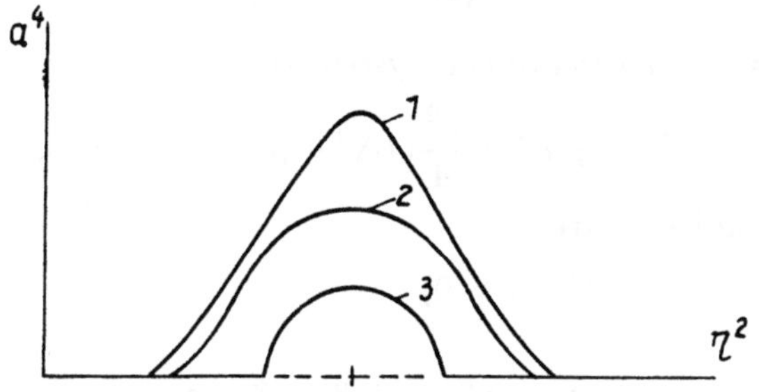


*Fig. 72.* Resonance curves for the case of linear friction.

To study the stability of the zero solution of equation (4.1) in the case $R(0,0) \equiv 0$, it is convenient to use the Cartesian coordinates $u$ and $v$ instead of the polar coordinates $a$ and $\varphi$ (4.3) as follows:

$$\begin{aligned} x &= u\cos\frac{\gamma}{2}t + v\sin\frac{\gamma}{2}t, \\ \dot{x} &= -\frac{\gamma}{2}u\sin\frac{\gamma}{2}t + \frac{\gamma}{2}v\cos\frac{\gamma}{2}t, \end{aligned} \tag{4.65}$$

with the additional condition

$$\dot{u}\cos\frac{\gamma}{2}t + \dot{v}\sin\frac{\gamma}{2}t = 0, \tag{4.66}$$

where $u$ and $v$ are new variables.

Substituting (4.65) into (4.1), we obtain

$$\frac{\gamma}{2}\dot{u}\sin\frac{\gamma}{2}t - \frac{\gamma}{2}\dot{v}\cos\frac{\gamma}{2}t = \varepsilon\big[\Delta x + \alpha x^3 + (cx + dx^3)\cos\gamma t + R(x,\dot{x})\big]. \tag{4.67}$$

Solving (4.66) and (4.67) with respect to the derivatives $\dot{u}$ and $\dot{v}$, we get:

$$\begin{aligned} \gamma\dot{u} &= 2\varepsilon\big[\Delta x + \alpha x^3 + (cx + dx^3)\cos\gamma t + R(x,\dot{x})\big]\sin\frac{\gamma}{2}t, \\ \gamma\dot{v} &= -2\varepsilon\big[\Delta x + \alpha x^3 + (cx + dx^3)\cos\gamma t + R(x,\dot{x})\big]\cos\frac{\gamma}{2}t. \end{aligned} \tag{4.68}$$

It is assumed that the function $R(x,\dot{x})$ may be expanded in the series

$$R(x,\dot{x}) = \mu\dot{x} + [\dots]_2, \tag{4.69}$$

where $\mu = \left(\frac{\partial R}{\partial \dot{x}}\right)_0$ and $[\ldots]_2$ denotes the higher order terms with respect to $x$ and $\dot{x}$. Then, in the first approximation, we have the following variational equations for the trivial solution $u = v = 0$ of (4.68):

$$\begin{aligned} \gamma\dot{u} &= \varepsilon\Big[-\frac{1}{2}\mu\gamma u + \Big(\Delta - \frac{c}{2}\Big)v\Big], \\ \gamma\dot{v} &= -\varepsilon\Big[\Big(\Delta + \frac{c}{2}\Big)u + \frac{1}{2}\mu\gamma v\Big]. \end{aligned} \tag{4.70}$$

The characteristic equation of this system is

$$\gamma^2\lambda^2 + \varepsilon\gamma^2\mu\lambda + \frac{\varepsilon^2}{4}\big[4\Delta^2 + \mu^2\gamma^2 - c^2\big] = 0. \tag{4.71}$$

So, the stability conditions are:

$$1) \quad \mu > 0, \tag{4.72}$$

$$2) \quad 4\Delta^2 + \mu^2\gamma^2 - c^2 > 0. \tag{4.73}$$

The following zone is unstable zone for the zero solution:

$$4\Delta^2 + \mu^2\gamma^2 - c^2 < 0$$

or with the notation (4.12):

$$1 - \sqrt{\frac{C^2}{4} - \frac{\varepsilon^2\mu^2}{\omega^2}} < \eta^2 < 1 + \sqrt{\frac{C^2}{4} - \frac{\varepsilon^2\mu^2}{\omega^2}}. \tag{4.74}$$

## 5. Parametric Oscillations of Mechanical Systems with Hysteresis

Experience shows that the energy is dissipated in all real systems as a result of the internal friction. In contrast to viscous damping, damping due to the internal friction does not depend on the velocity, but on the displacement. This type of damping is attributed to the hysteresis phenomenon, associated with the cyclic stress in elastic materials. The energy loss per cycle of the stress is equal to the area inside the hysteresis loop.

1. We consider a system with a single degree of freedom specified by the differential equation

$$\ddot{x} + \omega^2(1 + \varepsilon c\cos\gamma t)x + \varepsilon\alpha\omega^2x^3 + \varepsilon\omega\delta\dot{x} + \varepsilon\Phi(x) = 0, \tag{5.1}$$

where $\omega, c, \alpha, \delta$ are constants, $\varepsilon$ is a positive small parameter and $\Phi(x)$ is a function characterizing the dissipation of energy in the elastic material.

It is assumed that there is a resonance relation $\gamma \simeq 2\omega$. Then, the approximate solution of (5.1) is found in the form

$$x = a\cos\Big(\frac{\gamma}{2}t + \theta\Big), \tag{5.2}$$

where $a$ and $\theta$ satisfy the equations

$$\frac{da}{dt} = \varepsilon A_1(a,\theta),$$
$$a\frac{d\theta}{dt} = \left(\omega - \frac{\gamma}{2}\right)a + \varepsilon B_1(a,\theta). \tag{5.3}$$

Functions $A_1$ and $B_1$ are partial solutions of the equations:

$$\left(\omega - \frac{\gamma}{2}\right)\frac{\partial A_1}{\partial\theta} - 2\omega B_1 = -\left[\frac{3}{4}\alpha\omega^2 a^3 + \frac{c\omega^2}{2}a\cos 2\theta + 2\langle\cos\psi\ \Phi(a\cos\psi)\rangle\right],$$
$$\left(\omega - \frac{\gamma}{2}\right)\frac{\partial B_1}{\partial\theta} + 2\omega A_1 = -\left[\omega^2\delta a + \frac{c\omega^2}{2}a\sin 2\theta + 2\langle\sin\psi\ \Phi(a\cos\psi)\rangle\right], \tag{5.4}$$

where $\psi = \frac{1}{2}\gamma t + \theta$; and $\langle\ \cdot\rangle$ is an averaging operator on time.

We limit ourselves by considering the resonance zone, in which

$$\omega^2 = \frac{\gamma^2}{4} + \varepsilon\Delta. \tag{5.5}$$

Then, we have the following formulae for $A_1$ and $B_1$:

$$A_1 = \frac{1}{2\omega}\left[-\omega^2\delta a + 2\langle\sin\psi\ \Phi(a\cos\psi)\rangle\right] + \frac{c\omega}{4}a\sin 2\theta,$$
$$B_1 = \frac{1}{2\omega}\left[\frac{3\alpha}{4}\omega^2 a^3 + 2\langle\cos\psi\ \Phi(a\cos\psi)\rangle\right] + \frac{c\omega}{4}a\cos 2\theta. \tag{5.6}$$

Following the Davidenkov-Pisarenko hypothesis, the function $\Phi(x)$ may be written as [41]

$$\Phi(x) = -\beta\frac{\omega^2}{3}\left(x^3 \pm 3\xi x^2 + 3\xi^2 x \mp 3\xi a^2\right), \tag{5.7}$$

where $\beta$ is a constant determined by experiment; $\xi = g(a)$ is some function of the amplitude $a$ and $g(0) = 0$.

Substituting (5.6) and (5.7) into (5.3) and invoking the averaging operator, we obtain

$$\frac{da}{dt} = -\frac{\varepsilon\omega}{2}\left(\delta a + \frac{8\beta}{3\pi}\xi a^2 - \frac{c}{2}a\sin 2\theta\right),$$
$$a\frac{d\theta}{dt} = \frac{\varepsilon\Delta}{\gamma}a + \frac{\varepsilon\omega}{2}\left[\frac{3\alpha}{4}a^3 - \beta a\left(\frac{a^2}{4} + \xi^2\right) + \frac{ca}{2}\cos 2\theta\right]. \tag{5.8}$$

These equations have a trivial solution $a = 0$, and nontrivial stationary solutions $a = a_0 = \text{const}$, $\theta = \theta_0 = \text{const}$, determined by the relations

$$\frac{c}{2}\sin 2\theta_0 = \delta + \frac{8\beta}{3\pi}\xi_0 a_0,$$
$$\frac{c}{2}\cos 2\theta_0 = -\frac{\Delta}{\omega^2} - \frac{3\alpha}{4}a_0^2 + \beta\left(\frac{a_0^2}{4} + \xi_0^2\right), \tag{5.9}$$

where $\xi_0 = g(a_0)$.

Eliminating the phase $\theta_0$ from the equations (5.9) gives

$$W(a_0, \gamma) = 0, \tag{5.10}$$

where

$$W(a_0, \gamma) = -\frac{c^2\omega^4}{4} + \omega^2\Big(\delta + \frac{8\beta}{3\pi}\xi_0 a_0\Big)^2 + \Big[\Delta + \frac{3\alpha}{4}\omega^2 a_0^2 + \beta\omega^2\Big(\frac{a_0^2}{4} + \xi_0^2\Big)\Big]^2. \tag{5.11}$$

From this, one obtains

$$\eta^2 = 1 + \frac{3}{4}\varepsilon\alpha a_0^2 - \frac{\varepsilon\beta}{4}(a_0^2 + 4\xi_0^2) \pm \frac{\varepsilon}{2}\sqrt{c^2 - 4\Big(\delta + \frac{8\beta}{3\pi}\xi_0 a_0^2\Big)^2}. \tag{5.12}$$

The maximum amplitude is determined by

$$c^2 - 4\Big(\delta + \frac{8\beta}{3\pi}\xi_0 a_0^2\Big)^2 = 0.$$

For $\xi_0 = a_0$, we have

$$a_{0\,\max} = \sqrt{\frac{3\pi(c - 2\delta)}{16\beta}}. \tag{5.13}$$

In Figs. 73 and 74, the resonance curves are plotted for the case $\delta = 0$, $g(a) = a$ with various values of $\beta$.

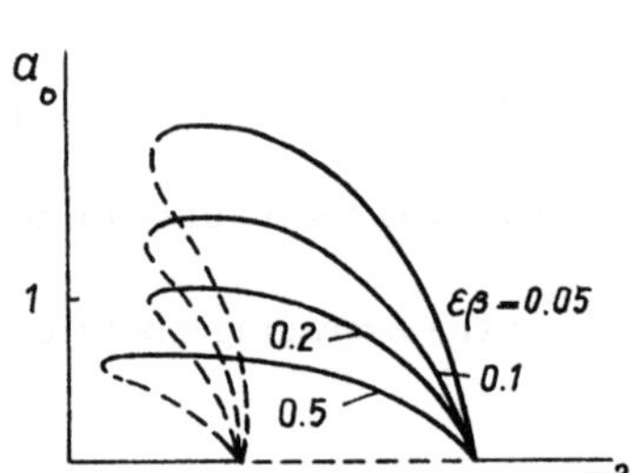


*Fig. 73.* Amplitude curves for a linear parametric oscillator ($\alpha = 0$) with various hysteresis parameter $\beta$.

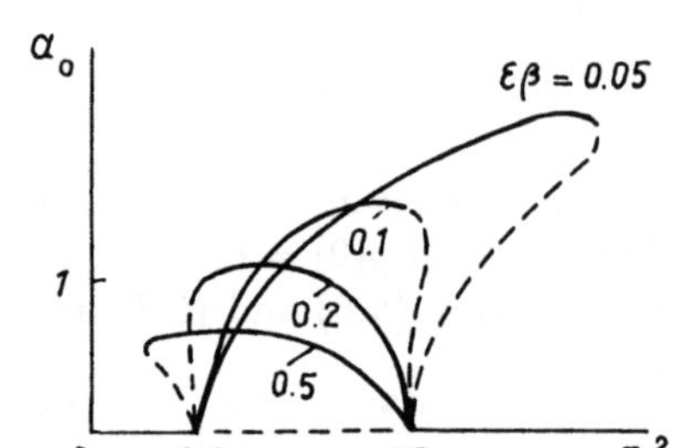


*Fig. 74.* Amplitude curves for a nonlinear parametric oscillator ($\alpha \neq 0$) with various hysteresis parameter $\beta$.

The variational equations for the stationary solution $a = a_0$, $\theta = \theta_0$ of the equations (5.8) are

$$\begin{aligned} \frac{d\delta a}{dt} &= -\varepsilon\omega a_0\Big\{\frac{4\beta}{3\pi}(\xi_0 a_0)'\delta a + + \Big[\frac{\Delta}{\omega^2} + \frac{3\alpha}{4}a_0^2 - \beta\Big(\frac{a_0^2}{4} + \xi_0^2\Big)\Big]\delta\theta\Big\}, \\ \frac{d\delta\theta}{dt} &= \frac{\varepsilon\omega}{2}\Big\{\Big[\frac{3\alpha}{2}a_0 - \beta\Big(\frac{a_0^2}{4} + \xi_0^2\Big)\Big]\delta a - 2\Big(\delta + \frac{8\beta}{3\pi}\xi_0 a_0\Big)\delta\theta\Big\}, \end{aligned} \tag{5.14}$$

where $\delta a = a - a_0$, $\delta\theta = \theta - \theta_0$, and prime denotes the derivative with respect to $a_0$. The characteristic equation of (5.14) has the form

$$\lambda^2 + \varepsilon\omega\left[\delta + \frac{8\beta}{3\pi}\xi_0 a_0 + \frac{4\beta}{3\pi}a_0(\xi_0 a_0)'\right]\lambda + \varepsilon^2 H = 0,$$

where

$$H = \frac{a_0}{4\omega^2}\cdot\frac{\partial W}{\partial a_0}\,.$$

Hence, the stability condition is

$$\begin{aligned} &\delta + \frac{8\beta}{3\pi}\xi_0 a_0 + \frac{4\beta}{3\pi}a_0(\xi_0 a_0)' > 0, \\ &\frac{\partial W}{\partial a_0} > 0. \end{aligned} \tag{5.15}$$

In Figs. 73 and 74, the heavy lines correspond to the stability and the dashed lines to the instability of the stationary oscillation.

By using the transformation of the type (3.26), we may easily verify the condition of instability of the zero solution a=0 of equations (5.8):

$$\delta^2 - \frac{c}{4} + \frac{\Delta^2}{\omega^4} > 0. \tag{5.16}$$

For the cases represented in Figs. 73 and 74, this means that the zero solution is unstable in the interval

$$1 - \frac{\varepsilon c}{2} \leq \eta^2 \leq 1 + \frac{\varepsilon c}{2} \tag{5.17}$$

(see dashed portion of the axis $\eta^2$).

2. Let us now study the parametric oscillations of a rod with hysteresis. The rod is subjected to a harmonic force, $P(t) = c\cos\gamma t$, directed longitudinally along the rod axis (Fig. 75).

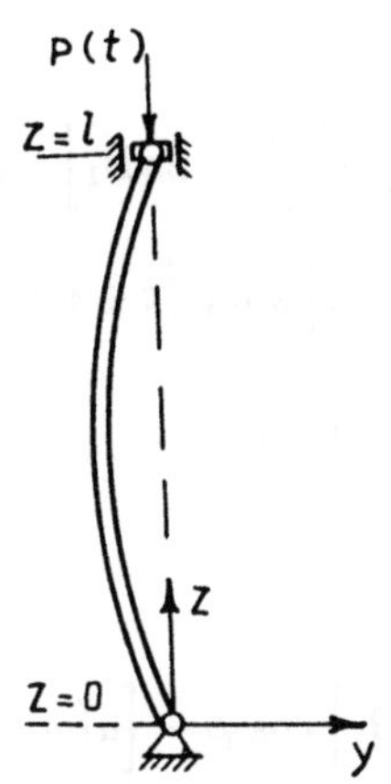


*Fig. 75.* An oscillating rod with hysteresis.

Taking into account the dissipation of the energy in the rod material, we have the following equation for the motion of the rod

$$EI\frac{\partial^4 y}{\partial z^4} + m\frac{\partial^2 y}{\partial t^2} + \varepsilon H\frac{\partial y}{\partial t} + \varepsilon\frac{\partial^2}{\partial z^2}\left[\Phi\left(\frac{\partial^2 y}{\partial z^2}\right)\right] + \varepsilon c\cos\gamma t\frac{\partial^2 y}{\partial z^2} = 0, \tag{5.18}$$

where $E$ is Young's modulus of the rod material, $I$ is the second moment of area of the cross-section about the neutral axis through its centroid, $m$ is the mass density, $y$ is the deflexion and $\frac{\partial^2}{\partial z^2}\left[\Phi\left(\frac{\partial^2 y}{\partial z^2}\right)\right]$ is a function characterizing the dissipation in rod material.

We will find the single-frequency oscillation of the rod (see paragraph 9, Chapter 3), supposing that some natural frequency of the rod, say, the lowest $\omega$, is close to $\gamma/2$, that is

$$\omega^2 = \frac{\gamma^2}{4} + \varepsilon\Delta$$

and the other natural frequencies of the rod are far from the resonance. In this case, we look for a solution to (5.18) in the form

$$y = S(z)a\cos\left(\frac{\gamma}{2}t + \psi\right), \tag{5.19}$$

where $S(z)$ is the normal function depending on the boundary conditions. The unknown functions $a$ and $\psi$ are determined from the equations

$$\begin{aligned}\frac{da}{dt} &= \varepsilon A_1(a, \psi),\\ \frac{d\psi}{dt} &= \omega - \frac{\gamma}{2} + \varepsilon B_2(a, \psi).\end{aligned} \tag{5.20}$$

We have

$$\begin{aligned}\frac{\partial^2 y}{\partial t^2} &= S(z)\Big\{\varepsilon\Big[\Big(\omega - \frac{\gamma}{2}\Big)\frac{\partial A_1}{\partial \psi} - 2aB_1\Big]\cos\Big(\frac{\gamma}{2}t + \psi\Big) - \varepsilon\Big[a\Big(\omega - \\ &\quad - \frac{\gamma}{2}\Big)\frac{\partial B_1}{\partial \psi} + 2\omega A_1\Big]\sin\Big(\frac{\gamma}{2}t + \psi\Big) - a\omega^2\cos\Big(\frac{\gamma}{2}t + \psi\Big)\Big\},\\ \frac{\partial^4 y}{\partial z^4} &= S^{IV}(z)a\cos\Big(\frac{\gamma}{2}t + \psi\Big).\end{aligned}$$

Hence, the equation (5.18) gives

$$\begin{aligned}&mS(z)\Big\{\Big[\Big(\omega - \frac{\gamma}{2}\Big)\frac{\partial A_1}{\partial \varphi} - 2aB_1\Big]\cos\varphi - \Big[a\Big(\omega - \frac{\gamma}{2}\Big)\frac{\partial B_1}{\partial \varphi} + 2\omega A_1\Big]\sin\varphi\Big\} = \\ &= \frac{1}{2}H\gamma aS(z)\sin\varphi - ca\cos\gamma t\cos\varphi S''(z) - \frac{\partial^2}{\partial z^2}\Big[\Phi\Big(\frac{\partial^2 y}{\partial z^2}\Big)\Big],\end{aligned}$$

where $\varphi = \frac{\gamma}{2}t + \psi$. Multiplying the last equation first by $S\cos\varphi$ and then by $S\sin\varphi$ and integrating on $z$ from zero to $\ell$ and on $\varphi$ from zero to $2\pi$, we obtain

$$\Big(\omega - \frac{\gamma}{2}\Big)\frac{\partial A_1}{\partial\psi} - 2a\omega B_1 = -\frac{ca}{2mL}\Big(\int_0^\ell S(z)S''(z)dz\Big)\cos 2\psi -$$

$$-\frac{2}{mL}\int_0^\ell \Big\langle \frac{\partial^2}{\partial z^2}\Big[\Phi\Big(\frac{\partial^2 y}{\partial z^2}\Big)\Big]\cos\varphi\Big\rangle S(z)dz,$$

$$a\Big(\omega - \frac{\gamma}{2}\Big)\frac{\partial B_1}{\partial\psi} + 2\omega A_1 = -\frac{H}{2m}\gamma a + \frac{ca}{2mL}\Big(\int_0^\ell S(z)S''(z)dz\Big)\sin 2\psi +$$

$$+\frac{2}{mL}\int_0^\ell \Big\langle \frac{\partial^2}{\partial z^2}\Big[\Phi\Big(\frac{\partial^2 y}{\partial z^2}\Big)\Big]\sin\varphi\Big\rangle S(z)dz, \tag{5.21}$$

where

$$L = \int_0^\ell S^2(z)dz, \tag{5.22}$$

and $\langle \cdot \rangle$ is an averaging operator on time.

Supposing that $\omega$ is close enough to $\frac{\gamma}{2}$, so that

$$\omega^2 = \frac{\gamma^2}{4} + \varepsilon\Delta, \tag{5.23}$$

then, we have the following approximate formulae

$$A_1 = -\frac{H}{2m}a + \frac{ca}{4\omega mL}\Big(\int_0^\ell S(z)S''(z)dz\Big)\sin 2\psi +$$

$$+\frac{1}{\omega mL}\int_0^\ell \Big\langle \frac{\partial^2}{\partial z^2}\Big[\Phi\Big(\frac{\partial^2 y}{\partial z^2}\Big)\Big]\sin\varphi\Big\rangle S(z)dz,$$

$$B_1 = \frac{c}{4\omega mL}\Big(\int_0^\ell S(z)S''(z)dz\Big)\cos 2\psi +$$

$$+\frac{1}{a\omega mL}\int_0^\ell \Big\langle \frac{\partial^2}{\partial z^2}\Big[\Phi\Big(\frac{\partial^2 y}{\partial z^2}\Big)\Big]\cos\varphi\Big\rangle S(z)dz. \tag{5.24}$$

Hence,

$$\frac{da}{dt}=\frac{\varepsilon}{\omega}\Big\{-\frac{\omega h}{2}a+\frac{c^*}{4}a\sin 2\psi+\frac{1}{mL}\Big\langle\sin\varphi\int_0^\ell S(z)\frac{\partial^2}{\partial z^2}\Big[\Phi\Big(\frac{\partial^2 y}{\partial z^2}\Big)\Big]dz\Big\rangle\Big\},$$

$$a\frac{d\psi}{dt}=\frac{\varepsilon\Delta}{2\omega}a+\frac{\varepsilon}{\omega}\Big\{\frac{c^*}{4}a\cos 2\psi+\frac{1}{mL}\Big\langle\cos\varphi\int_0^\ell S(z)\frac{\partial^2}{\partial z^2}\Big[\Phi\Big(\frac{\partial^2 y}{\partial z^2}\Big)\Big]dz\Big\rangle\Big\},\tag{5.25}$$

where

$$h=\frac{H}{m},\quad c^*=\frac{c}{mL}\int_0^\ell S(z)S''(z)dz.$$

Following Davidenkov-Pisarenko [41], the function

$$\Phi\Big(\frac{\partial^2 y}{\partial z^2}\Big)=\Phi(S''a\cos\varphi)$$

has the form

$$\Phi(S''a\cos\varphi)=\mp\frac{E\nu a^n}{n\ell^n}(S''(z))^n[(1\pm\cos\varphi)^n-2^{n-1}]\int_F x^{n+1}dF,\tag{5.26}$$

where $n$ and $\nu$ are constants determined by experiment, and $x$ is the distance of elementary area of cross section $dF$ from the neutral axis of the rod. For a rectangular rod with cross section $r\times R$, $dF=rdx$ and for odd $n$, we have

$$\Phi(S''a\cos\varphi)=\pm 2\frac{E\nu r}{n\ell^n}a^n(S'')^n[(1\pm\cos\varphi)^n-2^{n-1}]\int_0^{R/2}x^{n+1}dx=$$

$$=\mp\frac{E\nu ra^nR^{n+2}}{2^{n+1}n(n+2)\ell^n}[(1\pm\cos\theta)^n-2^{n-1}](S'')^n,$$

and hence

$$\frac{\partial^2}{\partial z^2}[\Phi(S''a\cos\varphi)]=\mp\frac{E\nu ra^nR^{n+2}}{2^{n+1}n(n+2)\ell^n}[(1\pm\cos\varphi)^n-2^{n-1}]\frac{d^2}{dz^2}(S'')^n,\tag{5.27}$$

$$\Big\langle\sin\varphi\int_0^\ell S(z)\frac{\partial^2}{\partial z^2}\Big[\Phi\Big(\frac{\partial^2 y}{\partial z^2}\Big)\Big]dz\Big\rangle=\frac{E\nu ra^nR^{n+2}}{2^{n+1}n(n+2)\ell^n}\times$$

$$\times\langle\mp\sin\varphi[(1\pm\cos\varphi)^n-2^{n-1}]\rangle\int_0^\ell S\frac{d^2}{dz^2}(S'')^n dz,$$

$$\left\langle \cos\varphi \int_0^\ell S(z) \frac{\partial^2}{\partial z^2}\left[\Phi\left(\frac{\partial^2 y}{\partial z^2}\right)\right] dz\right\rangle = \frac{E\nu r a^n R^{n+2}}{2^{n+1} n(n+2)\ell^n} \times$$

$$\times \left\langle \mp \cos\varphi\left[(1 \pm \cos\varphi)^n - 2^{n-1}\right]\right\rangle \int_0^\ell S \frac{d^2}{dz^2}(S'')^n dz,$$

It is easy to verify that

$$\left\langle \mp \sin\varphi\left[(1 \pm \cos\varphi)^n - 2^{n-1}\right]\right\rangle = \frac{1}{2\pi}\int_0^\pi \sin\varphi\left[(1-\cos\varphi)^n - 2^{n-1}\right] d\varphi -$$

$$- \frac{1}{2\pi}\int_\pi^{2\pi} \sin\varphi\left[(1+\cos\varphi)^n - 2^{n-1}\right] d\varphi = \frac{(1-n)2^n}{(n+1)\pi}, \tag{5.28}$$

$$\left\langle \mp \cos\varphi\left[(1 \pm \cos\varphi)^n - 2^{n-1}\right]\right\rangle = -\frac{2^{3n} \cdot n \cdot \left[\Gamma\left(n+\frac{1}{2}\right)\right]^2}{(n+1)\pi\Gamma(2n+1)}.$$

On the basis of the formulae (5.27) and (5.28), one can write the equations (5.25) in the form:

$$\begin{aligned} \frac{da}{dt} &= \frac{\varepsilon}{\omega}\left(-\frac{\omega}{2}ha + \frac{c^*}{4}a\sin 2\psi - Q_1 a^n\right), \\ a\frac{d\psi}{dt} &= \frac{\varepsilon}{\omega}\left(\frac{\Delta}{2}a + \frac{c^*}{4}a\cos 2\psi - Q_2 a^n\right), \end{aligned} \tag{5.29}$$

where

$$\begin{aligned} Q_1 &= \frac{E\nu r(n-1)R^{n+2}}{2\pi m n(n+1)(n+2)L \cdot \ell^n}\int_0^\ell S(z)\frac{d^2}{dz^2}\left[S''(z)\right]^n dz, \\ Q_2 &= \frac{n \cdot 2^{2n}\left[\Gamma\left(n+\frac{1}{2}\right)\right]^2}{(n-1)\Gamma(2n+1)} Q_1. \end{aligned} \tag{5.30}$$

Equations (5.29) have a trivial solution $a = 0$, and nontrivial stationary solutions $a = a_0 \neq 0$, $\psi = \psi_0 = \text{const}$, determined by

$$\begin{aligned} c^* \sin 2\psi_0 &= 2\omega h + 4Q_1 a_0^{n-1}, \\ c^* \cos 2\psi_0 &= -2\Delta + 4Q_2 a_0^{n-1}. \end{aligned}$$

Eliminating the phase $\psi_0$ gives

$$\frac{c^{*2}}{4} = \left(\omega h + 2Q_1 a_0^{n-1}\right)^2 + \left(\Delta - 2Q_2 a_0^{n-1}\right)^2. \tag{5.31}$$

From this, it follows that

$$\eta^2 = 1 - \frac{2\varepsilon}{\omega^2} Q_2 a_0^{n-1} \pm \frac{\varepsilon}{\omega^2} \sqrt{\frac{c^{*2}}{4} - \left(\omega h + 2Q_1 a_0^{n-1}\right)^2}\,,$$

$$\eta = \frac{\gamma}{2\omega}. \tag{5.32}$$

A typical resonance curve is represented in Fig. 76 for the case $h = 0$, $n = 5$, $c > 0$, $Q_1 > 0$, $Q_2 > 0$ and for a rod with both ends pinned: $S(z) = \sin \frac{\pi}{\ell} z$. In this figure, the heavy and dashed lines correspond to the stability and instability of stationary oscillation, respectively.

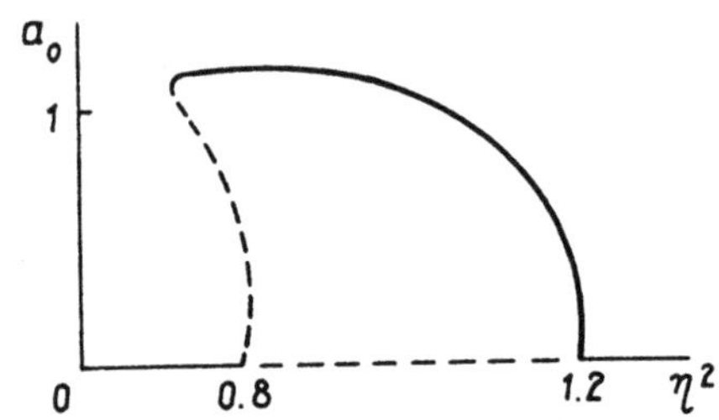


*Fig. 76.* Amplitude curve for a rod.

## 6. Indirectly-excited Parametric Oscillations

The importance of the theory of oscillations of nonlinear systems is now widely recognised but there are still phenomena, whose nature cannot be explained by well known models. The effects of intensive oscillations of a rigid body in the direction, which is not subjected directly to the external force, belong to these phenomena. Such oscillations are called indirect parametrically-excited oscillations, or coupled oscillations. The physical meaning of this term will become clear below.

The first attempt to study this problem was made by V. O. Kononenko. In 1953, he published the solution of the problem of the coupled bending - torsional oscillations, and then in 1963 the investigations of coupled oscillations of rigid bodies around their centre of gravity [14].

Let us consider an oscillating system described by the differential equations

$$\begin{aligned} &\ddot{x} + \omega_1^2 x = -\varepsilon F_1(\dot{x}, x, y), \\ &\ddot{y} + \omega_2^2 y = -\varepsilon F_2(\dot{y}, x, y) + q \sin \gamma t, \end{aligned} \tag{6.1}$$

$$\begin{aligned} F_1 &= h\dot{x} + \beta x^3 - c x y^2, \\ F_2 &= h\dot{y} + \beta y^3 - c x^2 y, \end{aligned} \tag{6.2}$$

where $\omega_1, \omega_2, h, c, \beta, q$ and $\gamma$ are constants; $\varepsilon$ is a small positive parameter.

The external harmonic force acts directly on the $y$ - coordinate, while the $x$ - coordinate is free from the external excitation. Under certain resonance conditions, the $y$ - coordinate can excite oscillation of the $x$ - coordinate. This excitation is called indirect parametric excitation, and the oscillation of the $x$ - coordinate is called indirectly-excited parametric oscillation.

It is easy to check that the system of Equations (6.1) possesses a solution:

$$x = 0, \qquad y \neq 0, \tag{6.3}$$

which corresponds to the oscillations of the system only in the $y$ - direction.

For the linear theory of oscillations, the solution (6.3) is always stable, and therefore no oscillations in the $x$ - direction occur. However, the nonlinearity of the equations of motion changes this situation. Under definite conditions, the solution (6.3) may become unstable and oscillation of the $x$ - coordinate arises.

The loss of stability of the motion (6.3) is expected in the zone of resonance. We will consider the case, in which $\omega_1$ and $\omega_2$ are linearly independent, and there is a resonance relation of the form

$$\omega_1^2 = \gamma^2 + \varepsilon\Delta, \tag{6.4}$$

where $\Delta$ is a detuning parameter.

We transform the system of equations (6.1) by means of the formulae:

$$\begin{aligned} x &= a\cos\gamma t + b\sin\gamma t, \\ \dot{x} &= -a\gamma\sin\gamma t + b\gamma\cos\gamma t, \\ y &= r\cos\varphi + q^*\sin\gamma t, \\ \dot{y} &= -r\omega_2\sin\varphi + q^*\gamma\cos\gamma t, \end{aligned} \tag{6.5}$$

where $q^* = \dfrac{q}{\omega_2^2 - \gamma^2}$ and $a, b, r$ and $\varphi$ are new variables. The transformed equations are:

$$\begin{aligned} \frac{da}{dt} &= \frac{\varepsilon}{\gamma}\big(\Delta x + F_1\big)\sin\gamma t, \\ \frac{db}{dt} &= \frac{-\varepsilon}{\gamma}\big(\Delta x + F_1\big)\cos\gamma t, \\ \frac{dr}{dt} &= \frac{\varepsilon}{\omega_2}F_2\sin\varphi, \\ \frac{d\psi}{dt} &= \frac{\varepsilon}{\omega_2 r}F_2\cos\varphi, \\ \varphi &= \omega_2 t + \psi. \end{aligned} \tag{6.6}$$

These equations belong to the standard form, to which the asymptotic method can be applied.

In the first approximation, one can replace the right-hand sides of (6.6) by their averaged values on time. So, we have

$$\begin{aligned}\frac{da}{dt} &= \frac{\varepsilon}{2\gamma}\Big\{-h\gamma a + b\big[\Delta - \frac{3}{4}cq^{*2} + \frac{3}{4}\beta(a^2+b^2)\big]\Big\} + \dots,\\ \frac{db}{dt} &= \frac{\varepsilon}{2\gamma}\Big\{-h\gamma b - a\big[\Delta - \frac{1}{4}cq^{*2} + \frac{3}{4}\beta(a^2+b^2)\big]\Big\} + \dots,\\ \frac{dr}{dt} &= -\frac{1}{2}\varepsilon h r, \quad \frac{d\psi}{dt} = \Psi(a,b,r,\psi),\end{aligned} \tag{6.7}$$

where the higher terms in (6.7) vanish when $r = 0$. The equation for $r$ is independent from the other equations. It has an asymptotically stable trivial solution $r = 0$. Hence, we are interested only in the first two equations of (6.7) with the omission of the higher terms, namely:

$$\begin{aligned}\frac{da}{dt} &= \frac{\varepsilon}{2\gamma}\Big\{-h\gamma a + b\big[\Delta - \frac{3}{4}cq^{*2} + \frac{3}{4}\beta(a^2+b^2)\big]\Big\},\\ \frac{db}{dt} &= \frac{\varepsilon}{2\gamma}\Big\{-h\gamma b - a\big[\Delta - \frac{1}{4}cq^{*2} + \frac{3}{4}\beta(a^2+b^2)\big]\Big\}.\end{aligned} \tag{6.8}$$

Equations (6.8) have a trivial solution

$$a = b = 0. \tag{6.9}$$

The non-trivial solution, $a = a_0 \neq 0$, $b = b_0 \neq 0$, is determined by

$$\begin{aligned}-h\gamma a_0 + b_0\big(\Delta - \frac{3}{4}cq^{*2} + \frac{3}{4}\beta A_0^2\big) &= 0,\\ h\gamma b_0 + a_0\big(\Delta - \frac{1}{4}cq^{*2} + \frac{3}{4}\beta A_0\big) &= 0,\end{aligned} \tag{6.10}$$

where $A_0^2 = a_0^2 + b_0^2$. Hence, we arrive at the amplitude of oscillation in the $x$ - direction:

$$h^2\gamma^2 + \big(\frac{3}{4}\beta A_0^2 + \Delta - \frac{3}{4}cq^{*2}\big)\big(\frac{3}{4}\beta A_0^2 + \Delta - \frac{1}{4}cq^{*2}\big) = 0. \tag{6.11}$$

From here, it follows:

$$A_0^2 = \frac{1}{3\beta}\Big[2(cq^{*2} - 2\Delta) \pm \sqrt{c^2q^{*4} - 16h^2\gamma^2}\Big] \tag{6.12}$$

or

$$A^2 = \nu^2 - 1 + Q \pm \sqrt{Q^2 - H^2\nu^2}\ , \tag{6.13}$$

where the case $\beta > 0$ is considered, and the following notation is introduced:

$$A^2 = \frac{3\varepsilon\beta}{4\omega_1^2}A_0^2\,, \quad Q = \frac{\varepsilon c}{2\omega_1^2}q^{*2},$$
$$H = \frac{\varepsilon h}{\omega_1}\,, \quad \nu = \frac{\gamma}{\omega_1}\,.$$

The relation (6.13) is represented graphically in Fig. 77 for the parameters: $Q = 2 \cdot 10^{-2}$, $H = 1.94 \cdot 10^{-2}$ (curve 1), and $H = 2 \cdot 10^{-2}$ (curve 2).

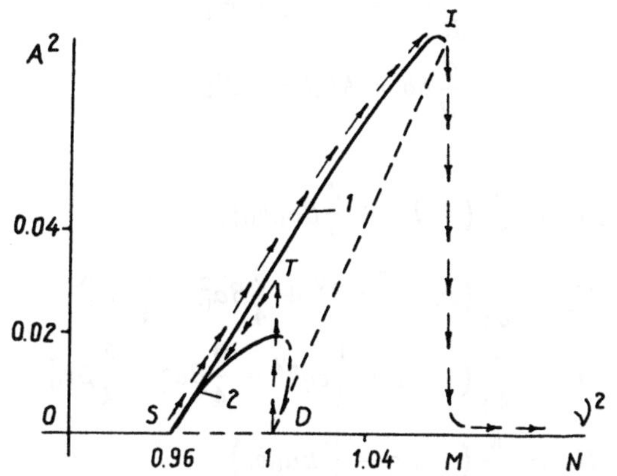


*Fig. 77.* Amplitude change with slowly varying frequency.

In a certain interval of $\nu^2$, we have obtained two values of the stationary amplitude of oscillation, which correspond to the plus and minus signs before the square root in (6.12). In order to explain which of these values corresponds to the real stationary process, it is necessary to study their stability.

First, we investigate the stability of the trivial solution (6.9) of equations (6.8). The variational equations for this solution are

$$\begin{aligned} \frac{d\delta a}{dt} &= \frac{\varepsilon}{2\gamma}\Big[-h\gamma\delta a + \big(\Delta - \frac{3}{4}cq^{*2}\big)\delta b\Big], \\ \frac{d\delta b}{dt} &= \frac{\varepsilon}{2\gamma}\Big[-\big(\Delta - \frac{1}{4}cq^{*2}\big)\delta a - h\gamma\delta b\Big]. \end{aligned} \tag{6.14}$$

The characteristic equation of (6.14) has the form

$$\lambda^2 + \varepsilon h\lambda + \frac{\varepsilon^2}{4\gamma^2}\Big[h^2\gamma^2 + \big(\Delta - \frac{1}{4}cq^{*2}\big)\big(\Delta - \frac{3}{4}cq^{*2}\big)\Big] = 0. \tag{6.15}$$

Hence, the stability condition for the trivial solution is

$$h^2\gamma^2 + \big(\Delta - \frac{1}{4}cq^{*2}\big)\big(\Delta - \frac{3}{4}cq^{*2}\big) > 0. \tag{6.16}$$

Inequality (6.16) requires that the free term in the quadratic equation relative to $A_0^2$ (6.11) must be positive. This means that the $\nu^2$ - axis (zero solution) is stable in the intervals, where equation (6.11) has either two positive roots, two negative roots or two complex-conjugate roots relative to $A_0^2$. So, for the case represented in Fig. 77, the zero solution is stable for $\nu^2$ outside the interval $\nu_S^2 \leq \nu^2 \leq \nu_D^2$, from which the resonance curve grows.

We now study the stability of the non-trivial solution of equations (6.8). Putting $a = a_0 + \delta a$, $b = b_0 + \delta b$ in (6.8) and using the stationary relations (6.10), we have the following variational equations:

$$\begin{aligned} \frac{d}{dt}\delta a &= M_1 \delta a + N_1 \delta b, \\ \frac{d}{dt}\delta b &= M_2 \delta a + N_2 \delta b, \end{aligned} \tag{6.17}$$

where

$$\begin{aligned} M_1 &= \frac{\varepsilon}{2\gamma}\Big( - h\gamma + \frac{3}{2}\beta a_0 b_0\Big), \\ N_1 &= \frac{\varepsilon}{2\gamma}\Big(\Delta - \frac{3}{4}cq^{*2} + \frac{3}{4}\beta a_0^2 + \frac{9}{4}\beta b_0^2\Big), \\ M_2 &= \frac{\varepsilon}{2\gamma}\Big( - \Delta + \frac{1}{4}cq^{*2} - \frac{9}{4}\beta a_0^2 - \frac{3}{4}\beta b_0^2\Big), \\ N_2 &= \frac{\varepsilon}{2\gamma}\Big( - h\gamma - \frac{3}{2}\beta a_0 b_0\Big). \end{aligned} \tag{6.18}$$

The characteristic equation of the system (6.17) is

$$\lambda^2 - (M_1 + N_2)\lambda + M_1 N_2 - M_2 N_1 = 0. \tag{6.19}$$

Since

$$M_1 + N_2 = -\varepsilon h < 0,$$

the stability condition becomes

$$M_1 N_2 - M_2 N_1 > 0.$$

Following (6.11) and (6.15), we can write this condition in the form:

$$\beta a_0 b_0 (a_0^2 - b_0^2) > 0$$

or

$$\beta a_0 b_0 (A_0^2 - 2b_0^2) > 0. \tag{6.20}$$

Multiplying successively the first equation of (6.10) by $-a_0$ and the second by $b_0$ and adding them, we obtain

$$\frac{1}{2}cq^{*2} a_0 b_0 = -h\gamma A_0^2.$$

Similarly, multiplying the first equation of (6.10) by $b_0$ and the second by $a_0$ and adding them, we get

$$\frac{1}{2}cq^{*2}b_0^2 = A_0^2(\Delta - \frac{1}{4}cq^{*2} + \frac{3}{4}\beta A_0^2).$$

Hence, the inequality (6.20) is equivalent to

$$\beta(\frac{3}{4}\beta A_0^2 + \Delta - \frac{1}{2}cq^{*2}) > 0 \tag{6.21}$$

or following (6.12):

$$\beta(\pm\sqrt{c^2q^{*4} - 16h^2\gamma^2}) > 0.$$

This means that for $\beta > 0$, the stability of stationary oscillation corresponds to the plus sign before the radical in (6.12); that is, the upper branch limited by the vertical tangent of the resonance curve is stable and the lower branch is unstable (see Fig. 77). For $\beta < 0$, the stable branch of the resonance curve corresponds to the minus sign before the radical in (6.12).

When $\nu$ increases from zero, the state of rest remains stable until the point S is reached. Beginning from this point, oscillation of the system in the $x$ - direction appears. By further increasing the frequency of the external force, the amplitude of oscillations grows at first along the curve STI. At the point I, a sudden drop of amplitude occurs. The value of the amplitude drops to zero at the point M, and the frequency of the external force is changed along the curve MN - that is, the amplitude tends to zero.

If we now begin to decrease the frequency of the external force, then the amplitude of oscillations changes along the straight line MD. On reaching the point D, the value of the amplitude jumps to the point T and is further changed along the upper branch of the resonance curve TS.

Note that in speaking about the change of frequency of the external force, we mean a very slow change, so that the system can be treated as stationary at each moment in time.

Thus, the presence of the nonlinear coupling term $kxy^2$ in the equations of motion indicates that the two coordinates $x$ and $y$ are dynamically coupled, and energy may under certain conditions, be transferred from the directly excited $y$ - coordinate to the coordinate initially at rest, the $x$ - coordinate. We will see an another example of this kind of nonlinear oscillations for an electromechanical system in the next paragraph.

## 7. Parametrically-excited Oscillations in an Electromechanical System [37,d]

Consider the oscillatory system shown in Fig. 78, in which the rigid body of mass m is attached to a fixed abutment through a massless elastic rod with restoring force

$$f = -kx - \beta_1 x^3, \tag{7.1}$$

where $x$ is the displacement of the mass from its equilibrium position. It is supposed that the mass is subjected to an electromagnetic force of a circuit, which consists of a capacitor with capacitance C, a coil with self-inductance L and a resistor of resistance R.

An input sine voltage is applied to the circuit: $E(t) = E_0 \sin \nu t$. The equations of motion of the system under consideration can be written in the form:

$$\begin{aligned} &\frac{d}{dt}\left[L(x)\dot{q}\right] + \frac{1}{C}q + R\dot{q} = E_0 \sin \nu t, \\ &m\ddot{x} + kx + \beta_1 x^3 + h_1 \dot{x} = \frac{1}{2}\dot{q}^2 \frac{dL(x)}{dx}, \end{aligned} \tag{7.2}$$

where $q$ is the charge of the capacitor, $q = \int i dt$ and $i$ is the current passing through the coil.

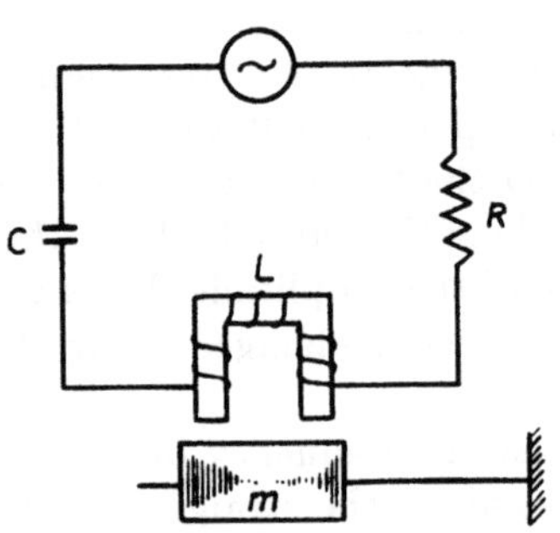


*Fig. 78.* An electromechanical oscillatory system.

Let us take into account the nonlinearity of the self-inductance by assuming that $L$ is related to $x$ by

$$L(x) = L_0(1 - \alpha_1 x + \alpha_2 x^2), \tag{7.3}$$

where $L_0, \alpha_1, \alpha_2$ are positive constants. Introducing the notation

$$\begin{aligned} &\omega^2 = \frac{k}{m}, \quad \Omega_0^2 = \frac{1}{L_0 C}, \quad \tau = \omega t, \\ &\beta = \frac{1}{k}\beta_1, \quad h = \frac{h_1}{m\omega}, \quad R_0 = \frac{R}{L\omega}, \\ &\gamma = \frac{\nu}{\omega}, \quad \Omega = \frac{\Omega_0}{\omega}, \quad e = \frac{E_0}{L_0\omega^2}, \end{aligned} \tag{7.4}$$

one can write equations (7.2) in the form:

$$\begin{aligned} &q'' + \Omega^2 q + R_0 q' + (\alpha_2 x - \alpha_1)x q'' + \\ &+ (2\alpha_2 x - \alpha_1)x' q' = e \sin \gamma\tau, \\ &x'' + x + hx' + \beta x^3 = \frac{1}{2m}(2\alpha_2 x - \alpha_1)q'^2, \end{aligned} \tag{7.5}$$

where prime denotes the derivative with respect to $\tau$.

We are specially interested in the appearance of the component of subharmonic oscillations with the frequency $\nu$ of the electrical circuit.

It is supposed that the nonlinear terms and the damping terms in (7.5) are small. This smallness can be characterized by using a small dimensionless parameter $\varepsilon$, which will be set equal to unity in the final results. Thus, we have

$$\begin{aligned} q'' + \Omega^2 q &= e \sin \gamma\tau - \varepsilon Q, \\ x'' + x &= -\frac{\alpha_1}{2m} q'^2 - \varepsilon X, \end{aligned} \tag{7.6}$$

where

$$\begin{aligned} \varepsilon Q &= R_0 q' + (\alpha_2 x - \alpha_1) x q'' + (2\alpha_2 x - \alpha_1) x' q', \\ \varepsilon X &= h x' + \beta x^3 - \frac{\alpha_2}{m} x q'^2. \end{aligned} \tag{7.7}$$

Equations (7.6) can be transformed into the standard form by means of the formulae

$$\begin{aligned} q &= e^* \sin \gamma t + B \sin \varphi, \\ q' &= \gamma e^* \cos \gamma\tau + \Omega B \cos \varphi, \end{aligned}$$

$$\begin{aligned} x &= -b_1 - \frac{b_1}{1-4\gamma^2} \cos 2\gamma\tau + a \sin \theta, \\ x' &= \frac{2\gamma b_1}{1-4\gamma^2} \sin 2\gamma\tau + \gamma a \cos \theta, \end{aligned} \tag{7.8}$$

where

$$\begin{aligned} e^* &= \frac{e}{\Omega^2 - \gamma^2}, \quad b_1 = \frac{\alpha e^{*2} \gamma^2}{4m}, \\ \varphi &= \Omega\tau + \Phi, \quad \theta = \gamma\tau + \psi \end{aligned} \tag{7.9}$$

and $B, a, \phi$ and $\psi$ are new variables.

Substituting (7.8) into (7.6), we obtain after some simple calculations

$$\begin{aligned} \Omega B' &= -\varepsilon Q \cos \varphi, \qquad \Omega B \phi' = \varepsilon Q \sin \varphi, \\ \gamma a' &= -a(1-\gamma^2) \sin \theta \cos \theta - \varepsilon X \cos \theta - \\ &\quad - \frac{\alpha_1}{2m} B\Omega (B\Omega \cos \varphi + 2\gamma e^* \cos \gamma\tau) \cos \varphi \cos \theta, \\ \gamma a \psi' &= a(1-\gamma^2) \sin^2 \theta + \varepsilon X \sin \theta + \\ &\quad + \frac{\alpha_1}{2m} B\Omega (B\Omega \cos \varphi + 2\gamma e^* \cos \gamma\tau) \cos \varphi \sin \theta. \end{aligned} \tag{7.10}$$

Assuming that $\gamma$ is close to unity and that $\gamma$ and $\Omega$ are linearly independent, and taking into account

$$\langle x\cos\theta\rangle = 0,\ \langle x\sin\theta\rangle = \frac{a}{2},\ \langle x\cos\theta\ \cos 2\gamma\tau\rangle = \frac{a}{4}\sin 2\psi,$$
$$\langle x'\cos\theta\rangle = \gamma\frac{a}{2},\ \langle x'\sin\theta\rangle = 0,\ \langle x\sin\theta\ \cos 2\gamma\tau\rangle = -\frac{a}{4}\cos 2\psi,$$
$$\langle x^3\sin\theta\rangle = \frac{3}{8}a^3 + \frac{3}{2}b_1^2 a + \frac{3}{4}\frac{b_1^2 a}{(1-4\gamma^2)^2} - \frac{3b_1^2 a}{2(1-4\gamma^2)}\cos 2\psi,$$

where $\langle\,\cdot\,\rangle$ is an averaging operator on time, we obtain, in the first approximation, the following averaged equations

$$\begin{aligned} B' &= -\frac{\varepsilon}{2}R_0 B, \quad \Omega B\phi' = -\frac{\varepsilon}{2}\alpha_1 b_1 \Omega^2 B,\\ \gamma a' &= -\frac{1}{2}\varepsilon h\gamma a + \frac{1}{2}\varepsilon c a\sin 2\psi + \ldots,\\ \gamma a\psi' &= \frac{1}{2}(1-\gamma^2+\varepsilon\sigma)a + \frac{3}{8}\varepsilon\beta a^3 + \frac{1}{2}\varepsilon c a\cos 2\psi + \ldots, \end{aligned} \tag{7.11}$$

where

$$c = \frac{1}{2}b_2 - \frac{3\beta b_1^2}{1-4\gamma^2}\,;\ \sigma = 3\beta b_1^2 + \frac{3\beta b_1^2}{2(1-4\gamma^2)^2} - b_2,$$

and the dots denote the terms which vanish when $B = 0$.

The first equation of (7.11) has an asymptotically stable solution, namely $B = 0$. Hence, the last two equations of (7.11) become

$$\begin{aligned} \gamma a' &= -\frac{1}{2}\varepsilon h\gamma a + \frac{1}{2}\varepsilon c a\sin 2\psi,\\ \gamma a\psi' &= \frac{1}{2}(1-\gamma^2+\varepsilon\sigma)a + \frac{3}{8}\varepsilon\beta a^3 + \frac{1}{2}\varepsilon c a\cos 2\psi. \end{aligned} \tag{7.12}$$

We have seen these equations in the paragraph 3. The following expressions will be found for the stationary solution of (7.12):

$$A^2 = -\delta + \gamma^2 - 1 \pm \sqrt{C^2 - \mathcal{H}^2\gamma^2}\,, \tag{7.13}$$

$$\sin 2\psi_0 = \frac{\mathcal{H}}{C}\gamma,\ \cos 2\psi_0 = \mp\frac{1}{C}\sqrt{C^2 - \mathcal{H}^2\gamma^2}\,, \tag{7.14}$$

where one should consistently use either the upper or lower signs in (7.13) and (7.14):

$$A^2 = \frac{3}{4}\varepsilon\beta a_0^2, \quad \delta = \varepsilon\sigma, \quad C = \varepsilon c, \quad \mathcal{H} = \varepsilon h.$$

The relation (7.13) is represented graphically in Fig. 79 for the parameters $\delta = 0.15$, $C^2 = 0.15$, $\mathcal{H}^2 = 0.1$, where the heavy and dashed lines correspond to the stability and instability of stationary oscillations, respectively. The arrows show how the amplitude changes when the frequency $\gamma$ increases or decreases.

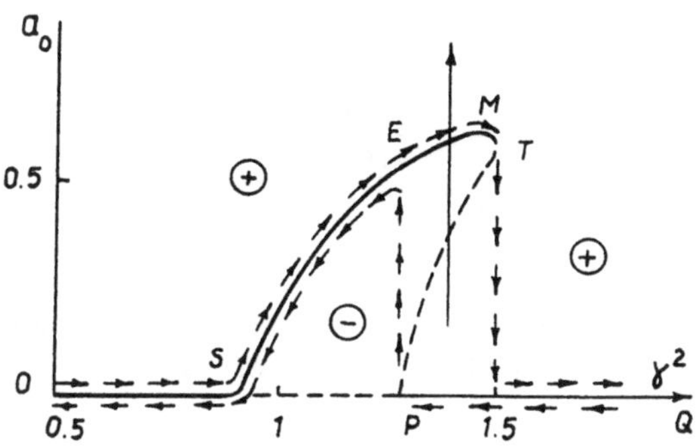


*Fig. 79.* The amplitude curve of subharmonic oscillations.

Following formulae (7.8) and (7.13), we have, in the first approximation,

$$\begin{aligned} x &= -b_1 - \frac{b_1}{1-4\gamma^2}\cos 2\gamma\tau + a_0 \sin(\gamma\tau + \psi_0), \\ q &= e^* \sin\gamma\tau, \end{aligned} \tag{7.15}$$

with $a_0$ and $\psi_0$ determined by (7.13) and (7.14). In the formula (7.15), there appears a subharmonic component of oscillation with frequency $\nu$ of the electrical circuit (since $\gamma\tau = \nu t$). The resulting oscillation of the mass m consists of two oscillations. The first is harmonic with the amplitude $\frac{b_1}{1-4\gamma^2}$ and frequency $2\nu$, which is twice the frequency of the circuit. The second is subharmonic with the amplitude $a_0$ and frequency $\nu$ of the circuit. The dependence of these amplitudes on the frequency $\nu$ is plotted in Fig. 80. When $\nu$ increases from zero, the mass $m$ oscillates with the frequency $2\nu$. At point B harmonic resonance occurs. Beginning from the point S, two oscillations, harmonic and subharmonic, coexist. As $\nu$ continues to increase the subharmonic oscillation becomes stronger. Between points P and D, the harmonic oscillation is very small and the subharmonic oscillation is largest. For $\nu > \nu_D$ there is again only small harmonic oscillation.

As $\nu$ decreases from $\nu_Q$, only a small harmonic oscillation is observed at first (see Fig. 80). At the point P, the subharmonic oscillation suddenly appears with a large amplitude corresponding to the point E. Then the subharmonic oscillation decreases and the two oscillations coexist. Beginning from the point S, when $\nu < \nu_S$, the subharmonic oscillation disappears and there is only the harmonic oscillation.

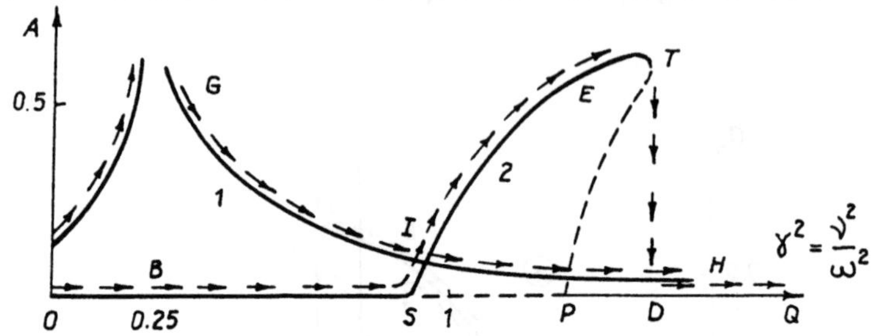


***Fig. 80.*** **The simultaneous appearance of both subharmonic and harmonic oscillations.**

CHAPTER 5

# INTERACTION OF NONLINEAR OSCILLATIONS

Linear oscillatory systems satisfy the superposition principle, that is the component oscillations of a linear system do not interact among themselves. In mathematical terms this translates into: a linear combination of particular solutions of a linear differential equation is a solution of this equation.

For nonlinear oscillatory systems, the component oscillations do not satisfy the superposition principle but always exhibit some kind of interaction.

In this chapter, we examine a nonlinear oscillatory system which consists of two nonlinear oscillators. Each of these oscillators undergoes definite self-sustained oscillation, comprising one or more oscillation modes, viz. self-excited, forced and parametrically excited. What will happen if the two oscillators are coupled in some manner? Will they oscillate, and will the oscillation be stable?

## 1. Forced Oscillations of Systems with Self-excitation. Synchronization Effect

We now consider forced oscillations of such a system, which could execute self-excited oscillation in the absence of an external force. Our aim is to discover what phenomena may be expected as a result of the externally imposed excitation of a self-excited oscillator. To this end, we consider a system described by the equation:

$$\ddot{x} + x = \varepsilon\big[(\alpha - \gamma x^2)\dot{x} + e\sin\nu t\big], \tag{1.1}$$

which is a classical Van der Pol equation with a small forcing term $\varepsilon e \sin\nu t$, where $\alpha$, $\gamma$, $e$ and $\nu$ are positive constants. It is supposed that $\nu$ is close to unity,namely:

$$\nu^2 = 1 + \varepsilon\Delta, \tag{1.2}$$

where $\Delta$ is a detuning parameter.

Applying the asymptotic method (1, Chapter 3) and using the amplitude and phase variables, $(a,\theta)$, given by

$$\begin{aligned} x &= a\cos(\nu t + \theta), \\ \dot{x} &= -a\nu\sin(\nu t + \theta), \\ \dot{a}\cos(\nu t &+ \theta) - a\dot{\theta}\sin(\nu t + \theta) = 0, \end{aligned} \tag{1.3}$$

we have

$$\begin{aligned}\nu\dot{a} &= -\varepsilon\big[\Delta x + (\alpha - \gamma x^2)\dot{x} + e\sin\nu t\big]\sin(\nu t + \theta),\\ \nu a\dot{\theta} &= -\varepsilon\big[\Delta x + (\alpha - \gamma x^2)\dot{x} + e\sin\nu t\big]\cos(\nu t + \theta).\end{aligned} \tag{1.4}$$

These equations were obtained by substituting equations (1.3) into the equation (1.1) and solving for $\dot{a}$ and $\dot{\theta}$. Since $a$ and $\theta$ are slowly varying functions of time, the change in their values during a time period $T = \dfrac{2\pi}{\nu}$ is very small. Hence, in the first approximation one may replace equations (1.4) by their time-averages over $(t, t+T)$, assuming $a$ and $\theta$ to be constant. Substituting (1.3) into (1.4), we get

$$\begin{aligned}\nu\dot{a} &= -\varepsilon\big[\Delta a\cos\varphi - a\nu(\alpha - \gamma a^2\cos^2\varphi)\sin\varphi + e\sin(\varphi - \theta)\big]\sin\varphi,\\ \nu a\dot{\theta} &= -\varepsilon\big[\Delta a\cos\varphi - a\nu(\alpha - \gamma a^2\cos^2\varphi)\sin\varphi + E\sin(\varphi - \theta)\big]\cos\varphi,\end{aligned}$$

where $\varphi = \nu t + \theta$. Expanding the right-hand sides of these equations into trigonometric series, we obtain:

$$\begin{aligned}\nu\dot{a} &= \frac{\varepsilon}{2}\big[a\nu(\alpha - \frac{1}{4}\gamma a^2) - e\cos\theta\big] + R,\\ \nu a\dot{\theta} &= \frac{\varepsilon}{2}(-\Delta a + e\sin\theta) + S,\end{aligned} \tag{1.5}$$

where $R$ and $S$ contain only harmonics $\sin k\varphi$ and $\cos k\varphi$, but no constant terms.

Averaging the right-hand sides of (1.5) over time, we have:

$$\begin{aligned}\nu\dot{a} &= \frac{\varepsilon}{2}\big[a\nu(\alpha - \frac{1}{4}\gamma a^2) - e\cos\theta\big],\\ \nu a\dot{\theta} &= \frac{\varepsilon}{2}(-\Delta a + e\sin\theta).\end{aligned} \tag{1.6}$$

To obtain information about steady-state oscillations, one may set $a = a_0$ and $\theta = \theta_0$, where $a_0$ and $\theta_0$ are constants. Hence, the steady-state equations are

$$\begin{aligned}a_0\nu(\alpha - \frac{1}{4}\nu a_0^2) &= e\cos\theta_0,\\ \Delta a_0 &= e\sin\theta_0.\end{aligned} \tag{1.7}$$

From these equations, we obtain:

$$\tan\theta_0 = \frac{\nu^2 - 1}{\varepsilon\nu(\alpha - \frac{\gamma}{4}a_0^2)}. \tag{1.8}$$

Eliminating $\theta_0$ from (1.7) gives

$$a_0^2\big[\varepsilon^2\nu^2(\alpha - \frac{1}{4}\gamma a_0^2)^2 + (\nu^2 - 1)^2\big] = \varepsilon^2 e^2. \tag{1.9}$$

We now return to the fundamental equation (1.1). We note that, when $e = 0$, we have the wellknown Van der Pol equation, which represents a self-excited oscillator with the amplitude $a_* = 2\sqrt{\frac{\alpha}{\gamma}}$ and frequency $\omega = 1$. The oscillation described by the equation (1.3) with stationary amplitude $a_0$ and phase $\theta_0$: $x = a_0 \cos(\nu t + \theta_0)$ has the frequency of the external force only, namely, $\nu$. The self-excited oscillation is entrained by the external excitation. The oscillation (1.3) is called *synchronized oscillation* which is characterized by the entrainment of the auto-periodic frequency by the external frequency. The synchronization effect is observed only when the exciting frequency is close enough to the natural frequency (i.e. when $\Delta$ is small enough).

The equation (1.9) is of the third degree in $a_0^2$, though only of the second degree in $\nu^2$. We arrange the terms according to powers of $\nu^2$ and obtain:

$$\nu^4 + \nu^2\big[\varepsilon^2(\alpha - \frac{1}{4}\nu a_0^2)^2 - 2\big] + 1 - \frac{\varepsilon^2 e^2}{a_0^2} = 0$$

with the solution

$$\nu_{1,2}^2 = 1 - \frac{\varepsilon^2}{2}\big(\alpha - \frac{1}{4}\gamma a_0^2\big)^2 \pm \Big\{\big[1 - \frac{\varepsilon^2}{2}\big(\alpha - \frac{1}{4}\gamma a_0^2\big)^2\big]^2 - \big(1 - \frac{\varepsilon^2 e^2}{a_0^2}\big)\Big\}^{1/2}. \quad (1.10)$$

We may determine from this the $\nu$ - value for each value of $a_0$ and so obtain the resonance curve $a_0(\nu)$.

One can simplify equation (1.9) by considering only the behaviour of forced oscillations with the frequency $\nu$ which is close to unity (see (1.2)). In this case, if we replace $\nu^2$ in the first term of (1.9) simply by unity, and neglect the quantity of smallness $\varepsilon^3$ in comparison with the quantity of $\varepsilon^2$, the approximate equation of the resonance curve will be

$$a_0^2\big[\varepsilon^2\big(\alpha - \frac{1}{4}\gamma a_0^2\big)^2 + (\nu^2 - 1)^2\big] = \varepsilon^2 e^2, \quad (1.11)$$

or

$$A\big[(1 - A)^2 + \sigma^2\big] = E^2, \quad (1.12)$$

where

$$A = \frac{1}{4\alpha}\gamma a_0^2 = \frac{a_0^2}{a_*^2}\,,\ E^2 = \frac{\gamma e^2}{4\alpha^3}\,,\ \sigma = \frac{\nu^2 - 1}{\alpha\varepsilon}\,, \quad (1.13)$$

and $a_* = 2\sqrt{\frac{\alpha}{\gamma}}$ is the amplitude of the purely self-excited Van der Pol oscillator, i.e. the amplitude of the solution of (1.1) when $e = 0$.

We proceed to discuss the curves given by equation (1.12) in the $(A, \sigma)$ plane for various values of the parameter $E$. This equation determines the family of cubic curves shown in Fig. 81. Clearly, we are interested only in positive values of $A$. The curves are symmetrical with respect to the $A$ - axis, $(\sigma = 0)$, because equation (1.12) is unchanged when $\sigma$ is replaced by $-\sigma$. For $E = 0$, i.e. for the zero external excitation, we find the results for the classical Van der Pol oscillator:

1) $A = 0$ with $\sigma$ arbitrary,
2) $\sigma = 0$, $A = 1$.

Therefore, the resonance curves degenerate into the line $A = 0$ ($\sigma$ - axis) and the point $\sigma = 0$, $A = 1$.

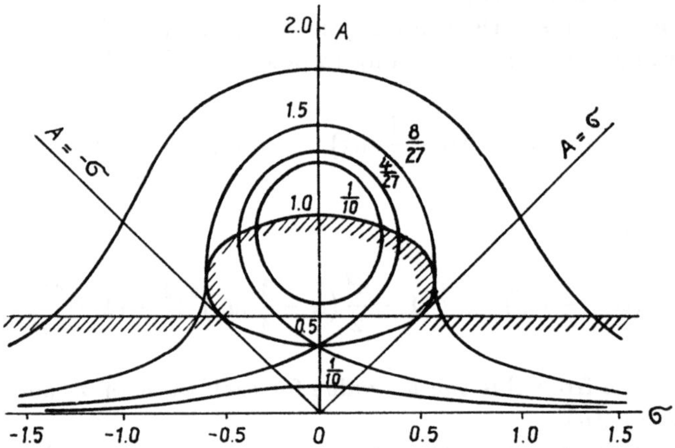


*Fig. 81.* Amplitude curves for the forced oscillations of the self-excited system with various values of external excitation.

If $E$ is small, but different from zero, we expect $A$ to be nearly 1 or nearly zero, so that one of the response curves would be oval which is approximately the circle

$$(A-1)^2 + \sigma^2 = E^2$$

with centre at $\sigma = 0$, $A = 1$. In addition, the other branch runs near the $\sigma$ - axis (Fig. 82). The oval expands with increasing $E$.

When $E$ increases, the resonance curves first consist of two branches, up to the critical value for which the two branches join. The resonance curve for this critical value of $E$ has a double point on the $A$ - axis with $A = 1/3$. The critical value is $E^2 = 4/27$. To verify this, we consider the geometric character of the resonance curves. For small values of $E$, the resonance curves have only one vertical tangent for $\sigma > 0$, while for larger values they have two; the transition takes place for the resonance curve which contains the vertex of the ellipse at $\sigma = 0$, $A = 1/3$, from which one finds from (1.12) the value $E^2 = 4/27$. One can easily verify that the resonance curve for $E^2 = 4/27$ has a singular point with crossing tangents at the point $\sigma = 0$, $A = 1/3$.

When $E$ is increased further, the resonance curves consist of a single branch, but $A$ is still not a single-valued function of $\sigma$ on these curves for values of $E$ near the critical value and for not too large values of $\sigma$.

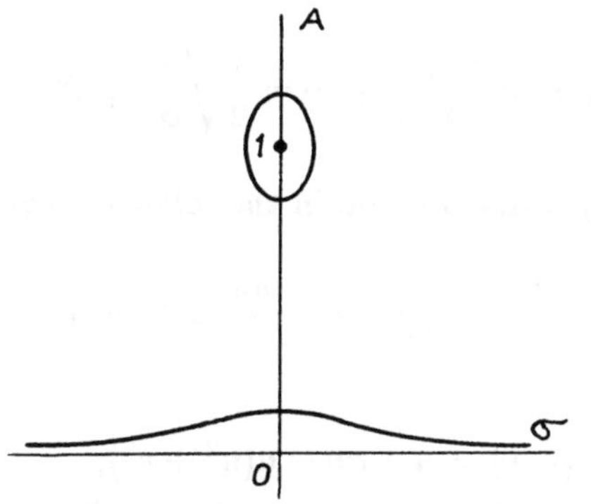


*Fig. 82.* Amplitude curve for small value of external excitation.

Fig. 81 shows a number of resonance curves for various values of $E$. Differentiating both sides of the equation (1.12) and setting $\frac{d\alpha}{dA} = 0$, we obtain the locus of the vertical tangents of the resonance curves:

$$(1 - A)(1 - 3A) + \sigma^2 = 0. \tag{1.14}$$

This is an ellipse with centre at $\sigma = 0$, $A = \frac{2}{3}$ and the two axes $\frac{2}{\sqrt{3}}$, $\frac{2}{3}$. This ellipse plays an important role in the further discussion of the stability of stationary oscillations.

It is easy to show that the resonance curve (1.12) is tangent to the ellipse (1.14) at its vertex when $E^2 = 8/27$. In fact, the transition to resonance curves with no vertical tangents occurs for the resonance curve passing through the vertex $\sigma = \frac{1}{\sqrt{3}}$, $A = \frac{2}{3}$ of the ellipse, and one finds from (1.12) the value $E^2 = 8/27$. Also indicated in Fig. 81 is the fact that the straight lines $A = \pm\sigma$ are tangent to the ellipse (1.14) at the points where $A = 1/2$. These features of the resonance curves are important for the later discussion.

In order to study the stability of the steady-state synchronized oscillations determined by equations (1.3) and (1.7), it is convenient to introduce new variables $a_1$ and $a_2$, connected with the amplitude $a$ and phase $\theta$ by the relations:

$$a_1 = -a \sin\theta, \quad a_2 = a\cos\theta. \tag{1.15}$$

Hence

$$a^2 = a_1^2 + a_2^2, \quad \tan\theta = -\frac{a_1}{a_2}\,. \tag{1.16}$$

Differentiating (1.15) with respect to time and taking into account (1.6), one has:

$$\begin{aligned}\dot{a}_1 &= \frac{\varepsilon}{2\nu}\big[\nu\alpha a_1 + \Delta a_2 - \frac{\nu\gamma}{4}a_1(a_1^2 + a_2^2)\big],\\ \dot{a}_2 &= \frac{\varepsilon}{2\nu}\big[-\Delta a_1 + \nu\alpha a_2 - \frac{\nu\gamma}{4}a_2(a_1^2 + a_2^2) - e\big].\end{aligned} \tag{1.17}$$

Using the notation (1.13) and also

$$\tau = \frac{2\varepsilon}{\alpha}t,\ u = \frac{1}{2}\sqrt{\frac{\gamma}{\alpha}}a_1,\ v = \frac{1}{2}\sqrt{\frac{\gamma}{\alpha}}a_2,\ E = \frac{e}{2\alpha}\sqrt{\frac{\gamma}{\alpha}} \tag{1.18}$$

and approximating $\nu$ by unity, we obtain the following equations:

$$\frac{du}{d\tau} = f_1(u,v),\quad \frac{dv}{d\tau} = f_2(u,v), \tag{1.19}$$

where

$$\begin{aligned}f_1(u,v) &= u + \sigma v - u(u^2+v^2),\\ f_2(u,v) &= -\sigma u + v - v(u^2+v^2) - E.\end{aligned} \tag{1.20}$$

The singular points of these equations $u_0$, $v_0$ are determined by

$$\frac{du}{d\tau} = \frac{dv}{d\tau} = 0$$

or

$$\begin{aligned}&u_0 + \sigma v_0 - u_0(u_0^2 + v_0^2) = 0,\\ &-\sigma u_0 + v_0 - v_0(u_0^2 + v_0^2) - E = 0,\end{aligned} \tag{1.21}$$

where

$$u_0^2 + v_0^2 = A. \tag{1.22}$$

In order to study the stability of the singular points obtained, first we replace $u$ and $v$ in (1.19) by $u_0 + \xi$ and $v_0 + \eta$, then develop the right-hand sides of (1.19) in powers of $\xi$ and $\eta$, and reject all but the linear terms in $\xi$ and $\eta$. The result is the following variational equations in $\xi$ and $\eta$:

$$\begin{aligned}\frac{d\xi}{d\tau} &= b_1\xi + b_2\eta,\\ \frac{d\eta}{d\tau} &= b_3\xi + b_4\eta,\end{aligned} \tag{1.23}$$

where

$$\begin{aligned}b_1 &= \frac{\partial f_1(u_0,v_0)}{\partial u_0},\quad b_2 = \frac{\partial f_1(u_0,v_0)}{\partial v_0},\\ b_3 &= \frac{\partial f_2(u_0,v_0)}{\partial u_0},\quad b_4 = \frac{\partial f_2(u_0,v_0)}{\partial v_0}.\end{aligned} \tag{1.24}$$

One finds

$$b_1 = 1 - 3u_0^2 - v_0^2, \quad b_2 = \sigma - 2u_0 v_0,$$
$$b_3 = -\sigma - 2u_0 v_0, \quad b_4 = 1 - u_0^2 - 3v_0^2.$$

In accordance with the discussion of paragraph 1, Chapter 2, we must calculate the quantities $p = b_1 + b_4$ and $q = b_1 b_4 - b_2 b_3$ and $p^2 - 4q = (b_1 - b_4)^2 + 4b_2 b_3$ in order to classify the singularities. We have:

$$\begin{aligned} p &= 2 - 4(u_0^2 + v_0^2) = 2(1 - 2A), \\ q &= \sigma^2 + (1 - A)(1 - 3A), \\ p^2 - 4q &= 4(A^2 - \sigma^2). \end{aligned} \tag{1.25}$$

The equations of the separatrices are

$(C_1)$ : $p = 0$ or $A = \dfrac{1}{2}$,

$(C_2)$ : $q = 0$ or $\sigma^2 + (1 - A)(1 - 3A) = 0$,

$(C_3)$ : $p^2 - 4q = 0$ or $A = \pm\sigma$.

These separatrices are shown in Fig. 81 in the $(A, \sigma)$ plane. The nature of the different areas which they bound is easily determined by applying the criteria developed in Chapter 2 to the relations (1.25). We may summarize the results as follows:

- With $q < 0$, that is in the interior of the ellipse (1.14), we have saddle points.
- With $q > 0$, that is at the exterior of the ellipse (1.14):

1) for the case $p > 0$ $\left(A < \dfrac{1}{2}\right)$, we have unstable nodes if $p^2 - 4q > 0$ $(A^2 > \sigma^2)$ and an unstable focus if $p^2 - 4q < 0$ $(A^2 < \sigma^2)$.

2) for the case $p < 0$ $\left(A > \dfrac{1}{2}\right)$, we have stable nodes if $p^2 - 4q > 0$ $(A^2 > \sigma^2)$ and *a* stable focus if $p^2 - 4q < 0$ $(A^2 < \sigma^2)$.

The division of the $(A, \sigma)$ - plane into the various regions of stability gives a convenient way of ascertaining at a glance what may be expected in a given oscillatory system under the different values of the detuning (parameter $\sigma$) as well as the external excitation (parameter A).

With reference to Fig. 81, we can summarize the situation with regard to harmonic oscillations of the equation (1.1) as follows:

1) If $E^2 > 8/27$, there is only one possible harmonic oscillation for a given value of $\sigma$ (i.e. for a given exciting frequency) and it is stable if $A > 1/2$, i.e. if the square of its amplitude $a_0$ is larger than half the square of the amplitude $a_*$ of the oscillation of a purely self-excited Van der Pol oscillator (since $A = a_0^2/a_*^2$ (see (1.13)) and is unstable for $A < 1/2$.

2) If $1/4 < E^2 < 8/27$, there is an interval of $\sigma$ - values, i.e. $1/2 < \sigma < 1/\sqrt{3}$, for which three different harmonic oscillations occur, of which either one or two will be stable depending upon whether $A > 1/2$ or not. The stable solution with larger amplitude corresponds to a node.

3) If $E^2 < 1/4$, there are three harmonic oscillations for all values of $\sigma$ between zero and a certain maximum (which depends on $E$), beyond which only one unstable harmonic oscillation occurs. When three periodic solutions occur only, the one with the largest amplitude is stable and it corresponds to a node.

From these results, one can reach the following conclusions. Under certain conditions the frequency of the free oscillation is canceled out, and is replaced by a synchronized oscillation, i.e. by an oscillation whose frequency is that of the external force, namely,

1) For a given amplitude of the exciting force ($E$), the synchronization effect is observed provided that the exciting frequency ($\nu$) is close enough to the natural frequency of the oscillator ($\sigma = 0$). The larger the amplitude of the exciting force, the greater is the frequency interval over which the synchronization occurs.

2) For a given exciting frequency, the oscillator is synchronized provided the exciting amplitude is large enough. The closer the exciting frequency is to unity ($\sigma \to 0$), the lower its threshold amplitude is.

**Combination oscillations for large detuning** (parameter $\sigma$)

As we have seen from Fig. 81, with a fixed value of $E$ (amplitude of external force) if the detuning $\sigma = \dfrac{\Delta}{\alpha}$ is sufficiently large compared with $E$, then no stable synchronized oscillation will be observed. The important question then arises: what can be predicted for large detuning.

When the detuning $\sigma$ is very large, we expect the oscillation of the system to be approximated closely by the purely self-excited oscillation ($e = 0$) with frequency $\omega = 1$. On the other hand, if the detuning is very small we observe from Fig. 81 that there is always one stable synchronized oscillation with the frequency $\nu$ of the external force. At the two extremes of small and large detuning, the resulting oscillations are combination ones which have the frequencies 1 and $\nu$ respectively.

One can verify the above statements by considering the solution of the equation (1.1) in the form:

$$x = a_0 \cos\varphi + b\cos t, \tag{1.26}$$

where, $\varphi = \nu t + \theta_0$ and $a_0$, $b$, $\theta_0$ are constants. The first term in (1.26) represents the forced oscillation, while the second term is the self-excited oscillation. There are three unknowns $a_0$, $b$ and $\theta_0$. To determine them, we substitute (1.26) into (1.1) and compare the coefficients of the harmonics in the resulting equations. Since we have only three unknowns, (1.26) can satisfy equation (1.1) approximately, requiring the coefficients. for example, of $\cos\varphi$, $\sin\varphi$ and $\sin t$ to vanish. If this is done, we obtain:

$$\text{for } \cos\varphi: \quad (1-\nu^2)a_0 = -\varepsilon e \sin\theta_0, \tag{1.27}$$

$$\text{for } \sin\varphi: \quad \varepsilon\nu\Big(\alpha - \frac{\gamma}{4}a_0^2 - \frac{\gamma}{2}b^2\Big)a_0 = \varepsilon e\cos\theta_0, \tag{1.28}$$

$$\text{for } \sin t: \quad \varepsilon b\Big(\alpha - \frac{\gamma}{2}a_0^2 - \frac{\gamma}{4}b^2\Big) = 0. \tag{1.29}$$

Here, it is necessary to assume that $1-\nu^2$ is not too small; because, for instance, if $\nu = 1$, i.e. if $\varphi = t + \theta_0$, the terms with $\sin(2\varphi - t)$ and $\sin(\varphi - 2t)$ contain also $\sin\varphi$ and $\cos\varphi$ and hence equations (1.27) and (1.28) will not be true.

Eliminating the phase $\theta_0$ from equations (1.27) and (1.28), we obtain:

$$\varepsilon^2 e^2 = a_0^2\Big[(1-\nu^2)^2 + \varepsilon^2\nu^2\Big(\alpha - \frac{\gamma}{2}a_0^2 - \frac{\gamma}{2}b^2\Big)^2\Big]. \tag{1.30}$$

From the equation (1.29), we have

a) The first subcase: $b = 0$. In this subcase, there is no self-excited oscillation and the equation (1.30) is exactly (1.9) which determines the amplitude $a_0$ of forced oscillation.

b) The second subcase $b^2 = \dfrac{4\alpha}{\gamma}(1-2A)$, where $A = \dfrac{\gamma}{4\alpha}a_0^2$. Since $\alpha > 0$ and $\gamma > 0$, the self-excited oscillation takes place only in the region $A < \dfrac{1}{2}$, or $a_0^2 < \dfrac{2\alpha}{\gamma}$. By setting $A = \dfrac{1}{2}$ $\left(a_0^2 = \dfrac{2\alpha}{\gamma},\ b = 0\right)$ in (1.30), we have the equation for determing of the corresponding two values of $\nu^2$:

$$\varepsilon^2 e^2 = \frac{2\alpha}{\gamma}\Big[(1-\nu^2)^2 + \frac{\varepsilon^2}{4}\nu^2\alpha^2\Big].$$

If we replace $\varepsilon^2\nu^2$ in the last term simply by $\varepsilon^2$, we neglect the quantity of smallness $\varepsilon^3$ in comparison with the quantities $\varepsilon^2$. This gives

$$\nu_{1,2}^2 \simeq 1 \pm \varepsilon\sqrt{\frac{\gamma}{2\alpha}e^2 - \frac{\alpha^2}{4}}\,. \tag{1.31}$$

Thus, in the zone $A < \dfrac{1}{2}$, the amplitude of self-excited oscillation $b$ and the amplitude of forced oscillation $a_0$ are determined by the formulae:

$$B = 1 - 2A, \tag{1.32}$$
$$E^2 = A\big[\sigma^2 + (3A-1)^2\big], \tag{1.33}$$

where the notation (1.13) is used, and $B = \dfrac{\gamma}{4\alpha}A$. By comparison with (1.12), we see that this portion of the amplitude is given by the same expression as for synchronization oscillation, except for the factor 3 multiplying $A$ inside the bracket. It is clear from (1.33) that $A \to 0$, as $\sigma \to \infty$ for any fixed value of $E$, so that $B \to 1$ from the lower side as $\sigma \to \infty$. This confirms the result we found earlier, when we discussed on the combination oscillations for large detuning, that the motion to be expected for very large $\sigma$ is pure self-excited oscillation. The amplitude curve $B = B(\sigma)$ and $A = A(\sigma)$ obtained from (1.32) and (1.33) are shown in Figure 83, together with the stable portion of the synchronized amplitude curve for $E > 8/27$. Fig. 83 shows how the oscillations will occur when the parameter $\sigma$ increases. With the increase of $\sigma$ from values less than $\sigma_1$ i.e. in the left part of Fig. 83, the amplitude of forced oscillation ($A$) increases, and reaches the value $\frac{1}{2}$ at $\sigma = \sigma_1$; the amplitude of self-excited oscillation ($B$) decreases and vanishes at $\sigma = \sigma_1$. In this interval of $\sigma$ ($\sigma < \sigma_1$), there are two simultaneous oscillations with close frequencies $\nu$ and 1, so there are beats.

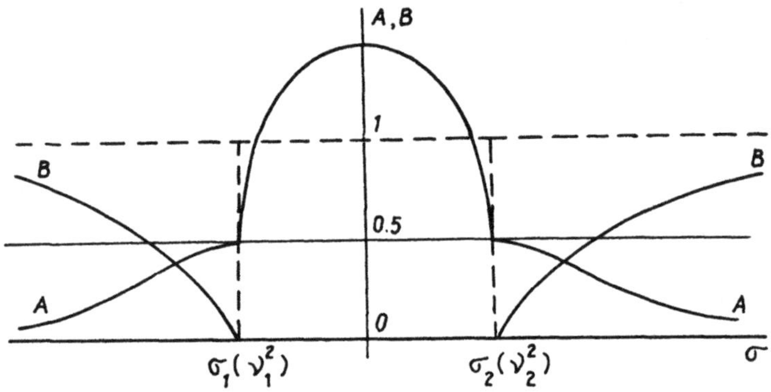


*Fig. 83.* Amplitude A of forced oscillation and amplitude B of self-excited oscillation.

Increasing $\sigma$ further, we observe that in the interval $[\sigma_1, \sigma_2]$ which corresponds to $\nu^2 = \nu_1^2$ and $\nu^2 = \nu_2^2$, there is only the synchronized oscillation with the frequency of the external force. In this interval, the self-excited frequency is entrained by the frequency of the external force.

Increasing $\sigma$ still further, we see that for $\sigma > \sigma_2$, the self-excited oscillation (B) appears again with increasing amplitude and reaches $B = 1$, while the amplitude of forced oscillation (A) decreases and approaches zero. Two oscillations with close frequencies exist simultaneously, and there are beats.

## 2. Interaction between Self-excited and Parametric Oscillations

First, we consider an oscillatory system described by the differential equation

$$\ddot{x} + \omega^2 x + \varepsilon R(\dot{x}) + \varepsilon \alpha x \cos \gamma t = 0, \tag{2.1}$$

where $\omega$, $\alpha$ and $\gamma$ are constants and $\varepsilon$ is a small positive parameter. The *negative* friction function $R(\dot{x})$ is assumed to be of the form:

$$R(\dot{x}) = -h_1 \dot{x} + h_3 \dot{x}^3, \tag{2.2}$$

where $h_1$ and $h_3$ are positive constants. When $\alpha = 0$, the equation (2.1) describes a self-excited oscillator and when $h_1 = h_3 = 0$ it describes a parametric oscillator. Each of them considered separately are self-sustained oscillators.

It is assumed that there exists a resonance relation

$$\omega^2 = \frac{\gamma^2}{4} + \varepsilon\Delta. \tag{2.3}$$

The solution of (2.1) is found in the following form

$$x = a\cos\theta, \quad \dot{x} = -\frac{a}{2}\gamma\sin\theta \tag{2.4}$$

with the additional condition

$$\dot{a}\cos\theta - a\dot{\psi}\sin\theta = 0, \tag{2.5}$$

where $\theta = \frac{\gamma}{2}t + \psi$. Substituting (2.4) into (2.1) and solving with respect to $\dot{a}$ and $\dot{\psi}$ with the help of (2.5), we obtain:

$$\begin{aligned} \dot{a} &= \frac{2\varepsilon}{\gamma}(\Delta x + R(\dot{x}) + \alpha x\cos\gamma t)\sin\theta, \\ a\dot{\psi} &= \frac{2\varepsilon}{\gamma}(\Delta x + R(\dot{x}) + \alpha x\cos\gamma t)\cos\theta. \end{aligned} \tag{2.6}$$

Since $a$ and $\psi$ are slowly varying in time one may, in the first approximation, replace the right-hand sides of (2.6) by their mean values. If this is done, we obtain:

$$\begin{aligned} \dot{a} &= \frac{\varepsilon}{\gamma}\left(\frac{1}{2}h_1\gamma a - \frac{3}{32}h_3\gamma^3 a^3 + \frac{\alpha}{2}a\sin 2\psi\right), \\ a\dot{\psi} &= \frac{\varepsilon}{\gamma}\left(\Delta a + \frac{\alpha}{2}a\cos 2\psi\right). \end{aligned} \tag{2.7}$$

The stationary solution of equations (2.7) is
1) $a_0 = 0$, $\psi_0$ is arbitrary, (2.8)
2) $a_0 \neq 0$ determined by:

$$\begin{aligned} &h_1\gamma - \frac{3}{16}h_3\gamma^3 a_0^2 + \alpha\sin 2\psi_0 = 0, \\ &2\Delta + \alpha\cos 2\psi_0 = 0. \end{aligned} \tag{2.9}$$

Eliminating the phase $\psi_0$ from these equations, we derive:

$$\gamma^2\left(h_1 - \frac{3}{16}h_3\gamma^2 a_0^2\right)^2 + 4\Delta^2 - \alpha^2 = 0 \tag{2.10}$$

or approximately

$$W(a_0^2, \eta^2) = 0, \tag{2.11}$$

where $\eta^2 = \gamma^2/4\omega^2$,

$$W(a_0^2, \eta^2) = \left(a_0^2 - \frac{4h_1}{3h_3\omega^2}\right)^2 + \frac{16(\eta^2-1)^2}{9\varepsilon^2\omega^2 h_3^2} - \frac{4\alpha^2}{9\omega^6 h_3^2}, \qquad (2.12)$$

that is, $\gamma^2$ in the first term of (2.10) is replaced by $4\omega^2$.

In the $(a_0^2, \eta^2)$ - plane the expression (2.11) represents an ellipse (Fig. 84) with the centre at

$$\eta_*^2 = 1, \quad a_{0*}^2 = \frac{4h_1}{3h_3\omega^2} \qquad (2.13)$$

and with two semiaxes:

$$\ell = \frac{\varepsilon|\alpha|}{2\omega^2}, \quad L = \frac{2|\alpha|}{3\omega^3 h_3}. \qquad (2.14)$$

Depending on the ratio $\delta = \dfrac{L}{a_{0*}^2} = \dfrac{\alpha}{2\omega h_1}$, the resonance curve either lies above the $\eta^2$ - axis ($\delta < 1$) or cuts this axis.

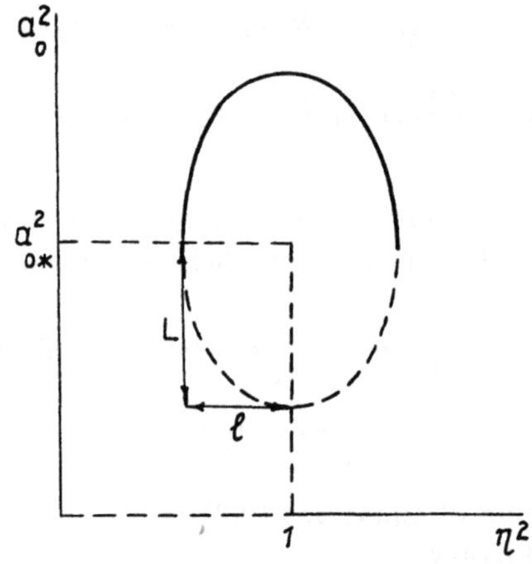


*Fig. 84.* Amplitude curve of a mixed self-excited and parametric oscillator.

Equation (1.1) is only a particular case of the equation (4.1) of Chapter 4. Using the stability criterions (4.13) and (4.14), Chapter 4, we obtain the following inequalities:

$$\frac{\partial}{\partial a_0}(a_0 S) > 0 \quad \text{or} \quad a_0^2 > \frac{2h_1}{3\omega^2 h_3} = \frac{1}{2}a_{0*}^2, \qquad (2.15)$$

$$\frac{\partial W}{\partial a_0^2} > 0, \qquad (2.16)$$

where

$$S(a_0, \gamma) = \frac{\gamma a_0}{4}\left(-h_1 + \frac{3}{16}h_3\gamma^2 a_0^2\right).$$

It is easy to verify that the function $W$ (2.12) is negative inside the ellipse (2.11) and positive outside it. Hence, the upper branch of the ellipse (2.11) limited by vertical tangents corresponds to the stability of stationary oscillations.

Comparing (4.69), Chapter 4 with (2.7) of this Chapter we obtain $\mu = -h_1 < 0$. So, following (4.72), Chapter 4, the zero solution $a = 0$ of the equations (2.7) is unstable.

## 3. Generalized Van Der Pol Equation

Another form of interaction between self-excited and parametric oscillations can be obtained in the generalized Van der Pol equation [37,g]:

$$\ddot{x} + \omega^2 x - \varepsilon\big[1 - (x + q\cos\nu t)^2\big]\dot{x} = 0, \tag{3.1}$$

where $\omega$, $\nu$, $q$ are constants, and $\varepsilon$ is a small parameter. When $q = 0$, we have the well-known Van der Pol equation, for which there exists only one stable stationary oscillation

$$x = 2\cos(\omega t + \alpha). \tag{3.2}$$

To investigate the oscillations described by the equation (3.1), we first transform it into the standard form by means of the formulae

$$x = a\cos\varphi, \quad \dot{x} = -a\omega\sin\varphi. \tag{3.3}$$

The transformed equations are

$$\begin{aligned} \frac{da}{dt} &= \varepsilon a\big[1 - (a\cos\varphi + q\cos\nu t)^2\big]\sin^2\varphi, \\ \frac{ad\psi}{dt} &= \varepsilon a\big[1 - (a\cos\varphi + q\cos\nu t)^2\big]\sin\varphi\cos\varphi, \end{aligned} \tag{3.4}$$

where $\psi = \varphi - \omega t$.

By employing the averaging method of nonlinear oscillations, we obtain in the first approximation the following equations for $a$ and $\psi$:

$$\begin{aligned} \frac{da}{dt} &= \varepsilon a f_1(a, \psi), \\ \frac{ad\psi}{dt} &= \varepsilon a f_2(a, \psi). \end{aligned} \tag{3.5}$$

Equations (3.5) have a zero solution $a = a_{**} = 0$ and the other non-trivial stationary solutions $\psi_*$, $a = a_* \neq 0$ determined by the system of equations:

$$f_1(a_*, \psi_*) = 0, \quad f_2(a_*, \psi_*) = 0. \tag{3.6}$$

The variational equation for the zero solution is

$$\frac{d\delta a}{dt} = \varepsilon f_1(0, \psi_{**})\delta a.$$

Hence, the zero solution is stable if ($\varepsilon > 0$):

$$f_1(0, \psi_{**}) < 0, \tag{3.7}$$

$\psi_{**}$ is arbitrary

The variational equations for the non-trivial stationary solutions have the form:

$$\begin{aligned} \frac{d\delta a}{dt} &= \varepsilon a_* \Big[\Big(\frac{\partial f_1}{\partial a}\Big)_* \delta a + \Big(\frac{\partial f_1}{\partial \psi}\Big)_* \delta\psi\Big], \\ \frac{d\delta\psi}{dt} &= \varepsilon\Big[\Big(\frac{\partial f_2}{\partial a}\Big)_* \delta a + \Big(\frac{\partial f_2}{\partial \psi}\Big)_* \delta\psi\Big]. \end{aligned} \tag{3.8}$$

The characteristic equation of this system is

$$\lambda^2 - \varepsilon\Big[a_*\Big(\frac{\partial f_1}{\partial a}\Big)_* + \Big(\frac{\partial f_2}{\partial \psi}\Big)_*\Big]\lambda + \varepsilon^2 a_*\Big[\Big(\frac{\partial f_1}{\partial a}\Big)_*\Big(\frac{\partial f_2}{\partial \psi}\Big)_* - \Big(\frac{\partial f_1}{\partial \psi}\Big)_*\Big(\frac{\partial f_2}{\partial a}\Big)_*\Big] = 0.$$

Therefore, the conditions of stability for nontrivial solutions are

$$\begin{aligned} a_*\Big(\frac{\partial f_1}{\partial a}\Big)_* + \Big(\frac{\partial f_2}{\partial \psi}\Big)_* &< 0, \\ \Big(\frac{\partial f_1}{\partial a}\Big)_*\Big(\frac{\partial f_2}{\partial \psi}\Big)_* - \Big(\frac{\partial f_1}{\partial \psi}\Big)_*\Big(\frac{\partial f_2}{\partial a}\Big)_* &> 0. \end{aligned} \tag{3.9}$$

We shall consider now various resonance cases that correspond to particular values of the frequencies $\omega$ and $\nu$.

**The resonance Case $\nu = 3\omega$**

In this case, the averaged equations (3.5) take the form

$$\begin{aligned} \frac{da}{dt} &= \frac{\varepsilon a}{4}\Big[2 - \frac{a^2}{2} - q^2 + aq\cos 3\psi\Big], \\ \frac{ad\psi}{dt} &= -\varepsilon\frac{a^2}{4}q\sin 3\psi, \end{aligned} \tag{3.10}$$

so

$$\begin{aligned} f_1(a, \psi) &= \frac{1}{2} - \frac{a^2}{8} - \frac{q^2}{4} + \frac{aq}{4}\cos 3\psi, \\ f_2(a, \psi) &= -\frac{aq}{4}\sin 3\psi. \end{aligned} \tag{3.11}$$

Since

$$f_1(0, \psi) = \frac{1}{2} - \frac{q^2}{4},$$

then the zero solution $a = 0$ of the system (3.10) is stable if

$$q < -\sqrt{2}, \quad \text{or} \quad q > \sqrt{2},$$

and unstable if

$$-\sqrt{2} < q < \sqrt{2}.$$

The non-trivial stationary solutions $(\psi_*, a_*)$ of the system (3.10) are determined from the equations:

1. *The first subcase*

$$\sin 3\psi_* = 0, \quad \cos 3\psi_* = +1,$$
$$a_*^2 - 2qa_* + 2q^2 - 4 = 0,$$

hence

$$3\psi_* = 2n\pi, \quad a_* = q \pm \sqrt{4 - q^2}. \tag{3.12}$$

As

$$\left(\frac{\partial f_1}{\partial \psi}\right)_* = \left(\frac{\partial f_2}{\partial a}\right)_* = 0$$

then the conditions of stability (3.9) become

$$\left(\frac{\partial f_1}{\partial a}\right)_* < 0, \quad \left(\frac{\partial f_2}{\partial \psi}\right)_* < 0. \tag{3.13}$$

We have

$$\left(\frac{\partial f_1}{\partial a}\right)_* = -\frac{a_*}{4} + \frac{q}{4} = \mp\sqrt{4 - q^2},$$
$$\left(\frac{\partial f_2}{\partial \psi}\right)_* = -\frac{3}{4} a_* q,$$

from which it follows that when $q > 0$, the branch $a_* = q - \sqrt{4 - q^2}$ is unstable, and the branch $a_* = q + \sqrt{4 - q^2}$ (branch EG, Fig. 85) is stable.

2. *The second subcase*

$$\sin 3\psi_* = 0, \quad \cos 3\psi_* = -1,$$
$$a_*^2 + 2qa_* + 2q^2 - 4 = 0,$$

hence,

$$3\psi_* = (2n + 1)\pi, \quad a_* = -q \pm \sqrt{4 - q^2}.$$

In this case, the conditions of stability are also given by the formulae (3.13). We have:

$$\left(\frac{\partial f_1}{\partial a}\right)_* = -\frac{a_*}{4} - \frac{q}{4} = \mp\frac{1}{4}\sqrt{4-q^2}\,,$$
$$\left(\frac{\partial f_2}{\partial \psi}\right)_* = \frac{3a_* q}{4}.$$

Consequently, when $q < 0$, the branch $a_* = -q - \sqrt{4-q^2}$ is unstable and the branch $a_* = -q + \sqrt{4-q^2}$ (branch DE, Fig. 85) is stable.

Fig. 85 shows the dependence of the amplitude of oscillations upon the parameter $q$ in the case $\nu = 3\omega$. The heavy lines correspond to the stable regimes of oscillations. With the increase of the parameter $q$, the amplitude of oscillations changes along the lines BCDEGHK. At the point C ($q = -\sqrt{2}$) and G ($q = 2$) there is a jump in the amplitude. A similar phenomena is also observed when decreasing the parameter $q$. Thus, in the case $\nu = 3\omega$, the amplitude of stationary oscillation of the generalized Van der Pol system is larger than that of the initial system. In the illustrations in this paragraph, the circle points are obtained by solving the corresponding equations on the analog computer. This is done to verify the validity of the theoretical investigations.

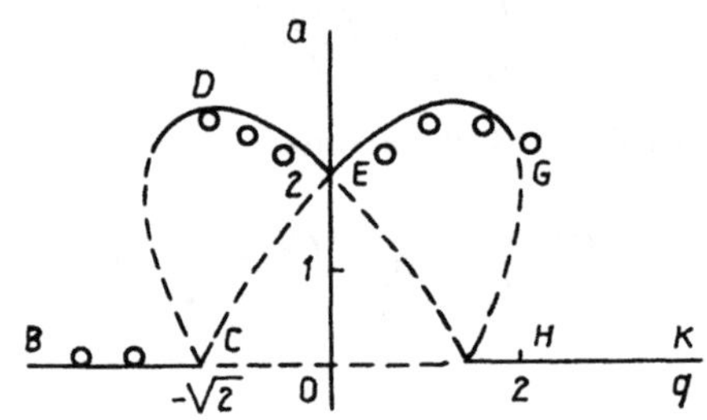


*Fig. 85.* The dependence of amplitude of oscillation of generalized Van der Pol oscillator on the parameter $q$.

**Oscillations in the case $\nu = \omega$**

Now, the averaged equations (3.5) are of the form:

$$\frac{da}{dt} = -\frac{\varepsilon a}{4}\Big(-2 + \frac{a^2}{2} + q^2 + aq\cos\psi - \frac{q^2}{2}\cos 2\psi\Big),$$
$$\frac{ad\psi}{dt} = -\frac{\varepsilon aq}{4}(a + q\cos\psi)\sin\psi, \tag{3.14}$$

therefore

$$f_1(a,\psi) = -\frac{1}{4}\Big(-2 + \frac{a^2}{2} + q^2 + aq\cos\psi - \frac{q^2}{2}\cos 2\psi\Big),$$
$$f_2(a,\psi) = -\frac{q}{4}(a + q\cos\psi)\sin\psi.$$

We have

$$f_1(0, \psi_*) = -\frac{1}{4}\left(-2 + q^2 - \frac{q^2}{2}\cos 2\psi_*\right),$$

where $\psi_*$ is an arbitrary number.

It will readily be seen that the zero solution $a = 0$ is stable if

$$f_1(0, \psi_*) < 0, \text{ or if } q < -2 \text{ or } q > 2.$$

Let us now consider the stability of the non-trivial stationary solutions $a_* \neq 0$. The first stationary solution $a_*$, $\psi_*$ is determined by the equations:

$$\sin\psi_* = 0, \quad a_*^2 + 2qa_* + q^2 - 4 = 0,$$
$$\cos\psi_* = 1, \quad \cos 2\psi_* = 1$$

from which we obtain

$$a_* = -q \pm 2, \quad \psi_* = 2n\pi,$$

where $n$ is an integer.

Since $\left(\frac{\partial f_1}{\partial \psi}\right)_* = \left(\frac{\partial f_2}{\partial a}\right)_* = 0$, the conditions of stability of the stationary solutions also take the form (3.13). We have:

$$\left(\frac{\partial f_1}{\partial a}\right)_* = -\frac{1}{4}(a_* + q\cos\psi_*) = \mp\frac{1}{2},$$
$$\left(\frac{\partial f_2}{\partial \psi}\right)_* = -\frac{q}{4}(a_*\cos\psi_* + q\cos 2\psi_*) = \mp\frac{q}{2},$$

where the upper sign corresponds to the straight line $a_* = -q + 2$, and the lower one to the straight line $a_* = -q - 2$. From it follows that for $q > 0$ the straight line $a_* = -q - 2$ is unstable, and the line $a_* = -q + 2$ is stable (Fig. 86).

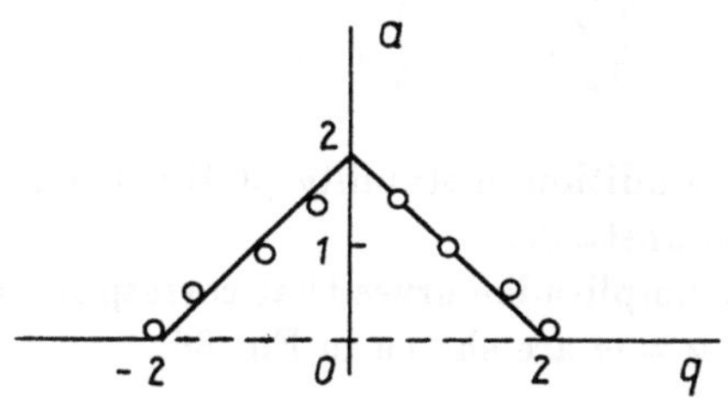


*Fig. 86.* The dependence of amplitude $a$ on the parameter $q$ in the case $\nu = \omega$.

The second stationary value $a_*$, $\psi_*$ is determined from the solution of the equations

$$\sin\psi_* = 0, \quad a_*^2 - 2qa_* + q^2 - 4 = 0,$$
$$\cos\psi_* = -1, \quad \cos 2\psi_* = 1,$$

hence,

$$\psi_* = (2n+1)\pi, \quad a_* = q \pm 2.$$

For the case under consideration, we have:

$$\left(\frac{\partial f_1}{\partial a}\right)_* = -\frac{1}{4}(a_* - q) = \mp\frac{1}{2},$$
$$\left(\frac{\partial f_2}{\partial \psi}\right)_* = -\frac{q}{4}(-a_* + q) = \pm\frac{q}{2},$$

where the upper sign corresponds to the straight line $a_* = q+2$, and the lower sign to the straight line $a_* = q-2$.

Following the inequalities (3.13), we conclude that the line $a_* = q-2$ is unstable, and the line $a_* = q+2$ is stable where $q < 0$.

The third stationary solution $a_*$, $\psi_*$ satisfies the equation

$$a_* + q\cos\psi_* = 0,$$

and consequently

$$\cos 2\psi_* = \frac{2a_*^2}{q^2} - 1.$$

Substituting these values into the equation $\dfrac{da}{dt} = 0$ yields:

$$a = \sqrt{q^2 - \frac{4}{3}}\,.$$

In this case, we have:

$$\left(\frac{\partial f_1}{\partial a}\right)_* - \frac{1}{4}(a_* + q\cos\psi_*) = 0,$$
$$\left(\frac{\partial f_2}{\partial \psi}\right)_* = -\frac{1}{3},\ldots$$

and therefore, the first condition of stability (3.9) is not satisfied, and the stationary solution obtained is unstable.

The branches of the amplitude-curves that correspond to the stable stationary oscillations in the case $\nu = \omega$ are shown in Fig. 86.

**Stationary oscillations in the remaining cases ($\nu \neq \omega, 3\omega$)**
For these cases, the averaged equations become quite simple

$$\frac{da}{dt} = \frac{\varepsilon a}{2}\left[1 - \left(\frac{a^2}{4} + \frac{q^2}{2}\right)\right],$$
$$\frac{d\psi}{dt} = 0.$$

It is easy to show that the zero solution $a = 0$ is stable if $q^2 \geq 2$, and unstable if $-\sqrt{2} < q < \sqrt{2}$.

The non-trivial stationary solution

$$a_*^2 = 4\left(1 - \frac{q^2}{2}\right)$$

is stable for the values $q^2 < 2$. Thus, for $q^2 < 2$, the system oscillates with the stationary amplitude $a_*$:

$$a_*^2 = 4 - 2q^2.$$

For $q^2 > 2$, the oscillations are damped (Fig. 87):

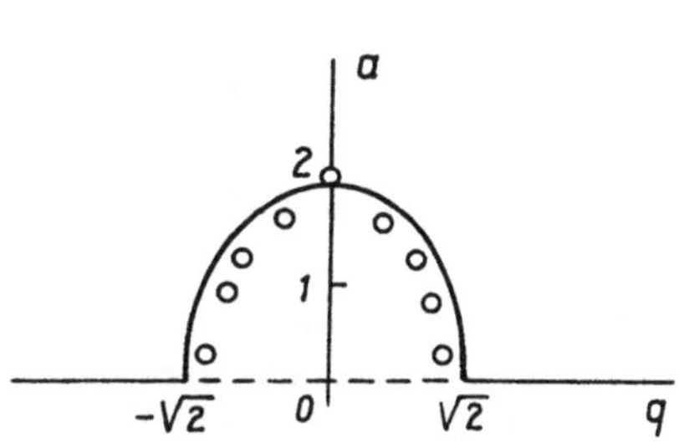


*Fig. 87.* The dependence of amplitude $a$ on the parameter $q$ in non resonance case.

## 4. Interaction of Subharmonic Oscillations

This paragraph deals with the interaction of subharmonic oscillations. As is generally known , a widely distributed species of nonlinear systems which exhibit subharmonic oscillations are:

$$\ddot{x} + \omega^2 x + \delta\dot{x} + c_3 x^3 = b_0 + b\cos(pt + \beta), \tag{4.1}$$

and

$$\ddot{x} + \omega^2 x + \delta\dot{x} + c_2 x^2 + c_3 x^3 = b\cos(pt + \beta). \tag{4.2}$$

These systems are characterized by the asymmetry of either the elastic force or of the external force. We will study the interaction of subharmonic oscillations in these systems by considering the combinative equation of the form [37,f]:

$$\ddot{x} + \omega^2 x + \delta\dot{x} + c_2 x^2 + c_3 x^3 = b_0 + b\cos(pt + \beta). \tag{4.3}$$

This equation becomes equation (4.1) when $c_2 = 0$, and equation (4.2) when $b_0 = 0$.

The interaction between the subharmonic and the parametric oscillations will be investigated by considering the equation (4.23) which coincides with (4.3) if $d = 0$, and exhibits parametric oscillations when $b_0 = b_2 = c_2 = 0$.

**Interaction of Subharmonic Oscillations of the 1/2-order**

Let us consider the case when $p = 2\nu$ and $\nu \sim 1$, namely

$$1 = \nu^2 + \varepsilon\sigma \tag{4.4}$$

and suppose that the friction force and the nonlinear terms are of order $\varepsilon$, so that we can write the equation (4.3) in the form:

$$\ddot{x} + \nu^2 x + \varepsilon(\sigma x + \delta\dot{x} + c_2 x^2 + c_3 x^3) = b_0 + b_2 \cos(2\nu t + \beta), \tag{4.5}$$

where $b_2 = b$ and $\varepsilon$ is a small parameter.

We shall find the periodic solution of the equation (4.5) in the form:

$$x_0 = B_0 + B_2 \cos(2\nu t + \beta) + A_0 \sin(\nu t + \psi_0), \quad B_2 = -\frac{b_2}{3\nu^2}. \tag{4.6}$$

Only the most important parts of the solution are considered here: the constant term $B_0$, the harmonic oscillation with the frequency $2\nu$ of the external force, and the subharmonic oscillation $A_0 \sin(\nu t + \psi_0)$, which occurs with a frequency one half that of the external force. We will concentrate our attention on the latter oscillation.

In order to obtain the unknown constants $B_0$, $A_0$, $\psi_0$, we substitute the expression (4.6) into (4.5) and compare the coefficients of $\sin(\nu t + \psi_0)$ and $\cos(\nu t + \psi_0)$ and the free terms. We have:

$$\nu^2 B_0 - b_0 + \varepsilon\Big\{\sigma B_0 + c_2\big(B_0^2 + \frac{1}{2}B_2^2 + \frac{1}{2}A_0^2\big) + c_3\Big[B_0\big(B_0^2 + \frac{3}{2}B_2^2 + \frac{3}{2}A_0^2\big) - \frac{3}{4K}B_2 A_0^2\big(\sigma + S + \frac{3}{4}c_3 A_0^2\big)\Big]\Big\} = 0,$$

$$A_0\big(\sigma + S + \frac{3}{4}c_3 A_0^2\big) = K A_0 \cos(\beta - 2\psi_0), \quad \delta\nu A_0 = K A_0 \sin(\beta - 2\psi_0), \tag{4.7}$$

where

$$K = (c_2 + 3c_3 B_0)B_2, \qquad S = 2c_2 B_0 + 3c_3\big(B_0^2 + \frac{1}{2}B_2^2\big). \tag{4.8}$$

From the last two equations of the system (4.7), we obtain a trivial stationary value $A_0 = 0$ and non-trivial values $A_0 \neq 0$ determined by the equation:

$$W = 0, \tag{4.9}$$

where

$$W = (\sigma + S + \rho^2)^2 + \delta^2\nu^2 - K^2, \qquad \rho^2 = \frac{3}{4}c_3 A_0^2. \tag{4.10}$$

Equation (4.9) gives the dependence of the frequency $\nu$ on the amplitude $A_0$ of subharmonic oscillation:

$$\rho^2 = -(\sigma + S) \pm \sqrt{K^2 - \delta^2\nu^2}\,. \tag{4.11}$$

Following this formula, the amplitude curves are presented graphically in Fig. 88. The dependence of $\rho$ on $c_2$ is given for various values of $B_0$ (which is taken approximately as $B_0 = b_0/\nu^2$) $B_0 = 0, 0.4, 0.6, 0.7, 0.8$ and $0.83$. The other parameters are $\delta^2 = 0.032$, $3c_3 = 0.1$, $B_2 = 1$, $\nu^2 = 1.25$. From Fig. 88, it is clear that for every value of $B_0$ there are two separated branches of the amplitude curves. As the value of $B_0$ increases, the right branch droops down and the left branch rises up.

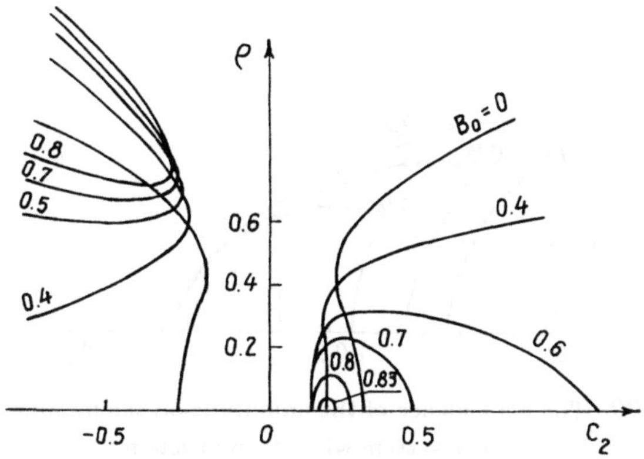


*Fig. 88.* The dependence of the amplitude on the parameter $c_2$.

Figure 89 gives the dependence of $\rho$ on $B_0$ for various values of $c_2$: 0, 0.1, 0.3, in the case $\delta^2 = 0.01$, $3c_3 = 0.1$, $B_2 = 1$, $\nu^2 = 1.4$. The amplitude curves are symmetrical with respect to the axis $B_0 = -10c_2$ (in Fig. 89 the symmetric branches are not presented). With an increase of the value $c_2$, the amplitude curves move to the left.

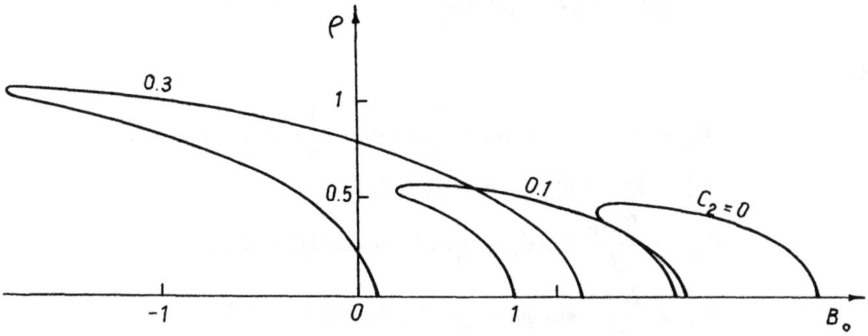


*Fig. 89.* The dependence of the amplitude on the parameter $B_0$.

The curves in Fig. 90 correspond to the system with stronger friction: $\delta^2 = 0.032$ ($3c_3 = 0.1$, $B_2 = 1$, $\nu^2 = 1.25$). From this figure, it is seen that the friction ($\delta$) strongly influences the location of the amplitude curves.

In order to study the stability of the subharmonic oscillation obtained, we put in the equation (4.5):

$$x = x_0 + \xi,$$

where $x_0$ is expressed by (4.6). In the first approximation, we have the following variational equation

$$\ddot{\xi} + \varepsilon\delta\dot{\xi} + \left[\nu^2 + \varepsilon(\sigma + 2c_2x_0 + 3c_3x_0^2)\right]\xi = 0.$$

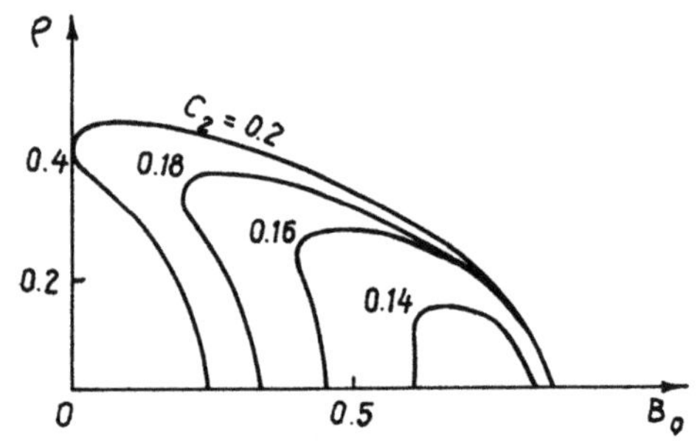


*Fig. 90.* The dependence of the amplitude on the parameter $B_0$ in the system with strong friction.

By using the transformation $\xi = \eta\exp(-\varepsilon\delta t/2)$, we obtain:

$$\ddot{\eta} + \left[\nu^2 + \varepsilon(\sigma + 2c_2x_0 + 3c_3x_0^2) - \frac{\varepsilon^2\delta^2}{4}\right]\eta = 0. \tag{4.12}$$

Substituting the expression $x_0$ from (4.6) yields:

$$\frac{d^2\eta}{d\tau^2} + \left[\frac{\theta_0}{\nu^2} + \frac{2\varepsilon}{\nu^2}\sum_{n=1}^{4}\theta_n\cos(n\tau - \varepsilon_n)\right]\eta = 0, \tag{4.13}$$

where

$$\begin{aligned}
\theta_0 &= \nu^2 + \varepsilon(\sigma + S) + \frac{3}{2}\varepsilon c_3 A_0^2 - \frac{1}{4}\varepsilon^2\delta^2, \quad \tau = \nu t,\\
\theta_n^2 &= \theta_{ns}^2 + \theta_{nc}^2, \quad n = 1, 2, 3, 4,\\
\theta_{1s} &= \frac{KA_0}{B_2}\cos\psi_0 - \frac{3}{2}c_3B_2A_0\cos(\beta - \psi_0),\\
\theta_{1c} &= \frac{KA_0}{B_2}\sin\psi_0 - \frac{3}{2}c_3B_2A_0\sin(\beta - \psi_0),\\
\theta_{2s} &= -K\sin\beta + \frac{3}{4}c_3A_0^2\sin 2\psi_0,
\end{aligned} \tag{4.14}$$

$$\theta_{2c} = K\cos\beta - \frac{3}{4}c_3A_0^2\cos 2\psi_0,$$

$$\theta_3 = \frac{3}{2}c_3B_2A_0, \quad \theta_4 = \frac{3}{2}c_3B_2^2.$$

The stability condition of the zero solution of the Hill equation (4.13) is:

$$\left[\theta_0 - \left(\frac{n\nu}{2}\right)^2\right]^2 + \frac{1}{2}\left[\theta_0 + \left(\frac{n\nu}{2}\right)^2\right]\varepsilon^2\delta^2 - \varepsilon^2\theta_n^2 + \frac{\varepsilon^4\delta^4}{16} > 0, \tag{4.15}$$

$n = 1, 2, 3, 4$.

For $n = 1, 3, 4$ the stability condition (4.15) has the form:

$$\left(\frac{4-n^2}{4}\right)^2\nu^4 + 0(\varepsilon) > 0,$$

which is always satisfied with a sufficiently small parameter $\varepsilon$.

For $n = 2$, we have

$$(\theta_0 - \nu^2)^2 + \frac{1}{2}(\theta_0 + \nu^2)\varepsilon^2\delta^2 - \varepsilon^2\theta_2^2 + \frac{\varepsilon^4\delta^4}{16} > 0. \tag{4.16}$$

Substituting here $\theta_0$, $\theta_2$ from (4.14), we find

$$\left(\sigma + S + \frac{3}{2}c_3A_0^2\right)^2 + \delta^2\nu^2 - \left[K^2 + \frac{9}{16}c_3^2A_0^4 - \frac{3}{2}c_3A_0^2K\cos(\beta - 2\psi_0)\right] > 0.$$

Taking into account the expressions (4.7) and (4.10), we obtain

$$3c_3\left(\sigma + S + \frac{3}{4}c_3A_0^2\right) > 0$$

or

$$\frac{\partial W}{\partial A_0} > 0. \tag{4.17}$$

The solution $A_0 = 0$ of the system (4.7) corresponds to the following solution of the system (4.5):

$$x_0 = B_0 + B_2\cos(2\nu t + \beta). \tag{4.18}$$

The variational equation (4.12) in this case has the form:

$$\frac{d^2\eta}{dt^2} + \left[\theta_0^* + 2\sum_{n=1}^{2}\varepsilon\theta_n^*\cos(2n\nu t - \varepsilon_n)\right]\eta = 0, \tag{4.19}$$

where

$$\begin{aligned} &\theta_0^* = \nu^2 + \varepsilon(\sigma + S) - \frac{\varepsilon^2\delta^2}{4}, \quad \theta_{1c}^* = K\cos\beta, \\ &\theta_{1s}^* = -K\sin\beta, \quad \theta_2^* = \frac{3}{2}c_3B_2^2. \end{aligned} \tag{4.20}$$

Then, the stability conditions of the solution (4.18) take the form:

$$(\theta_0 - n^2\nu^2)^2 + \frac{1}{2}(\theta_0 + n^2\nu^2)\varepsilon^2\delta^2 - \varepsilon^2\theta_n^2 + \frac{\varepsilon^4\delta^4}{16} > 0, \tag{4.21}$$

$n = 1, 2.$
In the case of $n = 2$, this equality has the form:

$$\nu^4 + 0(\varepsilon) > 0,$$

which is always satisfied for a sufficiently small value of $\varepsilon$

Substituting (4.20) into (4.21), we obtain for $n = 1$ the stability condition of the solution $A_0 = 0$:

$$(\sigma + S)^2 + \delta^2\nu^2 - K^2 > 0, \tag{4.22}$$

which shows that the free term of the quadratic equation in relation to $\rho^2$, (4.9) must be positive.

By using the rule given in Chapter 3, we can use the inequalities (4.7) and (4.22) to give the stable and unstable branches of the amplitude curves. The stability picture is given in Figs. 91, 92. The stable branches are presented by heavy lines, and the unstable ones by broken lines.

**Interaction between subharmonic and parametric oscillations**

Let us consider the equation

$$\begin{aligned} &\ddot{x} + \nu^2 x + \varepsilon(\sigma x + \delta\dot{x} + 2dx\cos 2\nu t + c_2x^2 + c_3x^3) = \\ &= b_0 + b_2\cos(2\nu t + \beta). \end{aligned} \tag{4.23}$$

In the system described by this equation, there exists, for $b_0 = b_2 = c_2 = 0$, a parametric oscillation, and for $d = 0$ a subharmonic oscillation.

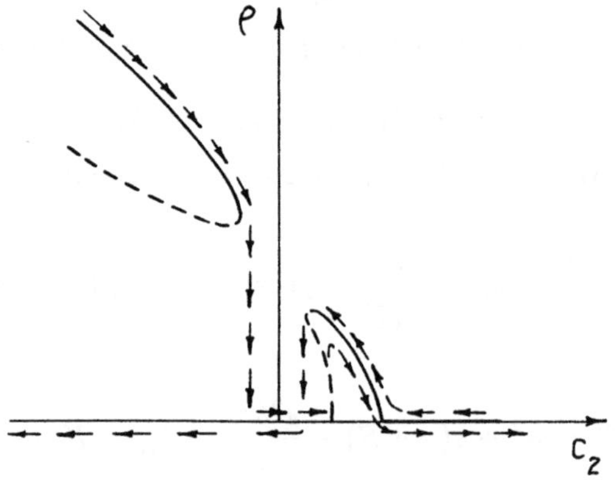


*Fig. 91.* The change of amplitude with slowly varying of the parameter $C_2$.

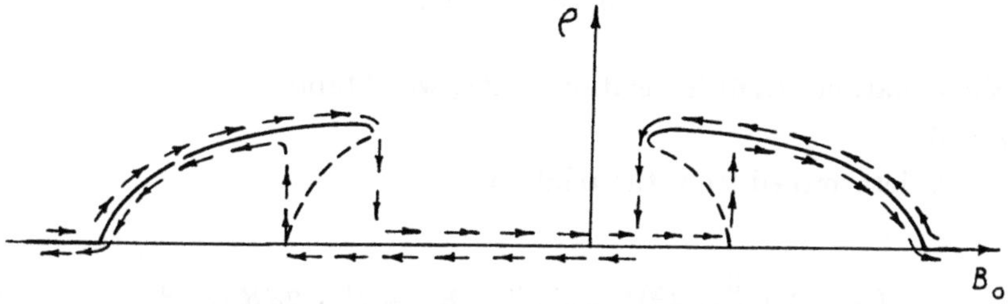


*Fig. 92.* The change of amplitude with slowly varying of the parameter $B_0$.

Although the manifestations of the subharmonic oscillation and the parametric one are different, both forms of oscillations have a common feature, and they can coexist under certain conditions.

The periodic solution of the equation (4.23) is found in the form:

$$x_0 = B_0 + B_2 \cos(2\nu t + \beta) + y \sin \nu t + z \cos \nu t, \tag{4.24}$$

where $B_2 = -b_2/3\nu^2$ and $B_0$, $y$, $z$ are constants. To determine the terms with frequency $\nu$, equal to half the frequency of the external force, we substitute (4.24) into (4.23) and compare the coefficients of $\sin \nu t$, $\cos \nu t$. We have:

$$\begin{aligned}
&(\varepsilon\sigma + \nu^2)B_0 + \varepsilon c_2\Big(B_0^2 + \frac{B_2^2}{2} + \frac{y^2+z^2}{2}\Big) + \\
&+ \frac{\varepsilon}{2} d B_2 \cos\beta + \varepsilon c_3 \Big\{ B_0 \Big[ B_0^2 + \frac{3}{2}B_2^2 + \frac{3}{2}(y^2+z^2)\Big] - \\
&- \frac{3}{4}B_2\big[\cos\beta(y^2-z^2) + 2yz\sin\beta\big]\Big\} = b_0, \\
&\big[\sigma + S + \frac{3}{4}c_3(y^2+z^2)\big]y = (K\cos\beta + d)y + (\delta\nu + K\sin\beta)z, \\
&\big[\sigma + S + \frac{3}{4}c_3(y^2+z^2)\big]z = (K\sin\beta - \delta\nu)y - (K\cos\beta + d)z,
\end{aligned} \tag{4.25}$$

where

$$K = (c_2 + 3c_3B_0)B_2, \quad S = 2c_2B_0 + 3c_3\big(B_0^2 + \frac{1}{2}B_2^2\big).$$

From the last two equations of (4.25), after some simple transformations, we obtain:

$$\begin{aligned}
&\big(\sigma + S + \frac{3}{4}c_3A_0^2\big)^2 A_0^2 = (K^2 + d^2 + 2dK\cos\beta)A_0^2 + \delta^2\nu^2A_0^2 + \\
&+ 2\delta\nu K\sin\beta(z^2 - y^2) + 4\delta\nu yz(K\cos\beta + d),
\end{aligned} \tag{4.26}$$

or

$$\Big[\big(\sigma + S + \frac{3}{4}c_3A_0^2\big)^2 + \delta^2\nu^2\Big]A_0^2 = (K^2 + d^2 + 2dK\cos\beta)A_0^2,$$

where

$$A_0^2 = y^2 + z^2.$$

Solving the equation (4.26) in relation to $A_0$, we obtain:

1) $A_0 = 0$,
2) $A_0 \neq 0$ determined from the relation:

$$\left(\sigma + S + \frac{3}{4}c_3 A_0^2\right)^2 + \delta^2\nu^2 = K^2 + d^2 + 2dK\cos\beta. \tag{4.27}$$

The dependence of $\rho = \frac{3}{4}c_3 A_0^2$ on the parameter $d$ is presented graphically in Fig. 93 for the case of $\delta^2 = 0.01$, $3c_3 = 0.1$, $B_2 = 1$, $\nu^2 = 1.4$, $\beta = \pi$, $B_0 = 1$, $c_2 = 0.2$. This figure shows that the component of the oscillation with frequency $\nu$ does not exist for all values of $d$, but only for those lying outside a certain interval.

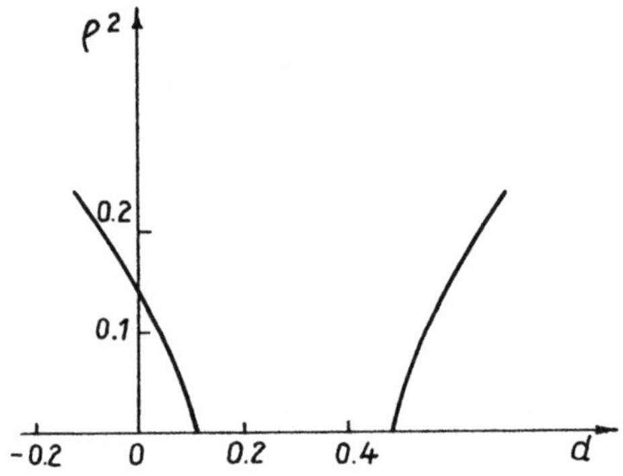


*Fig. 93.* The dependence of the amplitude of a mixed subharmonic and parametric oscillator on the parameter $d$.

In order to study the stability of the stationary oscillations, we use the variational equation

$$\ddot{\eta} + \left[\nu^2 + \varepsilon\left(\sigma + 2d\cos 2\nu t + 2c_2 x_0 + 3c_3 x_0^2\right) - \frac{\varepsilon^2\delta^2}{4}\right]\eta = 0. \tag{4.28}$$

Substituting the expression $x_0$ from (4.24), we can write the Equation (4.28) in the form:

$$\ddot{\eta} + \left[\theta_0 + 2\varepsilon\sum_{n=1}^{4}\theta_n\cos(n\nu t - \varepsilon_n)\right]\eta = 0, \tag{4.29}$$

where

$$
\begin{aligned}
\theta_0 &= \nu^2 + \varepsilon(\sigma + S) + \frac{3}{2}\varepsilon c_3 A_0^2 - \frac{\varepsilon^2\delta^2}{4}, \quad \theta_n^2 = \theta_{ns}^2 + \theta_{nc}^2, \\
\theta_{1s} &= Ky - \frac{3}{2}c_3 B_2(y\cos\beta + z\sin\beta), \\
\theta_{1c} &= Kz - \frac{3}{2}c_3 B_2(y\sin\beta - z\cos\beta), \\
\theta_{2s} &= -K\sin\beta + \frac{3}{2}c_3 yz, \\
\theta_{2c} &= d + K\cos\beta - \frac{3}{4}c_3(y^2 - z^2), \\
\theta_{3s} &= \frac{3}{2}c_3 B_2(y\cos\beta - z\sin\beta), \\
\theta_{3c} &= \frac{3}{2}c_3 B_2(y\sin\beta + z\cos\beta), \\
\theta_4 &= \frac{3}{2}c_3 B_2^2.
\end{aligned}
\tag{4.30}
$$

Now, the stability conditions are

$$
\left[\theta_0 - \left(\frac{n\nu}{2}\right)^2\right]^2 + 2\left[\theta_0 + \left(\frac{n\nu}{2}\right)^2\right]\frac{\varepsilon^2\delta^2}{4} + \frac{\varepsilon^4\delta^4}{16} > \varepsilon^2\theta_n^2, \tag{4.31}
$$

$n = 1, 2, 3, 4.$

For sufficiently small value of $\varepsilon$ the conditions (4.31) are satisfied for $n \neq 2$. In the case of $n = 2$, the inequality (4.31) yields:

$$
\begin{aligned}
&\left(\sigma + \frac{3}{4}c_3 A_0^2 + S\right)^2 + \delta^2\nu^2 > K^2 + d^2 + 2Kd\cos\beta + \frac{9}{16}c_3^2 A_0^4 - \\
&- \frac{3}{2}c_3 d(y^2 - z^2) - \frac{3}{2}Kc_3\left[2yz\sin\beta + (y^2 - z^2)\cos\beta\right].
\end{aligned}
\tag{4.32}
$$

Taking into account the formula (4.27) and

$$
\left(\sigma + S + \frac{3}{4}c_3 A_0^2\right)A_0^2 = 2Kyz\sin\beta + (K\cos\beta + d)(y^2 - z^2),
$$

obtained from the last two equations of the system (4.25), we can rewrite the condition (4.32) in the form:

$$
c_3\left(\sigma + S + \frac{3}{4}c_3 A_0^2\right) > 0 \tag{4.33}
$$

or

$$
\frac{\partial W_1}{\partial A_0} > 0, \tag{4.34}
$$

where

$$W_1 = 0 \tag{4.35}$$

is the equation of the amplitude curve,

$$W_1 = \left(\sigma + S + \frac{3}{4}c_3 A_0^2\right)^2 + \delta^2\nu^2 - K^2 - d^2 - 2dK\cos\beta.$$

When $A_0 = 0$, the solution (4.24) takes the form:

$$x_0 = B_0 + B_2\cos(2\nu t + \beta). \tag{4.36}$$

The stability (or instability) of the stationary value $A_0 = 0$ ($x_0 = y_0 = 0$) corresponds to the stability (or instability) of the solution (4.36).

The equation (4.28) in this case has the form:

$$\ddot{\eta} + \left\{\theta_0^0 + 2\varepsilon\left[\theta_2^0\cos(2\nu t - \varepsilon_2) + \theta_4^0\cos(4\nu t - \varepsilon_4)\right]\right\}\eta = 0, \tag{4.37}$$

where

$$\begin{aligned} &\theta_0^0 = \nu^2 + \varepsilon(\sigma + S) - \frac{\varepsilon^2\delta^2}{4}, \quad \theta_{2s}^0 = -K\sin\beta, \\ &\theta_{2c}^0 = d + K\cos\beta, \quad \theta_4^0 = \frac{3}{2}c_3B_2^2. \end{aligned} \tag{4.38}$$

As the right part of the equation (4.37) contains only cosines of even multiples of $t$, then the stability condition is:

$$\left(\theta_0^0 - n^2\nu^2\right)^2 + \frac{1}{2}\left(\theta_0^0 + n^2\nu^2\right)\varepsilon^2\delta^2 - \varepsilon^2\theta_{2n}^{02} + \frac{\varepsilon^4\delta^4}{16} > 0. \tag{4.39}$$

This condition is always satisfied for $n = 2$. For $n = 1$, it takes the form

$$(\sigma + S)^2 + \delta^2\nu^2 - K^2 - d^2 - 2Kd\cos\beta > 0. \tag{4.40}$$

The inequality (4.40) shows that the free term in the equation (4.35) must be positive. The geometric interpretation of the stability conditions (4.34) and (4.40) is similar to that presented above.

## 5. Interaction between Parametric and Forced Oscillations in Multidimensional Systems

This paragraph is devoted to the investigation of the interaction between parametric and forced nonlinear oscillations in multidimensional systems described by two nonlinear differential equations of the second order. The two modes $(x, y)$ of the system are excited by sinusoidal forces. These modes are coupled nonlinearly by means of the product of their coordinates. Under certain conditions, the oscillation of the first mode ($x$) parametrically excites the oscillation of the second one, and so the two oscillations of the second mode (parametric and forced oscillations) may coexiste and interact.

We shall now consider the stationary oscillations of the modes and their stability.

Let us consider oscillations of the system with two degrees of freedom described by a set of two differential equations of the type [37,e]:

$$\begin{aligned} &\ddot{x} + \lambda^2 x + \varepsilon\lambda^2(h_0\dot{x} + \alpha x^3 + cy^2x) = Q\sin\gamma t, \\ &\ddot{y} + \omega^2 y + \varepsilon\omega^2(h\dot{y} + \beta y^3 + bx^2y) = \varepsilon\omega^2 p\cos(\nu t + \delta), \end{aligned} \tag{5.1}$$

where $h_0 > 0$, $h > 0$, $\alpha$, $c$, $b$, $Q$, $p > 0$, $\beta$, $\delta$ are constants and $\varepsilon$ is a small parameter.

We assume the following relations between the frequencies $\omega$, $\nu$ and $\gamma$:

$$\omega^2 = \sigma^2\nu^2 + \varepsilon\omega^2\Delta, \quad \gamma = e\nu, \quad n\lambda \neq m\gamma, \tag{5.2}$$

where $\sigma$, $e$ are rational numbers, $\Delta$ is detuning of frequencies, and $m$, $n$ are integers.

First, we transform the equations (5.1) by means of the formulae

$$\begin{aligned} &x = q\sin\gamma t + a_1\cos\theta_1, \\ &\dot{x} = \gamma q\cos\gamma t - \lambda a_1\sin\theta_1, \\ &y = a\cos\theta, \\ &\dot{y} = -\sigma\nu a\sin\theta, \quad q = \frac{Q}{\lambda^2 - \gamma^2}, \end{aligned} \tag{5.3}$$

where $a_1, \theta_1, a, \theta$ are new variables which will be determined later.

Substituting (5.3) into (5.1) and transforming it, we obtain the following equations in the standard form:

$$\begin{aligned} \frac{da_1}{dt} &= \varepsilon\lambda(h_0\dot{x} + \alpha x^3 + cy^2x)\sin\theta_1, \\ a_1\frac{d\psi_1}{dt} &= \varepsilon\lambda(h_0\dot{x} + \alpha x^3 + cy^2x)\cos\theta_1, \\ \frac{da}{dt} &= \varepsilon\sigma\nu\Phi(x, y, \dot{y}, t)\sin\theta + O(\varepsilon^2), \\ \frac{ad\psi}{dt} &= \varepsilon\sigma\nu\Phi(x, y, \dot{y}, t)\cos\theta + O(\varepsilon^2), \end{aligned} \tag{5.4}$$

where

$$\begin{aligned} &\Phi(x, y, \dot{y}, t) = \Delta y + h\dot{y} + bx^2y + \beta y^3 - p\cos(\nu t + \delta), \\ &\psi_1 = \theta_1 - \lambda t, \quad \psi = \theta - \sigma\nu t. \end{aligned} \tag{5.5}$$

Averaging the right-hand parts of (5.4) over time, we have the equations of the first approximation for the unknowns $a$, $a_1$, $\psi$:

1) for $\sigma \neq e, \quad \sigma \neq 1$

$$\begin{aligned}\dot{a}_1 &= -\frac{\varepsilon\lambda h_0}{2}a_1,\\ \dot{a} &= -\frac{\varepsilon\sigma^2\nu^2}{2}ha + \dots,\\ a\dot{\psi} &= \varepsilon\sigma\nu\left[\frac{\Delta a}{2} + \frac{b}{4}q^2 - a + \frac{3}{8}\beta a^3\right] + \dots\end{aligned} \tag{5.6}$$

2) for $\sigma = 1, \quad e \neq 1$

$$\begin{aligned}\dot{a}_1 &= -\frac{\varepsilon\lambda}{2}h_0 a_1,\\ \dot{a} &= -\frac{\varepsilon\nu}{2}[\nu ha + p\sin(\psi - \delta)] + \dots,\\ a\dot{\psi} &= \frac{\varepsilon\nu}{2}[\Delta a + \frac{b}{2}q^2 a + \frac{3}{4}\beta a^3 - p cos(\psi - \delta)] + \dots\end{aligned} \tag{5.7}$$

3) for $e = \sigma \neq 1$

$$\begin{aligned}\dot{a}_1 &= -\frac{\varepsilon\lambda}{2}h_0 a_1,\\ \dot{a} &= -\frac{\varepsilon\sigma\nu}{2}[\sigma\nu ha + \frac{b}{4}q^2 a\sin 2\psi] + \dots,\\ a\dot{\psi} &= \frac{\varepsilon\sigma\nu}{2}\left[\Delta a + \frac{b}{2}q^2 a + \frac{3}{4}\beta a^3 - \frac{b}{4}q^2 a\cos 2\psi\right] + \dots\end{aligned} \tag{5.8}$$

4)for $e = \sigma = 1$

$$\begin{aligned}\dot{a}_1 &= -\frac{\varepsilon\lambda}{2}h_0 a_1,\\ \dot{a} &= -\frac{\varepsilon\nu}{2}[\nu ha + \frac{b}{4}q^2 a\sin 2\psi + p\sin(\psi - \delta)] + \dots,\\ a\dot{\psi} &= \frac{\varepsilon\nu}{2}\left[\Delta a + \frac{b}{2}q^2 a + \frac{3}{4}\beta a^3 - \frac{b}{4}q^2 a\cos 2\psi - p\cos(\psi - \delta)\right] + \dots,\end{aligned} \tag{5.9}$$

where the higher terms disappear when $a_1 = 0$. The first case is called the non-resonance case, the second - the harmonic resonance case, the third (5.8) - the parametric resonance case, and the last (5.9) - the harmonic and parametric resonance case. Obviously, the most interesting is the last resonance case. We shall investigate it in more detail.

The stationary solution of the system (5.9) is the one which is determined from the equations $\dot{a}_1 = \dot{a} = \dot{\psi} = 0$, or:

$$\begin{aligned}a_1 &= 0,\\ \nu ha + \frac{b}{4}q^2 a\sin 2\psi + p\sin(\psi - \delta) &= 0,\\ \Delta a + \frac{b}{2}q^2 a + \frac{3}{4}\beta a^3 - \frac{b}{4}q^2 a\cos 2\psi - p\cos(\psi - \delta) &= 0.\end{aligned} \tag{5.10}$$

Eliminating $\psi$ from the last two equations of (5.10), we obtain the following relation for the amplitude $a = a_0 = \text{const}$ of the stationary oscillation of the coordinate $y$:

$$M = 0, \tag{5.11}$$

where

$$\begin{aligned} M &\equiv (w^2 + u^2 - v^2)^2 - p^2\big[(w - v\cos 2\delta)^2 + (u + v\sin 2\delta)^2\big], \\ u &= \nu h a_0, \quad v = \frac{b}{4}q^2 a_0, \quad w = -a_0\Big(\Delta + \frac{b}{2}q^2 + \frac{3}{4}\beta a_0^2\Big). \end{aligned} \tag{5.12}$$

The relation (5.11) is expressed in the parameters of the initial system as:

$$\begin{aligned} &a_0^2\Big[\Big(\frac{1-k^2}{\varepsilon} + \frac{b}{2}q^2 + \frac{3}{4}\beta a_0^2\Big)^2 + \omega^2 h^2 - \frac{b^2}{16}q^4\Big]^2 - p^2\Big[\Big(\frac{1-k^2}{\varepsilon} + \\ &+ \frac{b}{2}q^2 + \frac{3}{4}\beta a_0^2 + \frac{b}{4}q^2\cos 2\delta\Big)^2 + \Big(\omega h + \frac{b}{4}q^2 \sin 2\delta\Big)^2\Big] = 0, \\ &k = \frac{\nu}{\omega}. \end{aligned} \tag{5.13}$$

Following this formula, the resonance curves are presented in Figs. 94-99.

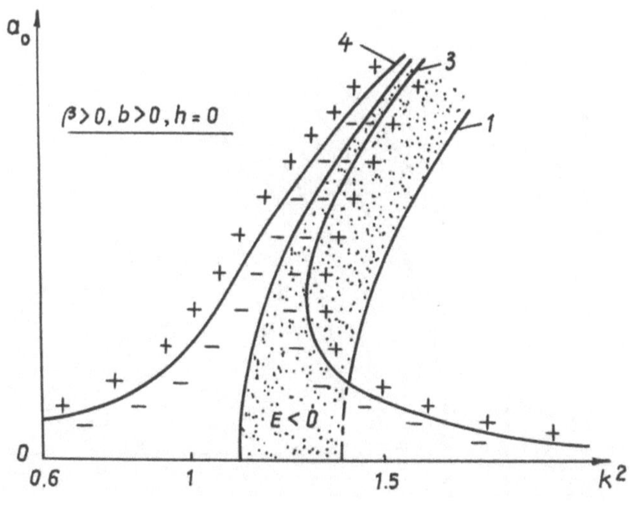


*Fig. 94.* Amplitude curves of mixed parametric and forced oscillation in absence of friction and for the case $\beta > 0$, $b > 0$.

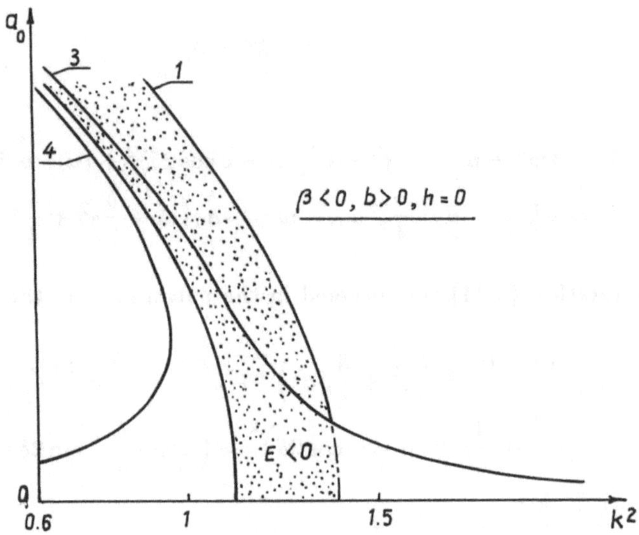


*Fig. 95.* **Amplitude curves of mixed parametric and forced oscillation in absence of friction and for the case $\beta < 0$, $b > 0$.**

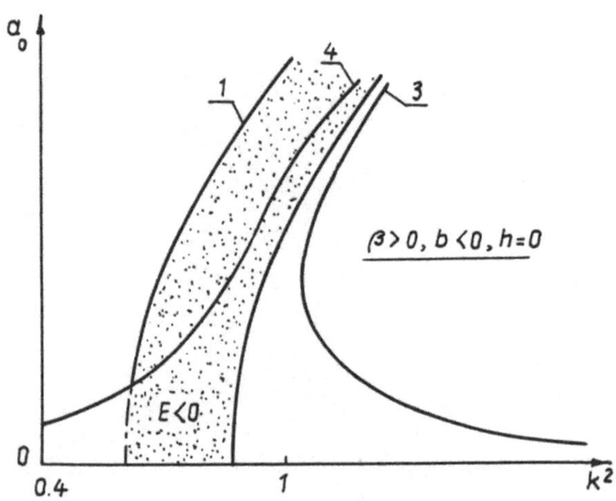


*Fig. 96.* **Amplitude curves of mixed parametric and forced oscillation in absence of friction and for the case $\beta > 0$, $b < 0$.**

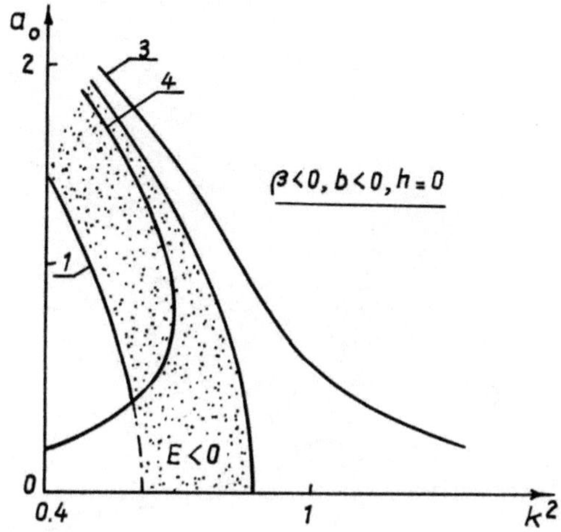


*Fig. 97.* Amplitude curves of mixed parametric and forced oscillation in the absence of friction and for the case $\beta < 0$, $b < 0$.

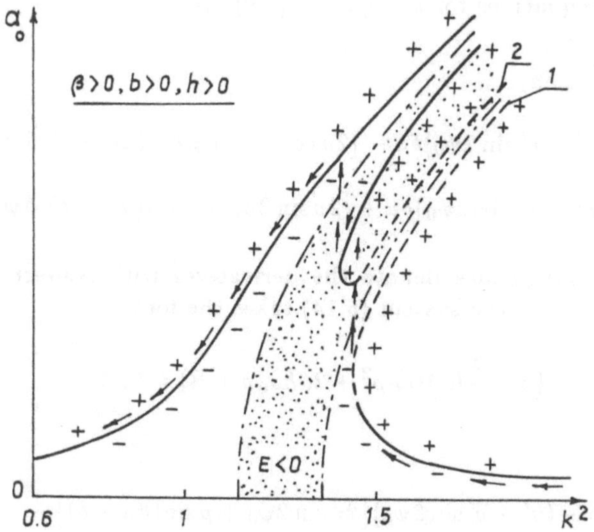


*Fig. 98.* Amplitude change with slowly varying frequency of external excitation in the case $\beta > 0$, $b > 0$, $h > 0$.

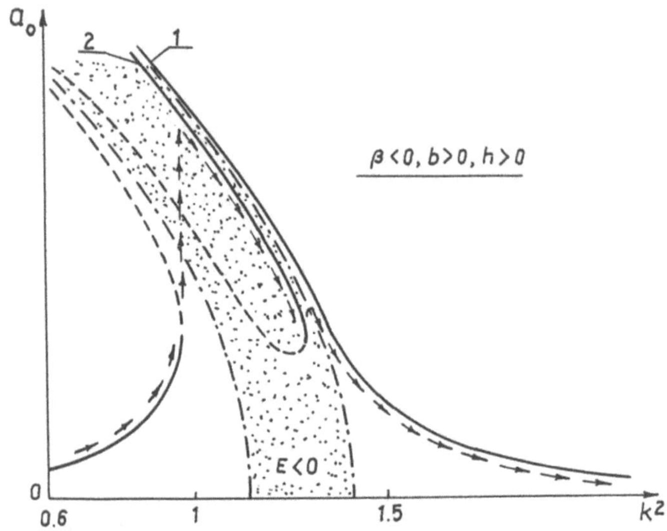


*Fig. 99.* **Amplitude change with slowly varying frequency of external excitation in the case $\beta < 0$, $b > 0$, $h > 0$.**

The variational equations for the system (5.9) are:

$$\begin{aligned}
\frac{d\delta a_1}{d\tau} &= -\frac{\lambda}{\nu} h_0 \delta a_1, \\
\frac{d\delta a}{d\tau} &= -(u' + v' \sin 2\psi_0)\delta a - \left[2v \cos 2\psi_0 + p\cos(\psi_0 - \delta)\right]\delta\psi, \\
a_0 \frac{d\delta\psi}{d\tau} &= -(w' + v' \cos 2\psi_0)\delta a + \left[2v \sin 2\psi_0 + p\sin(\psi_0 - \delta)\right]\delta\psi,
\end{aligned} \tag{5.14}$$

where $\tau = 2t/\varepsilon\nu$ and primes denote the derivatives with respect to $a_0$. The characteristic equation of the system (5.14) takes the form:

$$\left(\rho + \frac{\lambda}{\nu} h_0\right)\left(a_0\rho^2 + 2\nu h a_0 \rho + R\right) = 0,$$

where

$$\begin{aligned}
R = &-(u' + v' \sin 2\psi_0)\left[2v \sin 2\psi_0 + p\sin(\psi_0 - \delta)\right] - \\
&- (w' + v' \cos 2\psi_0)\left[2v \cos 2\psi_0 + p\cos(\psi_0 - \delta)\right].
\end{aligned} \tag{5.15}$$

By using the relations (5.10), (5.12), we may rewrite the expression $R$ in the form:

$$R = \frac{1}{2(w^2 + u^2 - v^2)} \frac{\partial M}{\partial a_0}. \tag{5.16}$$

As $h_0 > 0$, $h > 0$, $a_0 > 0$, the stability condition of the stationary solution is:

$$E\frac{\partial M}{\partial a_0} > 0, \tag{5.17}$$

where $E = w^2 + u^2 - v^2$. The resonance curves ($M = 0$) divide the plane $(a_0, k^2)$ into regions, in each of which the expression $M$ has a definite sign (+ or −). If moving up along the straight line parallel to the axis $a_0$, we pass from a region $M < 0$ to a region $M > 0$, then at the point of intersection between the straight line and the resonance curve the derivative $\partial M/\partial a_0$ is positive. So, this point corresponds to a stable state of oscillation if $E = w^2 + u^2 - v^2 > 0$ and to an unstable one if $E < 0$. On the contrary, if we pass from a region $M > 0$ to a region $M < 0$, then the point of intersection corresponds to a stable state of oscillation if $E < 0$ and to an unstable one if $E > 0$.

In a limiting case when $h = \delta = 0$ the equation for the stationary amplitude $a_0$ (5.13) may be written in the form:

$$\left(\frac{1-k^2}{\varepsilon} + \frac{3bq^2}{4} + \frac{3}{4}\beta a_0^2\right)^2\left[\left(\frac{1-k^2}{\varepsilon} + \frac{bq^2}{4} + \frac{3}{4}\beta a_0^2\right)^2 - \frac{p^2}{a^2}\right] = 0. \tag{5.18}$$

From here, we obtain a double root

$$k^2 = 1 + \frac{3}{4}\varepsilon bq^2 + \frac{3}{4}\varepsilon\beta a_0^2, \tag{5.19}$$

and two other roots

$$k^2 = 1 + \frac{\varepsilon}{4}bq^2 + \frac{3}{4}\varepsilon\beta a_0^2 + \frac{\varepsilon p}{a_0}\,, \tag{5.20}$$

$$k^2 = 1 + \frac{\varepsilon}{4}bq^2 + \frac{3}{4}\varepsilon\beta a_0^2 - \frac{\varepsilon p}{a_0}\,, \tag{5.21}$$

Following these formulae, the resonance curves are presented in Figs. 94-97, where the branch expressed by (5.19) is shown by the number 1, and the branches expressed by (5.20) and (5.21) are shown by the numbers 3 and 4, respectively. In the stippled region (the region of parametric resonance), the expression $E = w^2 + u^2 - v^2$ is negative. On the heavy lines, $E\partial M/\partial a_0$ is positive, so that they correspond to a stable state of oscillations. The broken lines correspond to the unstable state of oscillations. The signs + and − in the figures are those for the expression $M$. On the branch 1 of the resonance curves when $h = 0$, the stability of stationary oscillation is doubtful because $E = 0$ on this branch.

If we introduce a small value of friction $h$ ($h > 0$), the resonance curves have the form shown in Figs. 98, 99. These curves are obtained by solving the equation (5.13). The branch 1 in the case of $h = 0$ (Figs. 94-97) changes into two either stable (Fig. 99) or unstable branches (Fig. 98). With larger values of friction (h) the resonance curves take the forms presented in Fig. 100 and Fig. 101.

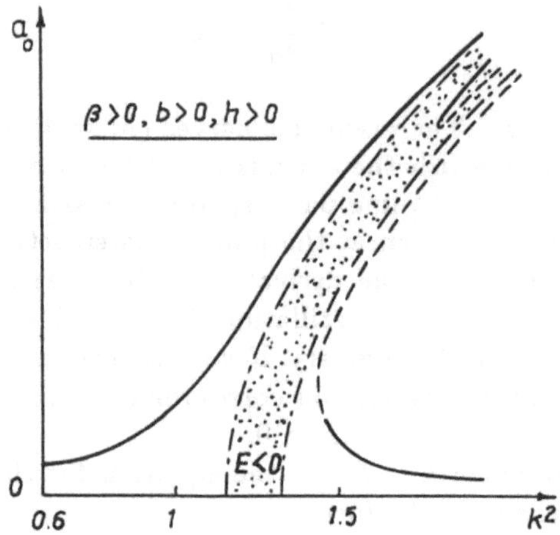


*Fig. 100.* Amplitude curve for the case of high friction and $\beta > 0$, $b > 0$, $h > 0$.

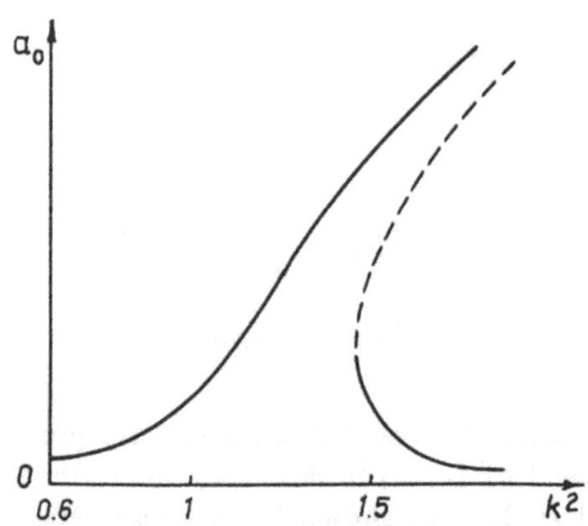


*Fig. 101.* Amplitude curve for the case of higher friction.

From these results, the following conclusions may be drawn:

1) Inside the region of parametric resonance, the parametric excitation caused by the first mode ($x$) strongly influences the stability of the stationary forced oscillations of the second mode ($y$). Some branches of the resonance curves of the second mode ($y$) which are unstable for $x = 0$ now become stable, and *vice versa.* Outside the region of parametric resonance, the mechanism of parametric excitation does not influence the stability of the forced oscillations.

2) The jump phenomenon of the amplitudes is observed for a hardening characteristic (Fig. 98), when the frequency $k$ decreases, and for a softening characteristic

(Figs. 99), when the frequency increases. The change of the stationary amplitudes follows the $M$ - form.

**Note.** To eliminate the phase $\psi$ between equations

$$v \sin 2\psi + p \sin(\psi - \delta) + u = 0, \tag{1}$$

$$v \cos 2\psi + p \cos(\psi - \delta) - w = 0, \tag{2}$$

we first multiply the equation (1) with $\sin(\psi-\delta)$ and the equation (2) with $\cos(\psi-\delta)$ then add them. Since

$$\sin(\psi + \delta) = \sin(\psi - \delta)\cos 2\delta + \cos(\psi - \delta)\sin 2\delta,$$
$$\cos(\psi + \delta) = \cos(\psi - \delta)\cos 2\delta - \sin(\psi - \delta)\sin 2\delta,$$

we have

$$(u - v\sin 2\delta)\sin(\psi - \delta) + (-w + v\cos 2\delta)\cos(\psi - \delta) = -p. \tag{3}$$

Similarly, multiplying the equation (1) with $\cos(\psi - \delta)$ and the equation (2) with $-\sin(\psi - \delta)$ then adding them, we have

$$(w + v\cos 2\delta)\sin(\psi - \delta) + (u + v\sin 2\delta)\cos(\psi - \delta) = 0. \tag{4}$$

Solving the equations (3) and (4) with respect to $\sin(\psi - \delta)$ and $\cos(\psi - \delta)$ and using the relation $\sin^2(\psi - \delta) + \cos^2(\psi - \delta) = 1$, we obtain the expression (5.12).

# CHAPTER 6

## AVERAGING METHOD

As is generally known, the averaging method first appeared and was developed in celestial mechanics. To solve these problems, various schemes of averaging were used, connected with the names Gauss, Fate, Delone-Hill and others. At this stage, the basic way of averaging was to replace the right-hand sides of complicated differential equations by averaged functions which did not explicitly contain either the time t or rapidly changing parameters of the system. The resulting equations were either exactly integrable or greatly simplified, and could be used to draw important qualitative and quantitative conclusions about the motion.

However, in the theory of nonlinear oscillations, the method of averaging was unknown for a long time. Only with the works of Van der Pol and the popularization by Mandelstam and Papaleksi, was the method used for study of nonlinear oscillatory processes in radio, electronics and mechanics begin.

In 1835, Ostrogradski wrote a note entitled "On a method of successive approximations", relating to a nonlinear equation with a cubic characteristic, and obtained the same result as that found by the method of averaging.

And still earlier, in 1682 Newton in his work "The mathematical principles of natural philosophies" found the formula governing the damping of small oscillations of a pendulum in an arbitrary resistive medium. In the first approximation, Newton's formula coincides with that obtained by the method of averaging.

In the beginning of the second decade of this century, the Dutch engineer Van der Pol gave a qualitatively new approach to the study of oscillatory motions and opened a new and very important domain in the theory of nonlinear oscillations.

In the first stage of the development of nonlinear mechanics, Van der Pol's method was used very successfully in practice. However, Van der Pol's method possesses considerable deficiencies. First, the approximations in the form given by Van der Pol were found intuitively. They are doubtful from a mathematical point of view. Van der Pol's method does not give the necessary information about its degree of exactitude, and about the limit of its application. Moreover, within this method, one cannot define more exact solutions, beside the first approximation. Another deficiency of Van der Pol's method is that it cannot be used for studying dynamic systems with several degrees of freedom, or systems with distributed parameters.

Considerable progress in the development of the method of averaging was made by Krylov and Bogoliubov. In 1937, they proved the applicability of the method of averaging, even for the case when the right-hand sides of the differential equations under consideration are quasi periodic functions of time.

The creation of the strict theory of the method of averaging belongs to Bogoliubov. He showed that this method is organically connected with the existence of a substitution of variables which allows the time $t$ to be eliminated from the right-hand sides of the equations with any degree of exactness relative to the small parameter $\varepsilon$.

By then, proceeding from delicate physical considerations, Bogoliubov showed how to build up not only the first order equations, but also the higher order equations, which give the solutions of the original equations with any given accuracy.

## 1. The Idea of Averaging by Bogoliubov

We give a brief description of the idea of averaging proposed by Bogoliubov which is connected with a special change of variables. More details of the averaging method can be found in Bogoliubov and Mitropolskii [6].

Let us consider a differential equation in the vectorial form

$$\frac{dx}{dt} = \varepsilon X(t, x), \tag{1.1}$$

where $\varepsilon$ is a small positive parameter, $t$ is time and $x$ is a point of $n$ - dimension Euclidean space $\mathbf{R}^n$: $x = (x_1, \dots, x_n)$, $X = (X_1, \dots, X_n)$. This structure is found frequently in the study of the process of oscillation, in particular, the equations for amplitude and phase of oscillations in quasi-linear systems. The $x$ - variable is slowly varying because its derivative is proportional to $\varepsilon$. This form of the equation of motion is known as the *standard* form.

We form a $m$-th approximation for the solution of equation (1.1):

$$x = \xi + \varepsilon F_1(t, \xi) + \varepsilon^2 F_2(t, \xi) + \cdots + \varepsilon^m F_m(t, \xi). \tag{1.2}$$

Using this equation as the formula for changing variables, one can reduce (1.1) into the following exact equation

$$\frac{d\xi}{dt} = \varepsilon X_0(\xi) + \varepsilon^2 P_2(\xi) + \cdots + \varepsilon^m P_m(\xi) + \varepsilon^{m+1} R(t, \xi, \varepsilon). \tag{1.3}$$

Neglecting the term $\varepsilon^{m+1} R(t, \xi, \varepsilon)$, we derive the *averaged* equation of the m-th approximation:

$$\frac{d\xi}{dt} = \varepsilon X_0(\xi) + \varepsilon^2 P_2(\xi) + \cdots + \varepsilon^m P_m(\xi), \tag{1.4}$$

which does not contain time as an explicit term and is simpler than the initial equation (1.1). The functions $X_0(\xi)$, $P_2(\xi)$, ..., $P_m(\xi)$ are determined as the results of averaging the right-hand side of (1.1) after substituting (1.2).

We now derive the physical interpretation of the averaging method, and its strict formulation.

The following notation will be used. Assuming that

$$X(t, x) = \sum_n e^{int} X_n(x), \tag{1.5}$$

we have

$$\underset{t}{M}\{X(t,x)\} = \langle X(t,x)\rangle = X_0(x),$$
$$\tilde{X}(t,x) = \sum_{n\neq 0} \frac{e^{int}}{in} X_n(x), \tag{1.6}$$
$$\approx{X}(t,x) = \sum_{n\neq 0} \frac{e^{int}}{(in)^2} X_n(x).$$

Obviously,

$$\frac{\partial \approx{X}}{\partial t} = \tilde{X}, \quad \frac{\partial \tilde{X}}{\partial t} = X(t,x) - \langle X(t,x)\rangle. \tag{1.7}$$

The wave operator ($\sim$) is called the integrating operator and $\underset{t}{M}$ or $\langle M\rangle$ is the averaging operator over explicit time, considering $x$ as constant.

The initial equation (1.1) is now

$$\frac{dx}{dt} = \varepsilon X(t,x) = \varepsilon \sum_n e^{int} X_n(x). \tag{1.8}$$

It is noted that the form of the approximate solution of (1.8) can be found, or more precisely can be guessed, by the aid of intuition. Namely, since $x$ is slowly varying, it can be represented as a superposition of a smoothly changing term $\xi$, and small oscillatory terms. In view of the smallness of the last terms, one can assume in the first approximation $x = \xi$ in the right-hand side of (1.8). Then, we have

$$\frac{dx}{dt} = \varepsilon X(t,\xi) = \varepsilon X_0 + \varepsilon \sum_{n\neq 0} e^{int} X_n(\xi), \tag{1.9}$$

which contains small sinusoidal oscillatory components. Considering these oscillatory terms to excite only small oscillations of $x$ close to $\xi$ and which do not influence the systematic change of $x$, one has the equation of the first approximation in the form

$$\frac{d\xi}{dt} = \varepsilon X_0(\xi) = \varepsilon\langle X(t,\xi)\rangle. \tag{1.10}$$

To obtain the second approximation for $x$, we must consider the oscillatory terms in the expression for $x$. Considering the term $\varepsilon e^{int} X_n(\xi)$ in (1.9) to excite an oscillation of the form

$$\frac{\varepsilon e^{int}}{in} X_n(\xi),$$

we obtain the following approximate expression

$$x = \xi + \varepsilon \sum_{n\neq 0} \frac{e^{int}}{in} X_n(\xi) = \xi + \varepsilon \tilde{X}(t,\xi). \tag{1.11}$$

Substituting (1.11) into equation (1.8), we have

$$\frac{dx}{dt} = \varepsilon X(t, \xi + \varepsilon \tilde{X}), \tag{1.12}$$

i.e.

$$\frac{dx}{dt} = \varepsilon \langle X(t, \xi + \varepsilon \tilde{X}) \rangle + \text{small sinusoidal oscillatory terms.}$$

Neglecting the influence of the sinusoidal oscillatory terms on the systematic change of $\xi$, we obtain the equation of the second approximation

$$\frac{d\xi}{dt} = \varepsilon \langle X(t, \xi + \varepsilon \tilde{X}) \rangle = \varepsilon \langle X(t, \xi) + \varepsilon \Big( \tilde{X} \frac{\partial}{\partial \xi} \Big) X(t, \xi) \rangle \tag{1.13}$$

and so on.

Obviously, the approximations in the above arguments affect their validity. For example, there could be an objection that in the derivation of the approximate equation (1.10), terms of the same degree of smallness as the retained term $\varepsilon X_0$ are neglected. However, one can give those approximations a more well-founded form and one can show that in the transformed equations the quantities of higher degree of smallness are neglected. With this end, we use in (1.1) the change of variables

$$x = \xi + \varepsilon \tilde{X}(t, \xi), \tag{1.14}$$

where $\xi$ is a new unknown.

Differentiating (1.14), we have

$$\frac{dx}{dt} = \frac{d\xi}{dt} + \varepsilon \frac{\partial \tilde{X}}{\partial \xi} \frac{d\xi}{dt} + \varepsilon \frac{\partial \tilde{X}}{\partial t} \,. \tag{1.15}$$

Taking into account the expression (1.7) for the integrating operator, we obtain:

$$\frac{\partial \tilde{X}}{\partial t} = X(t, \xi) - X_0(\xi).$$

Substituting (1.14) and (1.15) into (1.1), we get

$$\frac{d\xi}{dt} + \varepsilon \frac{\partial \tilde{X}}{\partial \xi} \frac{d\xi}{dt} + \varepsilon X(t, \xi) - \varepsilon X_0(t) = \varepsilon X\{t, \xi + \varepsilon \tilde{X}(t, \xi)\},$$

or

$$\Big\{ E + \varepsilon \frac{\partial \tilde{X}}{\partial \xi} \Big\} \frac{d\xi}{dt} = \varepsilon X_0(\xi) + \varepsilon \{ X(t, \xi + \varepsilon \tilde{X}) - X(t, \xi) \}, \tag{1.16}$$

where $E$ is the unit matrix.

Multiplying (1.16) from the left by the expression

$$\left\{E+\varepsilon\frac{\partial\tilde{X}}{\partial\xi}\right\}^{-1}, \tag{1.17}$$

we obtain the following equation for the new variable :

$$\frac{d\xi}{dt}=\varepsilon\left\{E+\varepsilon\frac{\partial\tilde{X}}{\partial\xi}\right\}^{-1}X_0(\xi)+ \\ +\varepsilon\left\{E+\varepsilon\frac{\partial\tilde{X}}{\partial\xi}\right\}^{-1}\{X(t,\xi+\varepsilon\tilde{X}(t,\xi))-X(t,\xi)\}. \tag{1.18}$$

Further, expanding (1.17) in a power series in $\varepsilon$, we have:

$$\left\{E+\varepsilon\frac{\partial\tilde{X}}{\partial\xi}\right\}^{-1}=E-\varepsilon\frac{\partial\tilde{X}}{\partial\xi}+O(\varepsilon^2).$$

Hence, equation (1.18) gives

$$\frac{d\xi}{dt}=\varepsilon X_0(\xi)+O(\varepsilon^2) \tag{1.19}$$

or in more detail

$$\frac{d\xi}{dt}=\varepsilon X_0(\xi)-\varepsilon^2\frac{\partial\tilde{X}}{\partial\xi}X_0(\xi)+\varepsilon^2\left(\tilde{X}\frac{\partial}{\partial\xi}\right)X(t,\xi)+O(\varepsilon^3).$$

Neglecting the terms of second degree of smallness, we obtain the so called equations of the first approximation

$$\frac{d\xi}{dt}=\varepsilon X_0(\xi). \tag{1.20}$$

Thus, if $\xi$ satisfies the equations (1.19) (exact equations), whose right-hand side differs from the right-hand side of (1.20) by second order terms, then the expression

$$x=\xi+\varepsilon\tilde{X}(t,\xi) \tag{1.21}$$

is the exact solution of the initial system of equations (1.1). Therefore, as the first approximation, one can take

$$x=\xi,$$

where $\xi$ is the solution of the equation of the first approximation (1.20).

The expression (1.21), in which $\xi$ satisfies the equation (1.20), is called the refinement of the first approximation.

Substituting (1.21) in the initial equation (1.1), one can verify that (1.21) satisfies (1.1) up to the second order.

Returning to the equation (1.20), we note that according to the definition of the averaging operator

$$X_0(\xi) = \langle X(t, \xi) \rangle,$$

we have the equation of the first approximation in the form

$$\frac{d\xi}{dt} = \varepsilon \langle X(t, \xi) \rangle. \tag{1.22}$$

Therefore, the equation of the first approximation is obtained from the initial equation (1.1) by averaging it over explicit time $t$. When averaging, we consider the quantity $\xi$ to be constant.

This formal process, consisting of replacing the exact equations by the averaged ones, is sometimes called the principle of averaging.

As we can verify further, the principle of averaging does not require that $X(t, \xi)$ be represented by the sum (1.5). Only the existence of the mean value

$$X_0(\xi) = \lim_{T \to \infty} \frac{1}{T} \int_0^T X(t, \xi) dt \tag{1.23}$$

is important.

It is worth mentioning that, in this or that form, the principle of averaging has been used to obtain approximate solutions in many applications. However, the mathematical foundation of this principle is due to Bogoliubov.

We now pass to the construction of the second approximation. We first note that in the construction of the first approximation, the equation (1.1) was transformed into (1.19) by means of the formula (1.21). To obtain the second approximation, we find an analogous formula for changing the variable $x$ to $\xi$, which satisfies an equation of the following form

$$\frac{d\xi}{dt} = \varepsilon X_0(\xi) + \varepsilon^2 P(\xi) + O(\varepsilon^3). \tag{1.24}$$

To obtain the change of variables in the most natural way, we find the expression

$$x = \Phi(t, \xi, \varepsilon), \tag{1.25}$$

which satisfies (1.1) with accuracy $O(\varepsilon^3)$, where $\xi$ is the solution of the equation of the type

$$\frac{d\xi}{dt} = \varepsilon X_0(\xi) + \varepsilon^2 P(\xi). \tag{1.26}$$

If, for $\xi$ determined by the equation of the first approximation

$$\frac{d\xi}{dt} = \varepsilon X_0(\xi),$$

the expression

$$x = \xi + \varepsilon \sum_{\nu \neq 0} \frac{e^{i\nu t}}{i\nu} X_\nu(\xi) = \xi + \varepsilon \tilde{X}(t, \xi),$$

satisfies the equation (1.1) with accuracy $O(\varepsilon^2)$, then the solution (1.25) will be found in the following form

$$x = \xi + \varepsilon \tilde{X}(t, \xi) + \varepsilon^2 F(t, \xi), \tag{1.27}$$

where $F$ is a summation of the form

$$F(t, \xi) = \sum_\nu e^{i\nu t} F_\nu(\xi). \tag{1.28}$$

But for (1.27)

$$\begin{aligned} \varepsilon X(t, x) &= \varepsilon X(t, \xi + \varepsilon \tilde{X}) + O(\varepsilon^2) = \\ &= \varepsilon X(t, \xi) + \varepsilon^2 \Big(\tilde{X} \frac{\partial}{\partial \xi}\Big) X(t, \xi) + O(\varepsilon^3). \end{aligned} \tag{1.29}$$

On the other hand, for $\xi$ determined by (1.26), by differentiation of (1.27), we obtain

$$\begin{aligned} \frac{dx}{dt} &= \frac{d\xi}{dt} + \varepsilon \frac{\partial \tilde{X}(t, \xi)}{\partial \xi} \frac{d\xi}{dt} + \varepsilon^2 \frac{\partial F(t, \xi)}{\partial \xi} \frac{d\xi}{dt} + \varepsilon \frac{\partial \tilde{X}(t, \xi)}{\partial t} + \varepsilon^2 \frac{\partial F(t, \xi)}{\partial t} = \\ &= \varepsilon X_0(\xi) + \varepsilon^2 P(\xi) + \varepsilon^2 \frac{\partial \tilde{X}(t, \xi)}{\partial \xi} X_0(\xi) + \varepsilon \frac{\partial \tilde{X}(t, \xi)}{\partial t} + \varepsilon^2 \frac{\partial F(t, \xi)}{\partial t} + O(\varepsilon^3). \end{aligned}$$

From here, we have

$$\frac{dx}{dt} = \varepsilon X(t, \xi) + \varepsilon^2 P(\xi) + \varepsilon^2 \frac{\partial \tilde{X}(t, \xi)}{\partial \xi} X_0(\xi) + \varepsilon^2 \frac{\partial F(t, \xi)}{\partial t} + O(\varepsilon^3), \tag{1.30}$$

since

$$\frac{\partial \tilde{X}(t, \xi)}{\partial t} = X(t, \xi) - X_0(\xi).$$

Thus, the expression (1.29) is equal to (1.30) with accuracy $O(\varepsilon^3)$ if the functions $P(\xi)$ and $F(t, \xi)$ are chosen in this way, that the following relationship is fulfilled:

$$\frac{\partial F(t, \xi)}{\partial t} = \Big(\tilde{X} \frac{\partial}{\partial \xi}\Big) X(t, \xi) - \frac{\partial \tilde{X}(t, \xi)}{\partial \xi} X_0(\xi) - P(\xi). \tag{1.31}$$

In view of

$$\begin{aligned} \tilde{X}(t, \xi) &= \sum_{\nu \neq 0} \frac{e^{i\nu t}}{i\nu} X_\nu(\xi), \\ X(t, \xi) &= \sum_\nu e^{i\nu t} X_\nu(\xi), \end{aligned} \tag{1.32}$$

we can write

$$\left(\tilde{X}\frac{\partial}{\partial\xi}\right)X(t,\xi)-\frac{\partial\tilde{X}(t,\xi)}{\partial\xi}X_0(\xi)=$$
$$=\sum_{\substack{\nu',\nu''\\ \nu'\neq 0}}e^{i(\nu'+\nu'')t}\frac{1}{i\nu'}\left(X_{\nu'}\frac{\partial}{\partial\xi}\right)X_{\nu''}(\xi)-\sum_{\nu\neq 0}\frac{e^{i\nu t}}{i\nu}\frac{\partial X_\nu(\xi)}{\partial\xi}X_0(\xi), \tag{1.33}$$

where the summation applies to all pairs $(\nu',\nu'')$ of frequencies which appear in (1.32). Hence, the expression (1.31) can be written

$$\left(\tilde{X}\frac{\partial}{\partial\xi}\right)X(t,\xi)-\frac{\partial\tilde{X}(t,\xi)}{\partial\xi}X_0(\xi)=\sum_\mu e^{i\mu t}\Phi_\mu(\xi),\quad (\mu=\nu,\ \nu'+\nu'')$$

and the relation (1.31) is fulfilled if

$$P(\xi)=\Phi_0(\xi)=\underset{t}{M}\left\{\left(\tilde{X}(t,\xi)\frac{\partial}{\partial\xi}\right)X(t,\xi)-\frac{\partial\tilde{X}}{\partial\xi}X_0(\xi)\right\}=$$
$$=\underset{t}{M}\left\{\left(\tilde{X}(t,\xi)\frac{\partial}{\partial\xi}\right)X(t,\xi)\right\}$$

and

$$F(t,\xi)=\sum_{\mu\neq 0}\frac{e^{i\mu t}}{i\mu}\Phi_\mu(\xi)=\overline{\left(\tilde{X}(t,\xi)\frac{\partial}{\partial\xi}\right)X(t,\xi)}-\frac{\partial\approx{X}}{\partial\xi}X_0(\xi). \tag{1.34}$$

Resuming, we can confirm that for $\xi$ determined by the equation

$$\frac{d\xi}{dt}=\varepsilon\underset{t}{M}\{X(t,\xi)\}+\varepsilon^2\underset{t}{M}\left\{\left(\tilde{X}(t,\xi)\frac{\partial}{\partial\xi}\right)X(t,\xi)\right\}, \tag{1.35}$$

the expression

$$x=\xi+\varepsilon\tilde{X}(t,\xi)+\varepsilon^2\overline{\left(\tilde{X}(t,\xi)\frac{\partial}{\partial\xi}\right)X(t,\xi)}-\varepsilon^2\frac{\partial\approx{X}}{\partial\xi}X_0(\xi) \tag{1.36}$$

satisfies the equation (1.1) with accuracy $0(\varepsilon^3)$, where $\widetilde{F(t,\xi)}=\int\limits_0^t F(t,\xi)dt$.

It is not difficult to show that, if the expression (1.36) is considered as the formula for changing the variables from $x$ to $\xi$, then $\xi$ satisfies the following equation

$$\frac{d\xi}{dt}=\varepsilon\underset{t}{M}\{X(t,\xi)\}+\varepsilon^2\underset{t}{M}\left\{\left(\tilde{X}\frac{\partial}{\partial\xi}\right)X(t,\xi)\right\}+O(\varepsilon^3), \tag{1.37}$$

which differs from the equation (1.35) by terms in $\varepsilon^3$.

Thus, if $\varepsilon$ satisfies the equation (1.37), then (1.36) is an exact solution of the equation (1.1).

As the second approximation, we take

$$x = \xi + \varepsilon \tilde{X}(t, \xi), \tag{1.38}$$

where $\xi$ is determined by the equation (1.35). In other words, for the second approximation, we take the form of the refinement of the first approximation in which $\xi$ satisfies the equation of the second approximation.

The expression (1.36), in which $\xi$ is determined from the equation (1.35), is the refinement of the second approximation. It satisfies the exact equation (1.1) with accuracy of order $\varepsilon^3$.

Between the exact solution (1.1) x(t) and the solution of the equation of the first approximation (1.10), there is a connection stated by Bogoliubov's theorem which is introduced here without proof.

**Bogoliubov's theorem**

If the function $X(t, x)$ satisfies the conditions

1) For some region $D$ of the space, $x, x', x''$, positive constants $M$ and $\lambda$ can be found such that for all real values $t > 0$ and for any points $x, x', x''$ of this region, the following inequalities

$$\begin{gathered} |X(t,x)| \leq M, \\ |X(t,x') - X(t,x'')| \leq \lambda |x' - x''| \end{gathered} \tag{1.39}$$

are satisfied.

2) There exists a limit

$$\lim_{T\to\infty} \frac{1}{T} \int_0^T X(t,x)dt = X_0(x),$$

which is uniform with respect to $x$ in the $D$ - region.

Then, corresponding to any small positive values of $\rho$ and $\eta$, and any large $L$, there is a positive number $\varepsilon_0$, such that if $\xi = \xi(t)$ is a solution of the equation (1.10) within the interval $0 < t < \infty$ and lying in the region $D$ with its $\rho$ - neighbourhood, the following inequality

$$|x(t) - \xi(t)| < \eta$$

is true for $0 < \varepsilon < \varepsilon_0$ in the interval $0 < t < L/\varepsilon$, where $x = x(t)$ is the solution of the equation (1.1), which coincides with $\xi(t)$ at $t = 0$ (for a proof, see [6]).

## 2. Averaging Differential Equations with Slowly Varying Parameters

Consider an oscillatory system with $N$ degrees of freedom specified by the kinetic energy T and potential energy $V$ as follows

$$T = \frac{1}{2}\sum_{i,j=1}^{N} a_{ij}(\tau)\dot{q}_i\dot{q}_j, \qquad V = \frac{1}{2}\sum_{i,j=1}^{N} b_{ij}(\tau)q_iq_j, \tag{2.1}$$

where $a_{ij}(\tau) = a_{ji}(\tau)$, $b_{ij}(\tau) = b_{ji}(\tau)$, $\tau = \varepsilon t$, and subject to the generalized forces

$$\varepsilon Q_j(\tau, \theta, q_1, \ldots, q_N, \dot{q}_1, \ldots, \dot{q}_N, \varepsilon), \quad (j = 1, 2, \ldots, N). \tag{2.2}$$

The equations of motion are then

$$\frac{d}{dt}\Big\{\sum_{i=1}^{N} a_{ij}(\tau)\dot{q}_i\Big\} + \sum_{i=1}^{N} b_{ij}(\tau)q_i = \varepsilon Q_j(\tau, \theta, q_1, \ldots, q_N, \dot{q}_1, \ldots, \dot{q}_N, \varepsilon). \tag{2.3}$$

Introducing the principal coordinates $x_1, x_2, \ldots, x_N$ according to the formulae (see appendix 1):

$$q_i = \sum_{k=1}^{N} \varphi_i^{(k)}(\tau)\xi_k, \quad (k = 1, 2, \ldots, N), \tag{2.4}$$

where $\varphi_i^{(k)}$ $(i, k = 1, 2, \ldots, N)$ are the principal functions depending on the slow variable $\tau$ and satisfying the orthogonality condition. The equations for principal coordinates have the form

$$\frac{d^2\xi_k}{dt^2} + \omega_k^2(\tau)\xi_k = \frac{\varepsilon}{m_k(\tau)} X_k(\tau, \theta, \xi_1, \ldots, \xi_N, \dot{\xi}_1, \ldots, \dot{\xi}_N, \varepsilon) \tag{2.5}$$

$(k = 1, 2, \ldots, N)$.
These equations may be transformed into the standard form, by using new conjugate complex variables $z_k$ and $z_{-k}$ as follows

$$\xi_k = z_k e^{i\omega_k t} + z_{-k}e^{-i\omega_k t}, \qquad \dot{\xi}_k = i\omega_k \dot{z}_k e^{i\omega_k t} - i\omega_k z_{-k}e^{-i\omega_k t}. \tag{2.6}$$

Equations for $z_k$, $z_{-k}$ are now:

$$\frac{dx}{dt} = \varepsilon X(\tau, \theta, x), \tag{2.7}$$

where $x = (z_1, \ldots, z_N, z_{-1}, \ldots, z_{-N})$ and $X$ are the points of an Euclidean space having $n$ dimensions, $n = 2N$.

Applying the averaging method one can form the averaged equations corresponding to (2.7) with the first, the second and the higher order approximations:

$$\frac{dx^{(m)}}{dt} = \varepsilon X^{(m)}(\tau, x^{(m)}). \tag{2.8}$$

We now formulate a theorem governing the magnitude of the accuracy $x(t) - x^{(m)}(t)$ of the $m$-th approximation in the interval $0 \leq t \leq T$; where $x_k(t)$ $(k = 1, 2, \ldots, N)$ is a partial solution of the initial equations (2.7), and $x^{(m)}(t)$ is a solution of the averaged equation (2.8) with the initial condition

$$x(0) = x^{(m)}(0); \quad |x(0)| \leq M.$$

**Theorem**

It is assumed that

1) The quadratic forms (2.1) are positive definite on any finite segment $0 \leq \tau \leq L$,

2) The functions $a_{ij}(\tau)$, $b_{ij}(\tau)$, $\nu(\tau)$ and $X_k(\tau, \theta, x_1, \ldots, x_N, \dot{x}_1, \ldots, \dot{x}_N, \varepsilon)$ $(i, j, k = 1, 2, \ldots, N)$ are infinitely differentiable for all finite values of their arguments and for sufficiently small values of $\varepsilon$.

3) The expressions $\left(\dfrac{\partial^m X_k}{\partial \varepsilon^m}\right)_{\varepsilon=0}$ are finite trigonometric polynomials of the angle $\theta$.

4) On all the segment $0 \leq \tau \leq L$, there exists an inequality

$$A_1(\tau, a, \psi) \leq Ca + C_1,$$

where $C$ and $C_1$ are constants, and

$$A_1(\tau, a, \psi) = \underset{\theta}{M}\Big\{X\big(\tau, \theta, a\cos(\theta + \psi)\big)\Big\}.$$

Then, for any large $L$, and for constant values of $M$, $S$, one can find positive numbers $\varepsilon_0$, $K_m$ such that for all $\varepsilon$ $(0 \leq \varepsilon \leq \varepsilon_0)$ in the interval $0 \leq t \leq L/\varepsilon$, the following inequality holds:

$$\left|x^{(m)}(t) - x(t)\right| < K_m \varepsilon^m, \quad \left|x^{(m)}(0) - x(0)\right| \leq S.\varepsilon^m, \quad |x(0)| \leq M,$$

where $x^{(m)}(t)$ and $x(t)$ are the solutions of the systems (2.8) and (2.7) respectively, (For proof, see [29,a]).

## 3. Averaging in Systems Excited by Impulsive Forces

The averaging method is used not only for the systems subjected to forces which are described by smooth functions, but also for systems excited by impulses.

Consider the differential equation

$$\frac{d^2x}{dt^2} + \omega^2 x = \varepsilon\left[f\left(x, \frac{dx}{dt}\right)\delta(x - x_0) + f_1\left(x, \frac{dx}{dt}\right)\right], \tag{3.1}$$

where $\delta(x)$ is a Dirac function determined by

$$\int\limits_{-0}^{0} \delta(x)dx = 1, \quad \text{and} \quad \delta(0) = 0 \quad \text{for} \quad x \neq 0. \tag{3.2}$$

In order to transform (3.1) into the standard form, we use the formula

$$\begin{aligned} x &= a\sin\psi, \\ \frac{dx}{dt} &= a\omega\cos\psi, \quad \psi = \omega t + \varphi. \end{aligned} \tag{3.3}$$

The equations for the new variables $a$ and $\varphi$ are

$$\begin{aligned} \frac{da}{dt} &= \frac{\varepsilon}{\omega} f(a\sin\psi, a\omega\cos\psi)\cos\psi\delta(a\sin\psi - x_0) + \\ &+ \frac{\varepsilon}{\omega} f_1(a\sin\psi, a\omega\cos\psi)\cos\psi, \\ \frac{d\varphi}{dt} &= -\frac{\varepsilon}{a\omega} f(a\sin\psi, a\omega\cos\psi)\sin\psi\delta(a\sin\psi - x_0) - \\ &- \frac{\varepsilon}{a\omega} f_1(a\sin\psi, a\omega\cos\psi)\sin\psi. \end{aligned} \tag{3.4}$$

As we know, the following properties are true for the delta function

$$\begin{aligned} f(x)\delta(x - x_0) &= f(x_0)\delta(x - x_0), \\ \delta(f(x)) &= \sum_i \frac{\delta(x - x_i)}{|f'(x_i)|}, \end{aligned} \tag{3.5}$$

where $f(x_i) = 0$.

Taking into account the relations (3.5), we have from (3.4):

$$\begin{aligned} \frac{da}{dt} &= \pm\frac{\varepsilon\sqrt{a^2 - x_0^2}}{\omega a} f\left(x_0, \pm\omega\sqrt{a^2 - x^2}\,\right)\delta(a\sin\psi - x_0) + \\ &+ \frac{\varepsilon}{\omega} f_1(a\sin\psi, a\omega\cos\psi)\cos\psi, \\ \frac{d\varphi}{dt} &= -\frac{\varepsilon x_0}{\omega a^2} f\left(x_0, \pm\omega\sqrt{a^2 - x^2}\,\right)\delta(a\sin\psi - x_0) - \\ &- \frac{\varepsilon}{\omega a} f_1(a\sin\psi, a\omega\cos\psi)\sin\psi, \end{aligned} \tag{3.6}$$

where the negative sign is taken for $\psi$ in the intervals:

$$\left(\frac{\pi}{2}, \frac{3}{2}\pi\right), \quad \left(-\frac{\pi}{2}, -\frac{3}{2}\pi\right), \quad \left(\frac{5\pi}{2}, \frac{7\pi}{2}\right), \ldots$$

For definitness, it is assumed that $x_0 \geq 0$. Since $\delta(a \sin\psi - x_0)$ is different from zero only when $a > x_0$, the study of the system (3.6) is interesting only in that case.

For further transformation of the system (3.6), it is necessary to find the roots of the equation

$$a \sin\psi = x_0. \tag{3.7}$$

Suppose that $\psi_a^{(1)}$ and $\psi_a^{(2)}$ are the roots of (3.7) lying in the intervals $\left(0, \frac{\pi}{2}\right)$ and $\left(\frac{\pi}{2}, \pi\right)$, respectively. Obviously, the roots of (3.7) are also

$$\psi_a^{(1)} + 2k\pi, \quad \psi_a^{(2)} + 2k\pi \quad (k = \pm 1, \pm 2, \ldots).$$

Taking this into account, following (3.5), we can write:

$$\delta(a \sin\psi - x_0) = \sum_{-\infty<k<\infty} \frac{\delta(\psi - \psi_a^{(1)} - 2k\pi) + \delta(\psi - \psi_a^{(2)} - 2k\pi)}{\sqrt{a^2 - x^2}}. \tag{3.8}$$

Equations (3.6) can be represented as

$$\begin{aligned}\frac{da}{dt} = \frac{\varepsilon}{\omega a}\Big\{ & f\left(x_0, \omega\sqrt{a^2 - x_0^2}\,\right) \sum_k \delta(\psi - \psi_a^{(1)} - 2k\pi) - \\ & - f\left(x_0, -\omega\sqrt{a^2 - x_0^2}\,\right) \sum_k \delta(\psi - \psi_a^{(2)} - 2k\pi)\Big\} + \\ & + \frac{\varepsilon}{\omega} f_1(a \sin\psi, a\omega \cos\psi) \cos\psi,\end{aligned} \tag{3.9}$$

$$\begin{aligned}\frac{d\varphi}{dt} = -\frac{\varepsilon x_0}{\omega a^2 \sqrt{a^2 - x_0^2}}\Big\{ & f\left(x_0, \omega\sqrt{a^2 - x_0^2}\,\right) \sum_k \delta(\psi - \psi_a^{(1)} - \\ & - 2k\pi) + f\left(x_0, \omega\sqrt{a^2 - x_0^2}\,\right) \sum_k \delta(\psi - \psi_a^{(2)} - 2k\pi)\Big\} - \\ & - \frac{\varepsilon}{a\omega} f_1(a \sin\psi, a\omega \cos\psi) \sin\psi.\end{aligned}$$

Using the well-known equality

$$\sum_{-\infty<k<\infty} \delta(x + 2k\pi) = \frac{1}{\pi}\left\{\frac{1}{2} + \sum_{k=1}^{\infty} \cos kx\right\},$$

we can transform (3.9) into the form

$$\frac{da}{dt} = \frac{\varepsilon}{\pi\omega a}\Big\{f\Big(x_0, \omega\sqrt{a^2 - x_0^2}\,\Big)\Big[\frac{1}{2} + \sum_{k=1}^{\infty}\cos k(\psi - \psi_a^{(1)})\Big] -$$
$$- f\Big(x_0, -\omega\sqrt{a^2 - x_0^2}\,\Big)\Big[\frac{1}{2} + \sum_{k=1}^{\infty}\cos k(\psi - \psi_a^{(2)})\Big]\Big\} +$$
$$+ \frac{\varepsilon}{\omega} f_1(a\sin\psi, a\omega\cos\psi)\cos\psi, \tag{3.10}$$

$$\frac{d\varphi}{dt} = -\frac{\varepsilon x_0}{\pi\omega a^2\sqrt{a^2 - x_0^2}}\Big\{f\Big(x_0, \omega\sqrt{a^2 - x_0^2}\,\Big)\Big[\frac{1}{2} + \sum_{k=1}^{\infty}\cos k(\psi - \psi_a^{(1)})\Big] +$$
$$+ f\Big(x_0, -\omega\sqrt{a^2 - x_0^2}\,\Big)\Big[\frac{1}{2} + \sum_{k=1}^{\infty}\cos k(\psi - \psi_a^{(2)})\Big]\Big\} -$$
$$- \frac{\varepsilon}{\omega a} f_1(a\sin\psi, a\omega\cos\psi)\sin\psi.$$

In order to write the equations of the first approximation, we average (3.10) over time, or equivalently, over $\psi$ (because $\psi = \omega t + \varphi$):

$$\frac{da}{dt} = \frac{\varepsilon}{2\pi a\omega}\Big[f\Big(x_0, \omega\sqrt{a^2 - x_0^2}\,\Big) - f\Big(x_0, -\omega\sqrt{a^2 - x_0^2}\,\Big)\Big] + \frac{\varepsilon}{\omega}a_{01}(a),$$
$$\frac{d\varphi}{dt} = -\frac{\varepsilon x_0}{2\pi a^2\omega\sqrt{a^2 - x_0^2}}\Big[f\Big(x_0, \omega\sqrt{a^2 - x_0^2}\,\Big) + \tag{3.11}$$
$$+ f\Big(x_0, -\omega\sqrt{a^2 - x_0^2}\,\Big)\Big] - \frac{\varepsilon}{a\omega}a_{02}(a),$$

where

$$a_{01}(a) = \frac{1}{2\pi}\int_0^{2\pi} f_1(a\sin\psi, a\omega\cos\psi)\cos\psi d\psi,$$
$$a_{02}(a) = \frac{1}{2\pi}\int_0^{2\pi} f_1(a\sin\psi, a\omega\cos\psi)\sin\psi d\psi. \tag{3.12}$$

After elementary calculations, we obtain the following refinement of the first approximation

$$a_1 = a + \frac{\varepsilon}{\pi\omega^2 a}\Big\{f\Big(x_0, \omega\sqrt{a^2 - x_0^2}\,\Big)\sum_{k=1}^{\infty}\frac{\sin k(\psi - \psi_a^{(2)})}{k} -$$
$$- f\Big(x_0, -\omega\sqrt{a^2 - x_0^2}\,\Big)\sum_{k=1}^{\infty}\frac{\sin k(\psi - \psi_a^{(2)})}{k}\Big\} +$$
$$+ \frac{\varepsilon}{\omega^2}\sum_{k=1}^{\infty}\frac{a_{k_1}\sin k\psi - b_{k_1}\cos k\psi}{k},$$

$$\varphi_1 = \varphi - \frac{\varepsilon x_0}{\pi a^2 \omega^2 \sqrt{a^2 - x_0^2}} \Big\{ f\big(x_0, \omega\sqrt{a^2 - x_0^2}\,\big) \sum_{k=1}^{\infty} \frac{\sin k(\psi - \psi_a^{(1)})}{k} +$$

$$+ f\big(x_0, -\omega\sqrt{a^2 - x_0^2}\,\big) \sum_{k=1}^{\infty} \frac{\sin k(\psi - \psi_a^{(2)})}{k} \Big\} -$$

$$- \frac{\varepsilon}{a\omega^2} \sum_{k=1}^{\infty} \frac{a_{k_2} \sin k\psi - b_{k_2} \cos k\psi}{k}, \qquad (3.13)$$

where $a_{k_1}$, $b_{k_1}$, $a_{k_2}$ and $b_{k_2}$ $(k = 1, 2, \ldots)$ are the Fourier coefficients of the functions

$$f_1(a \sin\psi, a\omega\cos\psi)\cos\psi, \quad f_1(a\sin\psi, a\omega\cos\psi)\sin\psi,$$

and $a$ and $\varphi$ are solutions of (3.11).

Since the summation $\sum\limits_{k=1}^{\infty} \frac{\sin k\psi}{k}$ is a Fourier summation for the function $\frac{\pi - \psi}{2}$, $0, \psi < 2\pi$, the formulae (3.13) determine the expressions of the refinement of the first approximations for $a$ and $\psi$ with the help of discontinuous functions with discontinuities at the points $\psi - \psi_a^{(1)} = 2k\pi$, $\psi - \psi_a^{(2)} = 2k\pi$, $k = 0.1, 2, \ldots$

We now analyze the stationary regimes. To this end, we set $\frac{da}{dt} = 0$ in (3.11):

$$a_{01}(a) + \frac{1}{2\pi a}\Big[f\big(x_0, \omega\sqrt{a^2 - x_0^2}\,\big) - f\big(x_0, -\omega\sqrt{a^2 - x_0^2}\,\big)\Big] = 0. \qquad (3.14)$$

Denoting the stationary amplitude by $a_0$ - the root of the equation (3.14), we have the corresponding total phase

$$\psi = \big[\omega + \varepsilon B(a_0)\big]t + \varphi_0.$$

Then, the expression $a_1$ (3.13) is a periodic discontinuous function with the period

$$T(a_0) = \frac{2\pi}{\omega + \varepsilon B(a_0)}, \qquad (3.15)$$

where $\varepsilon B(a_0)$ denotes the right-hand side of the second equation of (3.11).

The functions

$$x = a_1 \sin\psi_1, \quad \frac{dx}{dt} = a_1\omega\cos\psi_1, \qquad (3.16)$$

where $a_1$ and $\psi_1$ are determined by (3.13) in which $a$ is changed into $a_0$, are periodic with the period $T(a_0)$. The derivative $\frac{dx}{dt}$ is a periodic discontinuous function with the points of discontinuity being the roots of the equation

$$a_0 \sin\Big\{\big[\omega + \varepsilon B(a_0)\big]t + \varphi_0\Big\} = x_0. \qquad (3.17)$$

As an example, let us consider the equation of motion of a clock pendulum excited by impulses:

$$J\frac{d^2x}{dt^2} + \left\{\lambda\frac{dx}{dt} - I\frac{\frac{dx}{dt} + \left|\frac{dx}{dt}\right|}{2}\delta(x - x_0)\right\} + kx = 0, \tag{3.18}$$

where $x_0$ is the angle of deviation of the pendulum, at which the pendulum is subjected to a impulse $I$; $\delta(x)$ is a Dirac function.

After some calculations, one obtains the following averaged equations for the amplitude $a$:

$$\frac{da}{dt} = -\frac{\lambda}{2J}a \quad \text{if} \quad a \le x_0, \tag{3.19}$$

and

$$\frac{da}{dt} = -\frac{\lambda}{2J}a + \frac{1}{2\pi J}\sqrt{1 - \frac{x_0^2}{a^2}} \text{ if } a > x_0, \tag{3.20}$$

and for the total phase

$$\frac{d\psi}{dt} = \omega(a), \tag{3.21}$$

where

$$\omega(a) = \begin{cases} \omega & \text{if} \quad a \le x_0, \\ \omega - \dfrac{Ix_0}{2\pi J a^2} & \text{if} \quad a > x_0. \end{cases}$$

From (3.19), it follows that with small enough initial values $a_0$, for example $a_0 \le x_0$, the amplitude $a(t)$ decreases:

$$a(t) \to 0 \quad \text{when} \quad t \to \infty,$$

and the oscillations are damped with time.

## 4. Conditions for Uniformity in the Averaging Method

In the previous section, we considered the problem of construction of approximations. In this section, we give an analysis of the conditions for uniformity of the application of the averaging method.

Let us consider a differential equation in the standard form:

$$\frac{dx}{dt} = \varepsilon X(t, x) + \varepsilon^2 X_1(t, x) + \cdots + \varepsilon^m X_{m-1}(t, x), \tag{4.1}$$

where $X_i(t, x)$ $(i = 0, \ldots, m-1)$ are periodic functions with period $2\pi$.

In order to form the $m$-th approximation, we consider the expression

$$x = \xi + \varepsilon F_1(t, \xi) + \varepsilon^2 F_2(t, \xi) + \cdots + \varepsilon^m F_m(t, \xi), \tag{4.2}$$

in which $\xi$ is the solution of the averaged equation:

$$\frac{d\xi}{dt} = \varepsilon P_1(\xi) + \varepsilon^2 P_2(\xi) + \cdots + \varepsilon^m P_m(\xi) \tag{4.3}$$

and the functions $F_i(t, \xi)$ $(i = 1, 2, \ldots, m)$ are periodic with respect to t with period $2\pi$.

Earlier, for determination of the functions $F_j(t, \xi)$ and $P_j(\xi)$ $(j = 1, 2, \ldots, m)$, we required that they are represented by a sum of the form

$$\sum_{\mu \neq 0} e^{i\mu t} F_{j\mu}(\xi) \tag{4.4}$$

and therefore, have the zero mean value. Now, these functions will be determined in a different way.

Substituting (4.2) into (4.1) and taking into account (4.3), we obtain:

$$\begin{aligned} &\varepsilon\Big\{P_1(\xi) - \frac{\partial F_1}{\partial t}\Big\} + \varepsilon^2\{P_2(\xi) + \frac{\partial F_2}{\partial \xi} + \frac{\partial F_1}{\partial \xi} P_1(\xi)\Big\} + \cdots = \\ &= \varepsilon X(t, \xi) + \varepsilon^2\Big\{X_1(t, \xi) + X'_\xi(t, \xi) F_1(t, \xi)\Big\} + \ldots \end{aligned} \tag{4.5}$$

Comparing the coefficients of like powers of $\varepsilon$ in (4.5), we have:

$$\begin{aligned} &P_1(\xi) + \frac{\partial F_1}{\partial t} = X(t, \xi), \\ &P_2(\xi) + \frac{\partial F_2}{\partial t} + \frac{\partial F_1}{\partial \xi} P_1(\xi) = \\ &= X_1(t, \xi) + X'_\xi(t, \xi) F_1(t, \xi), \\ &\cdots\cdots\cdots\cdots\cdots\cdots \end{aligned} \tag{4.6}$$

Since the function $F_1(t, \xi)$ is periodic with respect to $t$ with period $2\pi$, then from (4.5), one obtains

$$P_1(\xi) = \frac{1}{2\pi} \int_0^{2\pi} X(t, \xi) dt = X_0(\xi). \tag{4.7}$$

Then, from (4.5) and (4.7), we find

$$F_1(t, \xi) = u_1(t, \xi) + q_1(\xi), \tag{4.8}$$

where $q_1(\xi)$ is an arbitrary function of $\xi$, and $u_1(t, \xi)$ is determined from the equation

$$\frac{du_1}{dt} = X(t, \xi) - P_1(\xi), \quad \int_0^{2\pi} u_1(t, \xi) dt = 0. \tag{4.9}$$

Similarly, from (4.6) and from the periodicity of $F_2$, we obtain

$$P_2(\xi) = \frac{1}{2\pi}\int_0^{2\pi}\Big[X_1(t,\xi) + X'_\xi(t,\xi)F_1(t,\xi) - \frac{\partial F_1}{\partial \xi}P_1(\xi)\Big]dt. \tag{4.10}$$

Denoting

$$\frac{1}{2\pi}\int_0^{2\pi}\Big[X_1(t,\xi) + X'_\xi(t,\xi)u_1(t,\xi)\Big]dt = X_{10}(\xi) \tag{4.11}$$

we have from (4.7) and (4.10):

$$P_2(\xi) = X_{10}(\xi) + \frac{\partial X_0(\xi)}{\partial \xi}q_1(\xi) - X_0(\xi)\frac{dq_1}{d\xi}\,. \tag{4.12}$$

From the equations (4.6), one can find $F_2(t,\xi)$ with accuracy to an arbitrary function $q_2(\xi)$.

Earlier, the determination of the averaging method for the equation (4.1) required that the functions $F_j(t,\xi)$ in (4.2 ) have a zero mean value on t, and therefore, $q_1(\xi) = 0$, $q_2(\xi) = 0, \ldots$

Equations (4.3), in which $P_1(\xi)$, $P_2(\xi), \ldots$ are determined by the expressions (4.7), (4.10), and $X_{10}(\xi)$ by (4.11), and $q_1(\xi)$ is an arbitrary function of $\xi$, are called the modified averaged equations.

The application of the Krylov-Bogoliubov method for the construction of asymptotic expansions demonstrated the non-uniformity of the averaged equations. Following that indication, we may obtain the modified averaged equations as follows.

It is assumed that for equations in the standard form (1.1) with the help of the method used earlier, the solution is found as

$$x = \eta + \varepsilon F_1(t,\eta) + \varepsilon^2 F_2(t,\eta) + \ldots, \tag{4.13}$$

where the functions $F_1(t,\eta)$, $F_2(t,\eta), \ldots$ have a zero mean value over $t$, and $\eta$ is determined from the averaged equation of the type (4.3)

$$\frac{d\eta}{dt} = \varepsilon P_1(\eta) + \varepsilon^2 P_2(\eta) + \ldots \tag{4.14}$$

By introduction of a change of variables

$$\eta = \xi + \varepsilon q_1(\xi) + \varepsilon^2 q_2(\xi) + \ldots \tag{4.15}$$

and by inversion, we have

$$\xi = \eta - \varepsilon q_1(\eta) + \ldots$$

Equation (4.14) in the new variable $\xi$ will be

$$\begin{aligned}\frac{d\xi}{dt} &= \varepsilon P_1(\eta) + \varepsilon^2 P_2(\eta) + \varepsilon^2 \frac{dq_1}{d\eta} P_1(\eta) + O(\varepsilon^3) = \\ &= \varepsilon P_1(\xi) + \varepsilon^2 \Big[P_2(\xi) + \frac{\partial P_1}{\partial \xi} q_1(\xi) - P_1(\xi)\frac{dq_1}{d\xi}\Big] + O(\varepsilon^3).\end{aligned} \tag{4.16}$$

Thus, we have obtained the equation which coincides with the second equation of (1.8), Chapter 1, whose right-hand side coincides with the expressions (4.7) and (4.12).

To eliminate non-uniformity in the construction of the averaged equations, it is necessary to require some additional conditions. One of them may be, for example, the condition of the zero mean value of the functions $F_i(t, \xi)$. Together with this, the non-uniformity of the construction of averaged equations in many cases gives them a form compatible for analysis by the corresponding choice of the functions $q_1(\xi), q_2(\xi), \ldots$.

## 5. Averaging in Systems Containing Slow and Rapid Motions

In many oscillatory systems, only a part of the equations of motion can be transformed into standard form. An example is a system in which some elements rapid motions and the others execute slow ones.

On the basis of the ideas of averaging the equations in the standard form and the equations with slowly varying parameters, V. M. Volosov worked out a general scheme for averaging for systems of the differential equations which contain slow and rapid motions, of the form [53]:

$$\begin{aligned}\frac{dx}{dt} &= X(t, x, y, \varepsilon), \\ \frac{dy}{dt} &= \varepsilon Y(t, x, y, \varepsilon),\end{aligned} \tag{5.1}$$

where $x$ and $y$ are $k$ and $m$-dimension vectors, respectively, in $n$-dimensional Euclidean space $E_n$ $(k+m = n)$, and $X(t, x, y, \varepsilon)$ and $Y(t, x, y, \varepsilon)$ are $k$ and $m$-dimension vector functions, respectively; $\varepsilon$ is a small positive parameter; the $x$ variables are rapid while the $y$ variables are slow.

Expanding the right-hand sides of the system (5.1) into power series in $\varepsilon$, we have

$$\begin{aligned}\frac{dx}{dt} &= X_0(t, x, y) + \varepsilon X_1(t, x, y) + \varepsilon^2 \ldots, \\ \frac{dy}{dt} &= \varepsilon Y_1(t, x, y) + \varepsilon^2 Y_2(t, x, y) + \varepsilon^3 \ldots\end{aligned} \tag{5.2}$$

Together with (5.2), we consider the degenerate, or so-called unexcited, system

$$\frac{dx}{dt} = X_0(t, x, y), \qquad \frac{dy}{dt} = 0, \quad y = \text{const}, \tag{5.3}$$

obtained form (5.2) by setting $\varepsilon = 0$.

In the system (5.1), the variables $y$ change with small velocity $\varepsilon Y$. The influence of the rapid motions $x$ on the character of $y$ leads to the fact that the rapid variables $x$ are superimposed on slow velocity of the change of variables $y$: $\varepsilon Y(t, x, y, \varepsilon)$, because $Y = Y(t, x, y, \varepsilon)$. We suppose that this influence of rapid motions on the velocity of slow motions can be averaged in the long term over time, i.e. that, along the integral curves of the rapid motions there exist mean values of the velocity $\varepsilon Y$. Then we consider closeness of the solutions of the system (5.1) to the solutions of the averaged system.

Thus, we suppose that along every integral curve $x = x(t, y)$ ($y = \text{const}$) of the system (5.3), there exists a limit

$$\overline{Y}_1(y) = \lim_{T\to\infty} \frac{1}{T} \int_0^T Y[t, x(t, y), y, 0]dt. \tag{5.4}$$

Note that usually, the averaging along an integral curve is simply averaging over explicit time. For averaging the system (5.2), there are two approaches.

*First approach*: Since in (5.2) the $y$ - variables change slowly and $x$ - variables rapidly, then averaging the rapid action of the $x$ - variables over a long time interval leads to the following averaged equation instead of the second equation of the system (5.1):

$$\frac{d\overline{y}}{dt} = \varepsilon Y_1. \tag{5.5}$$

If the right-hand sides of (5.5) depend only on y and therefore not on the trajectory $x = x(t, y)$ of an unexcited system (5.3), then such averaging simplifies the problem considerably, because the system (5.2) simplifies into independent systems of $k$ - th and $m$-th degrees.

The problem about approximate solutions of the system (5.2) is reduced to the problem of approximate solutions of the averaged system (5.5).

In this approach, the essential limit is the condition of independence of the right-hand side of the equation (5.5) on the trajectory of the unexcited system (5.3). However, note that in many important cases, one can give methods, consisting in a special change of variables. In this way, we can reduce the general case, in which the mean value (5.4) depends on the integral curve of the unexcited system (5.3), to the special case we are considering here.

This first approach is simple and is based on the application of the usual ideas of the averaging method for equations (5.2).

*The second approach* is more general and consists of the construction of higher approximations for $y$ and for rapid motions $x$. It is based on the Bogoliubov's ideas about averaging and leads to the construction of the following averaged system for (8.2):

$$\begin{aligned} \frac{d\overline{x}}{dt} &= X_0(t, \overline{x}, \overline{y}) + \varepsilon A_1(\overline{y}) + \varepsilon^2 A_2(\overline{y}) + \dots, \\ \frac{d\overline{y}}{dt} &= \varepsilon \overline{Y}_1(\overline{y}) + \varepsilon^2 B_2(\overline{y}) + \dots \end{aligned} \tag{5.6}$$

The solutions of (5.6) approximate the solutions of (5.2) with any degree of exactness in the time interval $\frac{1}{\varepsilon}$. The unknown functions $A_1(\overline{y})$, $A_2(\overline{y})$, ..., $B_2(\overline{y})$, ... are determined with the help of averaging along the trajectory of the unexcited system (5.3), and the function $\overline{Y}_1(\overline{y})$ is determined by (5.4).

The equations (5.6) can be integrated more simply than the original equations (5.2), if we limit ourselves by considering only a finite number of terms in the right-hand sides, because in (5.6) the rapid and the slow variables $x$ and $y$ are separated.

We give here the basic results for the construction of the transformation from the system (5.2) to the averaged one.

Inasmuch as the basic idea is the averaging along the integral curves $x = x(t, y)$ of the unexcited system (5.3), we suppose that in the region under consideration there exists a general solution of the unexcited system (5.3)

$$x = \varphi(t, y, x_0, t_0), \tag{5.7}$$

where $x_0$ is a constant vector and $\varphi(t_0, y, x_0, t_0) \equiv x_0$. Suppose that vector-function (5.7) has sufficient derivatives and that there is only one integral curve (5.7) passing through every point of the region of the variables $x, y, t$ under consideration. Since the function $\varphi(t, y, x_0, t_0)$ is a general solution of (5.2), then, the rank of the matrix $\left\| \frac{\partial \varphi}{\partial x_0}, \frac{\partial \varphi}{\partial t_0} \right\|$ is to the number of $x$ - variables, i.e. to $k$. We assume that for the functions (5.7) and for those on the right-hand sides of (5.2), all necessary conditions for further calculations are fulfilled.

With these assumptions, we make a change of variables in the form

$$\begin{aligned} x &= \overline{x} + \varepsilon u_1(t, \overline{x}, \overline{y}) + \varepsilon^2 u_2(t, \overline{x}, \overline{y}) + \dots, \\ y &= \overline{y} + \varepsilon v_1(t, \overline{x}, \overline{y}) + \varepsilon^2 v_2(t, \overline{x}, \overline{y}) + \dots, \end{aligned} \tag{5.8}$$

where $u_1(t, \overline{x}, \overline{y})$ and $u_2(t, \overline{x}, \overline{y})$ are $k$-dimension vector functions; $v_1(t, \overline{x}, \overline{y})$ and $v_2(t, \overline{x}, \overline{y})$ are $m$ - dimension vector functions. They are determined in such a way that the change (5.8) leads (5.2) to the averaged equations (5.6). When $\varepsilon = 0$, the systems (5.2) and (5.6) become the unexcited system (5.3). At the same time, from (5.8) it follows that $x = \overline{x}$ and $y = \overline{y}$, i.e. the solutions of (5.2) and (5.6), coincide when they have the same initial conditions.

In order to solve the problem stated, it is necessary to find the vector-function

$$\begin{aligned} &u_1(t, \overline{x}, \overline{y}), u_2(t, \overline{x}, \overline{u}), \dots, v_1(t, \overline{x}, \overline{y}), \\ &v_2(t, \overline{x}, \overline{y}), \dots, A_1(\overline{y}), A_2(\overline{y}), \dots, B_2(\overline{y}), \dots \end{aligned} \tag{5.9}$$

To this end, we first differentiate (5.8) and substitute the derivatives into (5.2) after representing all quantities through the new variables $\overline{x}$ and $\overline{y}$ following (5.8). By comparison of the coefficients of like powers of $\varepsilon$, we have:

$$\frac{\partial u_1}{\partial t} + \frac{\partial u_1}{\partial x} X_0(t, \overline{x}, \overline{y}) = X_1(t, \overline{x}, \overline{y}) + \frac{\partial X_0}{\partial x} u_1 + \frac{\partial X_0}{\partial y} v_1 - A_1(\overline{y}), \tag{5.10}$$

$$\frac{\partial u_2}{\partial t} + \frac{\partial u_2}{\partial x} X_0(t,\overline{x},\overline{y}) = X_2(t,\overline{x},\overline{y}) + \frac{1}{2}\left\{\frac{\partial^2 X_0}{\partial x^2} u_1^2 + \frac{\partial^2 X_0}{\partial y^2} v_1^2 + \right.$$
$$\left. + 2\frac{\partial^2 X_0}{\partial x \partial y} u_1 v_1\right\} + \frac{\partial X_0}{\partial x} u_2 + \frac{\partial X_0}{\partial y} v_2 + \frac{\partial X_1}{\partial x} u_1 +$$
$$+ \frac{\partial X_1}{\partial y} v_1 - \frac{\partial u_1}{\partial x} A_1(\overline{y}) - \frac{\partial u_1}{\partial y} y_1(t,\overline{x},\overline{y}) - A_2(\overline{y}), \tag{5.11}$$

........................

$$\frac{\partial v_1}{\partial t} + \frac{\partial v_1}{\partial x} X_0(t,\overline{x},\overline{y}) = y_1(t,\overline{x},\overline{y}) - \overline{y}_1(y), \tag{5.12}$$
$$\frac{\partial v_2}{\partial t} + \frac{\partial v_2}{\partial x} X_0(t,\overline{x},\overline{y}) = y_2(t,\overline{x},\overline{y}) - B_2(\overline{y}) +$$
$$+ \frac{\partial y_1}{\partial x} u_1 + \frac{\partial y_1}{\partial y} v_1 - \frac{\partial v_1}{\partial x} A_1(\overline{y}) - \frac{\partial v_1}{\partial y} \overline{y}_1(\overline{y}),$$

.....................

and from these equations the unknown functions (5.9) will be determined.

We can now find the change of variables (5.8). In the general case, we obtain the following equations in the $n$-th approximation

$$\frac{d\overline{x}}{dt} = X_0(t,\overline{x},\overline{y}) + \varepsilon A_1(\overline{y}) + \varepsilon^2 A_2(\overline{y}) + \cdots + \varepsilon^{n-1} A_{n-1}(\overline{y}),$$
$$\frac{d\overline{y}}{dt} = \varepsilon \overline{Y}_1(\overline{y}) + \varepsilon^2 B_2(\overline{y}) + \cdots + \varepsilon^n B_n(\overline{y}). \tag{5.13}$$

Although we have not achieved the full elimination of the rapid motion or the full separation of variables, the averaged equations (5.13) are considerably simple, than the original ones (5.2), and the second group of equations for the slow variables $\overline{y}$ is independent of the rapid variables $\overline{x}$. Determining $\overline{y}_n = y_n(t)$ from (5.13) and substituting them into the equations for the rapid variables $\overline{x}$ and integrating, we obtain $\overline{x} = \overline{x}_{n-1}(t)$. Further, with the help of the formulae (5.8), we find the following approximations for the solutions of (5.2):

$$x = \overline{x} + \varepsilon u_1(t,\overline{x}_{n-1},\overline{y}_n) + \varepsilon^2 u_2(t,\overline{x}_{n-1},\overline{y}_n) + \cdots + \varepsilon^{n-2} u_{n-2}(t,x_{n-1},y_n),$$
$$y = \overline{y} + \varepsilon v_1(t,\overline{x}_{n-1},\overline{y}_n) + \varepsilon^2 v_2(t,\overline{x}_{n-1},\overline{y}_n) + \cdots + \varepsilon^{n-1} v_{n-1}(t,x_{n-1},y_n),$$

where, as before, the slow motion $y$ is calculated to one degree more exact than the rapid motion.

The mathematical foundation of this averaging method and more detailed calculations can be found in the book [53].

## 6. Averaging in Systems Containing Rotation. Motion of Satellites

Up to now it was always been supposed that the unexcited system of differential equations describes oscillatory processes. However, there are a number of cases in which this approach is inadequate.

The problems connected with studying rotational regimes can be solved with the help of the corresponding development of the averaging method.

Consider a system, which we will call a pendulum, whose motion is described by the equation:

$$\frac{d^2x}{dt^2} + f(x) = 0, \tag{6.1}$$

where $f(x)$ is a periodic function with period $2\pi$; and $f(0) = 0$.

Assume, that the mean value $\overline{f}(x)$ is

$$\overline{f}(x) = \int_0^{2\pi} f(x)dx = 0. \tag{6.2}$$

The phase-plane of equation (6.1) has the form shown in Fig. 102. The closed branches limit the zone of $x$, $\dot{x}$, corresponding to periodic motions. Non-closed phase trajectories describe the rotational motions of the system under consideration and are determined by a large initial energy.

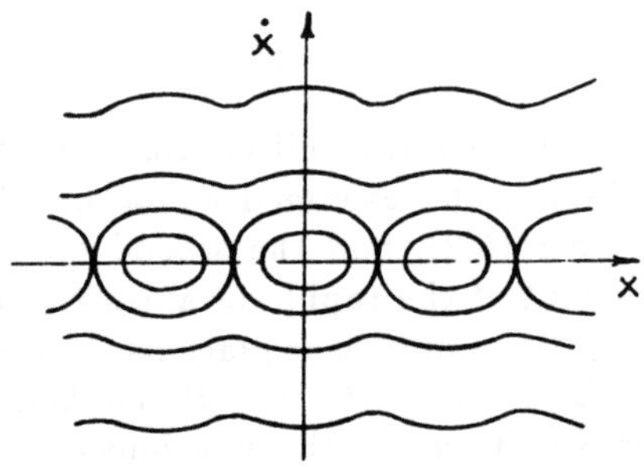


*Fig. 102.* Phase trajectories of a rotational system.

We consider the motions described by the excited equation

$$\frac{d^2x}{dt^2} + f(x) = \varepsilon F\left(t, x, \frac{dx}{dt}\right) \tag{6.3}$$

and also the equation with slowly varying variables

$$\frac{d^2x}{dt^2} + f(\tau, x) = \varepsilon F\left(\tau, t, x, \frac{dx}{dt}\right), \tag{6.4}$$

where $\varepsilon$ is a small positive parameter.

Setting $\varepsilon = 0$ in (6.3) and (6.4), we obtain the unexcited equations of the form (6.1). In order to study equations (6.3) and (6.4), we use the method proposed by Moiseev [35], and Volosov and Morgunov [53], which is development of the Bogoliubov's averaging method.

We first consider the case of a large initial energy of the unexcited system (6.1) when the motion is represented by non-closed phase trajectories. Consider a curve describing rotational motion in Fig. 102. This curve cuts the axis $O\dot{x}$ at the point $\dot{x} = \Omega$. The phase trajectory corresponding to that level of energy, is a wavy curve lying in the neighbourhood of the straight line $\dot{x} = \Omega$.

Hence, in equation (6.1), it is natural to introduce the change of variables

$$\frac{dx}{dt} = \Omega + y. \tag{6.5}$$

Letting $t = \varepsilon t_1$, where $\varepsilon = \dfrac{1}{\Omega}$, we obtain

$$\begin{aligned} \frac{dx}{dt_1} &= 1 + \varepsilon y, \\ \frac{dy}{dt_1} &= -\varepsilon f(x). \end{aligned} \tag{6.6}$$

The system (6.6) belongs to the type considered before. Therefore, the solution of the *unexcited* system (6.1) or equivalently (6.6), will be found in the form of a power series

$$\begin{aligned} x &= \overline{x} + \varepsilon u_1(\overline{y}, \overline{x}) + \varepsilon^2 u_2(\overline{y}, \overline{x}) + \dots, \\ y &= \overline{y} + \varepsilon v_1(\overline{y}, \overline{x}) + \varepsilon^2 v_2(\overline{y}, \overline{x}) + \dots, \end{aligned} \tag{6.7}$$

where the new variables $\overline{x}$ and $\overline{y}$ satisfy the equations

$$\begin{aligned} \frac{d\overline{x}}{dt} &= 1 + \varepsilon B_1(\overline{y}) + \varepsilon^2 B_2(\overline{y}) + \dots, \\ \frac{d\overline{y}}{dt} &= \varepsilon A_1(\overline{y}) + \varepsilon^2 A_2(\overline{y}) + \dots \end{aligned} \tag{6.8}$$

It is required that $u_i(\overline{y}, \overline{x})$, $v_i(\overline{y}, \overline{x})$ are bounded and satisfy the conditions

$$u_i(\overline{y}, 0) = v_i(\overline{y}, 0) = 0. \tag{6.9}$$

Substituting the series (6.7) into (6.6), taking into account (6.8) and comparing the coefficients of like powers of $\varepsilon$, we obtain

$$\begin{aligned} &\frac{\partial v_1}{\partial \overline{x}} = -f(\overline{x}) - A_1(\overline{y}), \quad \frac{\partial u_1}{\partial \overline{x}} = \overline{y} - B_1(\overline{y}), \\ &\frac{\partial v_2}{\partial \overline{x}} = -\frac{\partial f}{\partial \overline{x}} u_1 - \frac{\partial v_1}{\partial \overline{y}} A_1(\overline{y}) - \frac{\partial v_1}{\partial \overline{x}} B_1(\overline{y}) - A_2(\overline{y}), \\ &\frac{\partial u_2}{\partial \overline{x}} = v_1 - \frac{\partial u_1}{\partial \overline{y}} A_1(\overline{y}) - \frac{\partial u_1}{\partial \overline{x}} B_1(\overline{y}) - B_2(\overline{y}), \\ &\dots\dots\dots\dots\dots\dots \end{aligned} \tag{6.10}$$

Consider the first equation of (6.10). For the boundedness of $v_1(\overline{y}, \overline{x})$, it is necessary and sufficient that

$$A_1(\overline{y}) = -\overline{f}(x). \tag{6.11}$$

But, following (6.2) $\overline{f}(x) = 0$. Then,

$$A_1(\overline{y}) = 0. \tag{6.12}$$

Further, we find

$$v_1(\overline{y}, \overline{x}) = -\int_0^{\overline{x}} f(x)dx + C(\overline{y}), \tag{6.13}$$

where $C(\overline{y})$ is an arbitrary function of $\overline{y}$ and according to (6.9) $C(\overline{y}) \equiv 0$.
Introducing the notation

$$\int_0^x f(x)dx = \Psi(x), \quad \int_0^x \psi(x)dx = \Phi(x) \tag{6.14}$$

we have

$$v_1(\overline{x}) = -\Psi(\overline{x}). \tag{6.15}$$

From the next two equations of (6.10), we obtain

$$B_1(\overline{y}) = \overline{y}, \quad u_1(\overline{y}, \overline{x}) = 0, \tag{6.16}$$

$$A_2(\overline{y}) = 0, \quad v_2(\overline{y}, \overline{x}) = \overline{y}\Psi(\overline{x}), \tag{6.17}$$

$$B_2(\overline{y}) = -\overline{\Psi}(\overline{x}), \quad u_2(\overline{y}, \overline{x}) = -\phi(\overline{x}) + \overline{\Psi}(\overline{x})\overline{x}. \tag{6.18}$$

With accuracy $O(\varepsilon^2)$, the expressions (6.7) and (6.8) become

$$\begin{aligned} x &= \overline{x} - \varepsilon^2\Big[\phi(\overline{x}) - \overline{\Psi}(\overline{x})\overline{x}\Big], \\ y &= \overline{y} - \varepsilon\Psi(\overline{x}) + \varepsilon^2\overline{y}\Psi(\overline{x}). \end{aligned} \tag{6.19}$$

$$\begin{aligned} \frac{d\overline{x}}{dt_1} &= 1 + \varepsilon\overline{y} - \varepsilon^2\overline{\Psi}(\overline{x}), \\ \frac{d\overline{y}}{dt_1} &= 0. \end{aligned} \tag{6.20}$$

Taking into account the initial values

$$t = 0, \quad y = 0, \quad x = 0,$$

we find from (6.8)

$$\overline{y} = 0.$$

Then, the expressions (6.7) and the first equation of (6.8) can be represented in the form

$$x = \overline{x} + \varepsilon^2\Big[\overline{x}\overline{\Psi}(\overline{x}) - \phi(\overline{x})\Big], \tag{6.21}$$

$$y = -\varepsilon\Psi(\overline{x}), \quad \frac{d\overline{x}}{dt_1} = 1 - \varepsilon^2\overline{\Psi}(\overline{x}). \tag{6.22}$$

Integrating (6.22) and substituting the value obtained $\overline{x} = \overline{x}(t)$ into the expressions (6.21), we have:

$$\begin{aligned} x &= \Big(\Omega - \frac{1}{\Omega}\overline{\Psi}\Big)t + \frac{1}{\Omega^2}\Big\{\overline{\Psi}\Big(\Omega - \frac{1}{\Omega}\overline{\Psi}\Big)t - \phi\Big[\Big(\Omega - \frac{1}{\Omega}\overline{\Psi}\Big)t\Big]\Big\}, \\ y &= -\frac{1}{\Omega}\Psi\Big[\Big(\Omega - \frac{1}{\Omega}\overline{\Psi}\Big)t\Big], \end{aligned} \tag{6.23}$$

where $\varepsilon = \dfrac{1}{\Omega}$.

Introducing a new constant $\lambda$:

$$\lambda = \Omega - \frac{1}{\Omega}\overline{\Psi}, \tag{6.24}$$

one can represent the general solution of (6.1) in the case of large energy in the following form

$$x = \lambda(t + t_0) + \frac{1}{\lambda^2}\Big\{\overline{\Psi}\lambda(t + t_0) - \phi\big[\lambda(t + t_0)\big]\Big\} + O\Big(\frac{1}{\lambda^3}\Big)\ ; \tag{6.25}$$

which contains two arbitrary constants $\lambda$ and $t_0$.

**Example 23**

For illustration, we consider the rotational motion of the mathematical pendulum

$$\frac{d^2x}{dt^2} + \sin x = 0. \tag{6.26}$$

We have

$$f(x) = \sin x$$

and hence,

$$\Psi(x) = 1 - \cos x, \quad \overline{\Psi} = 1, \quad \phi(x) = x - \sin x.$$

According to (6.25), we obtain

$$x = \lambda t + \frac{1}{\lambda^2}\sin\lambda t + O\Big(\frac{1}{\lambda^3}\Big)\ \cdot \tag{6.27}$$

The method presented for finding the solution of the degenerate system in the case of large energy can be applied with success for more general equations. For example, the equation

$$\frac{d^2x}{dt^2} + f(x,\xi) = 0, \tag{6.28}$$

where $\xi$ is a vector satisfying the equation

$$\frac{d\xi}{dt} = R(x,\xi), \tag{6.29}$$

with the assumption that the function $f(x,\xi)$ is periodic in $x$ with period $2\pi$ for all $\xi$. Then by use of the formula (6.5), equation (6.28) can be reduced to the system

$$\begin{aligned} \frac{dx}{dt_1} &= 1 + \varepsilon y, \\ \frac{dy}{dt_1} &= -\varepsilon f(x,\xi), \\ \frac{d\xi}{dt_1} &= \varepsilon R(x,\xi). \end{aligned} \tag{6.30}$$

Using the method presented above, one can find the asymptotic solution of this system.

**Example 24**

Consider oscillations of a pendulum with a variable restoring force, described by the equation

$$\frac{d^2x}{dt^2} + f(x,t) = 0 \tag{6.31}$$

with a large initial energy.

Equation (6.31) is transformed into (6.28) and (6.29) by the change of variables

$$t = \xi, \qquad \frac{d\xi}{dt} = 1. \tag{6.32}$$

Using the formula (6.5), one can reduce (6.31) to

$$\begin{aligned} \frac{dx}{dt_1} &= 1 + \varepsilon y, \\ \frac{dy}{dt_1} &= -\varepsilon f(x,\xi), \\ \frac{d\xi}{dt_1} &= \varepsilon. \end{aligned} \tag{6.33}$$

After that, we find the asymptotic representation for the general solution of equation (6.31)

$$x(t) = \lambda t + \frac{1}{\lambda^2}\Big\{\overline{\Psi}(t)\lambda t - \phi(t,\lambda t)\Big\}, \tag{6.34}$$

where $\lambda = \Omega - \frac{1}{\Omega}\overline{\Psi}(t)$.

**Excitation of rapid rotation**

We consider the equation (6.4) :

$$\frac{d^2x}{dt^2} + f(\tau, x) = \varepsilon F\left(\tau, t, x, \frac{dx}{dt}\right) . \tag{6.35}$$

Suppose that the initial energy of the system is large enough; i.e. when $\varepsilon = 0$ the degenerate equation

$$\frac{d^2x}{dt^2} + f(\tau, x) = 0 \quad (\tau = \text{const}) \tag{6.36}$$

has rotational motion.

Further, suppose that

$$\overline{f}(\tau) = \frac{1}{2\pi}\int_0^{2\pi} f(\tau, x)dx = 0, \tag{6.37}$$

and the solution of the degenerated equation (6.36) has the form

$$\begin{aligned} x &= z(\tau, \psi, a), \\ \frac{dx}{dt} &= \omega(\tau, a)z'_\psi(\tau, \psi, a), \end{aligned} \tag{6.38}$$

where

$$z(\tau, \psi, a) = \psi + aZ(\tau, \psi, a). \tag{6.39}$$

Here $a$ is an arbitrary constant, and

$$\psi = \omega(\tau, a)(t - t_0). \tag{6.40}$$

Comparing the expression (6.39) with the asymptotic solution (6.25), we obtain

$$\omega = \lambda, \quad a = \frac{1}{\lambda^2}, \quad \omega = \frac{1}{\sqrt{a}}, \tag{6.41}$$

$$z(\tau, \psi, a) = \overline{\Psi}(\tau)\psi - \phi(\psi, \tau) + O\left(\frac{1}{\lambda^3}\right) = Z(\tau, \psi) + O\left(\frac{1}{\lambda^3}\right) . \tag{6.42}$$

Thus, with accuracy to $O\left(\frac{1}{\lambda^3}\right)$ the function Z does not depend on $a$.

We transform the equation (6.35) by means of the formulae (6.38). Taking into account:

$$\omega^2(\tau, a)z''_{\psi^2} + f(\tau, z) \equiv 0,$$

$$z'_a \frac{da}{dt} + z'_\psi \frac{d\psi}{dt} = \omega(\tau, a) z'_\psi - \varepsilon z'_\tau,$$

$$(\omega z'_\psi)'_a \frac{da}{dt} + \omega z''_{\psi^2} \frac{d\psi}{dt} = -\omega^2(\tau, a) z''_{\psi^2} + \varepsilon\left[F(\tau, z, \omega z'_\psi) - (\omega z'_\psi)'_\tau\right],$$

and therefore

$$\begin{aligned} \frac{da}{dt} &= \frac{\varepsilon\{-F(\tau, t, z, \omega z'_\psi) z'_\psi + (\omega z'_\psi)'_\tau z'_\psi - \omega z''_{\psi^2} z'_\tau\}}{\omega z'_a z''_{\psi^2} - z'_\psi (\omega z'_\psi)'_a}, \\ \frac{d\psi}{dt} &= \omega(\tau, a) + \frac{\varepsilon\{F(\tau, t, z, \omega z'_\psi) z'_a - (\omega z'_\psi)'_\tau z'_a + (\omega z'_\psi)'_a z'_\tau\}}{\omega z'_a z''_{\psi^2} - z'_\psi (\omega z'_\psi)'_a}, \end{aligned} \tag{6.43}$$

and also

$$\begin{aligned} & z(\tau, \psi, a) = \psi + aZ(\tau, \psi), \\ & z'^2_\psi = 1 + 2aZ'_\psi + a^2 Z'^2_\psi, \\ & z'_\psi = 1 + aZ'_\psi, \quad z''_{a\psi} = Z'_\psi, \quad \omega'_\tau = 0, \\ & z''_{\psi^2} = aZ''_{\psi^2}, \quad z''_{\psi\tau} = aZ''_{\psi\tau}, \quad z'_\tau = aZ'_\tau, \end{aligned} \tag{6.44}$$

we obtain the following equations

$$\begin{aligned} \frac{da}{dt} &= -\frac{\varepsilon}{\mathcal{D}_0}\left\{F z'_\psi + \sqrt{a}\left[-Z''_{\psi\tau} + a\left(Z'_\tau Z''_{\psi^2} - Z''_{\psi\tau} Z'_\psi\right)\right]\right\}, \\ \frac{d\psi}{dt} &= \omega(a) - \frac{\varepsilon}{\mathcal{D}_0}\left\{F z'_a - \frac{Z'_\tau}{2\sqrt{a}} + \sqrt{a}\left(\frac{Z'_\psi Z'_\tau}{2} - Z Z''_{\psi\tau}\right)\right\}, \end{aligned} \tag{6.45}$$

where

$$\mathcal{D}_0 = \frac{1}{2}\omega^3(a) + \frac{1}{\omega}\left(Z Z''_{\psi^2} - \frac{1}{2} Z'^2_\psi\right), \tag{6.46}$$

$$z'_\psi = 1 + aZ'_\psi, \quad z'_a = Z(\tau, \psi), \quad \omega = \frac{1}{\sqrt{a}} \cdot$$

Further, the system obtained can be simplified according to the successive changes of variables, as before.

**The resonance phenomena in rotational systems**

Resonance phenomena in systems with rotational elements have particular characteristics which have not been fully investigated .

We consider a particular case of the equation (6.35)

$$\frac{d^2x}{dt^2} + f(x) = \varepsilon F\left(\nu t, x, \frac{dx}{dt}\right), \tag{6.47}$$

where $F\left(\nu t, x, \dfrac{dx}{dt}\right)$ is a periodic function in $t$ with period $\dfrac{2\pi}{\nu}$.

We suppose that the unexcited equation

$$\frac{d^2x}{dt^2} + f(x) = 0 \tag{6.48}$$

has the solution

$$x = z(\psi, a) \quad (\psi = \omega(a)(t - t_0)), \tag{6.49}$$

where

$$z(\psi + 2\pi, a) = z(\psi, a) \tag{6.50}$$

in the case of an oscillatory regime and

$$z(\psi + 2\pi, a) = z(\psi, a) + 2\pi, \tag{6.51}$$

in the case of a rotational regime.

It is assumed that the frequency of the free oscillations $\omega(a)$ is close to $\frac{p}{q}\nu$, where $p$ and $q$ are simple natural numbers. Instead of $\psi$, we introduce the new variables $\vartheta$ (the phase lag) as follows

$$\psi = \frac{p}{q}\nu t + \vartheta, \tag{6.52}$$

and use in (6.47) the transformation to new variables $a$, $\vartheta$ in (6.47):

$$\begin{aligned} x &= z\Big(\frac{p}{q}\theta + \vartheta, a\Big), \\ \frac{dx}{dt} &= \omega(a) z'_\psi\Big(\frac{p}{q}\theta + \vartheta, a\Big) . \end{aligned} \tag{6.53}$$

We obtain the equations in the standard form

$$\frac{da}{dt} = \frac{-\varepsilon}{D(a)} F\Big(\theta, z\big(\frac{p}{q}\theta + \vartheta, a\big), \omega(a) z'_\psi\big(\frac{p}{q}\theta + \vartheta, a\big)\Big) z'_\psi\big(\frac{p}{q}\theta + \vartheta, a\big), \tag{6.54}$$

$$\frac{d\vartheta}{dt} = \omega(a) - \frac{p}{q}\nu + \frac{\varepsilon}{D(a)} F\Big(\theta, z\big(\frac{p}{q}\theta + \vartheta, a\big), \omega(a) z'_\psi\big(\frac{p}{q}\theta + \vartheta, a\big)\Big) z'_a\big(\frac{p}{q}\theta + \vartheta, a\big),$$

where $\nu t = \theta$, $\psi = \frac{p}{q}\theta + \vartheta$.

We denote the solution of the equation

$$\omega(a) = \frac{p}{q}\nu \tag{6.55}$$

by $a_0$, and we consider the resonance solutions of (6.54) when

$$\omega(a) - \frac{p}{q}\nu = O(\varepsilon). \tag{6.56}$$

In this case

$$a - a_0 = O(\varepsilon)$$

and we can write

$$\omega(a) - \frac{p}{q}\nu = \omega'(a_0)(a - a_0) + O(\varepsilon^k), \ k > 1. \tag{6.57}$$

Taking (6.57) into account, and averaging the right-hand sides of (6.54), we obtain

$$\begin{aligned} \frac{da}{dt} &= -\frac{\varepsilon}{2\pi p \mathcal{D}(a_0)} \int_0^{2\pi p} F(\theta, \frac{p}{q}\theta + \vartheta, a) z'_\psi d\theta, \\ \frac{d\vartheta}{dt} &= \omega'(a_0)(a - a_0) - \frac{\varepsilon}{2\pi p \mathcal{D}(a_0)} \int_0^{2\pi p} F(\theta, \frac{p}{q}\theta + \vartheta, a) z'_a d\theta. \end{aligned} \tag{6.58}$$

For the case of rapid rotations, following the formulae (6.39) and (6.25), we have

$$z(\psi, a) = \psi + aZ(\psi, a), \tag{6.59}$$

where

$$\begin{aligned} \omega = \frac{1}{\sqrt{a}}, \quad a = \frac{1}{\lambda^2}, \\ Z(\psi, a) = Z(\psi) + O(\frac{1}{\lambda^3}) . \end{aligned} \tag{6.60}$$

Hence, the averaged system (6.58) can be represented in the form

$$\begin{aligned} \frac{da}{dt} &= -\frac{\varepsilon}{2\pi p \mathcal{D}(a_0)} \int_0^{2\pi p} F(\theta, \frac{p}{q}\theta + \vartheta, a)\left(1 + aZ'_\psi(\psi)\right) d\theta, \\ \frac{d\vartheta}{dt} &= \omega'(a_0)(a - a_0) + \frac{\varepsilon}{2\pi p \mathcal{D}(a_0)} \int_0^{2\pi p} F(\theta, \frac{p}{q}\theta + \vartheta, a) Z(\psi) d\theta, \end{aligned} \tag{6.61}$$

where, $\psi = \frac{p}{q}\theta + \vartheta$ and according to (6.46):

$$\mathcal{D}(a_0) = \frac{1}{2}\sqrt{a_0^3} + \sqrt{a_0}(ZZ''_{\psi^2} - \frac{1}{2}{Z'_\psi}^2). \tag{6.62}$$

With the same accuracy, we can represent the system (6.61) in the form:

$$\begin{aligned} \frac{da}{dt} &= -\frac{2\varepsilon\sqrt{a_0^3}}{2\pi p} \int_0^{2\pi p} F(\theta, \frac{p}{q}\theta + \vartheta)\left(1 + a_0 Z'_\psi(\psi)\right) d\theta, \\ \frac{d\vartheta}{dt} &= -\frac{a - a_0}{2\sqrt{a_0^3}} - \frac{2\varepsilon\sqrt{a_0^3}}{2\pi p} \int_0^{2\pi p} F(\theta, \frac{p}{q}\theta + \vartheta, a) Z(\psi) d\theta. \end{aligned} \tag{6.63}$$

The stationary values $a_0$ and $\vartheta_0$ corresponding to the uniform rotation are obtained from (6.63) by setting $\frac{da}{dt} = \frac{d\vartheta}{dt} = 0$. Their stability is investigated with the help of variational equations.

As an example, we consider a particular case of the equation (6.47):

$$\frac{d^2x}{dt^2} + f(x) = \varepsilon\left[f(\nu t) - 2h\frac{dx}{dt}\right], \tag{6.64}$$

where $h > 0$ is a constant and $\overline{f(\nu t)} = 0$.

We limit ourselves by only investigating the principal resonance $p = q = 1$. The averaged equations (6.63) are now

$$\begin{aligned} \frac{da}{dt} &= -2\varepsilon\sqrt{a_0^3}\left\{a_0\overline{f(\theta)Z'_\psi} - \frac{2h}{\sqrt{a_0}}\right\}, \\ \frac{d\vartheta}{dt} &= -\frac{a - a_0}{2\sqrt{a_0^3}} + 2\varepsilon\sqrt{a_0^3}\,\overline{f(\theta)Z}. \end{aligned} \tag{6.65}$$

The stationary regimes are determined from the equations

$$\overline{f(\theta)Z'_\psi} = \frac{2h}{\sqrt{a_0^3}}, \quad 2\varepsilon\overline{f(\theta)Z} = \frac{a - a_0}{a_0^3}. \tag{6.66}$$

From the first equation of (6.66), we obtain $\vartheta = \vartheta_0$, and from the second $a = a_0$. To study the stability of the solutions $a_0$, $\theta_0$, we use the variational equations

$$\frac{d\delta a}{dt} = -2\varepsilon\sqrt{a_0^5}\,\overline{f(\theta)Z''_{\psi^2}}\,\delta\vartheta, \tag{6.67}$$

$$\frac{d\delta\vartheta}{dt} = -\frac{\delta a}{2\sqrt{a_0^3}} + 2\varepsilon\sqrt{a_0^3}\,\overline{f(\theta)Z'_\psi}\,\delta\vartheta. \tag{6.68}$$

Equation (6.68) can be simplified by using the first equation of (6.66). We have

$$\frac{d\delta\vartheta}{dt} = -\frac{\delta a}{2\sqrt{a_0^3}} + 4\varepsilon h\delta\vartheta. \tag{6.69}$$

The characteristic equation of the system (6.67) is

$$\lambda^2 - 2\varepsilon h\lambda - \varepsilon a_0\overline{f(\theta)Z''_{\psi^2}} = 0. \tag{6.70}$$

For $\varepsilon > 0$, $h > 0$, this equation always has a root with positive real part. Therefore, any resonance regime in the system described by equation (6.64) for the rapid rotation is unstable if there is a damping force.

The averaging method can be used to study rotational motions in the resonance case for more complicated systems than (6.4), for example:

$$\begin{aligned}
&\frac{d}{dt}\Big[m(y)\frac{dx}{dt}\Big] + f(y,x) = \varepsilon F\Big(y,\theta,x,\frac{dx}{dt},\varepsilon\Big),\\
&\frac{dy}{dt} = \varepsilon \mathcal{Y}\Big(y,\theta,x,\frac{dx}{dt},\varepsilon\Big),\\
&\frac{d\theta}{dt} = \nu(y) + \varepsilon\Theta\Big(y,\theta,x,\frac{dx}{dt},\varepsilon\Big),
\end{aligned} \tag{6.71}$$

where $x$ is an one-dimensional coordinate, $y = \{y_1, \dots, y_n\}$ are slowly varying parameters and $F\big(y,\theta,x,\frac{dx}{dt},\varepsilon\big)$, $\mathcal{Y}\big(y,\theta,x,\frac{dx}{dt},\varepsilon\big)$ and $\Theta\big(y,\theta,x,\frac{dx}{dt},\varepsilon\big)$ are nonlinear excited functions which are periodic on $\theta$.

**Motion of satellite about its centre of mass**

As an illustrative example for using the asymptotic method to study the resonance phenomena in the presence of rotation, we consider the plane motion of a satellite about its centre of mass which moves on an elliptic orbit in the gravitational field.

It is assumed that the principal central axis of inertia of the satellite is always perpendicular to the orbit-plane. Following Beletxki, we write the equation of motion of the satellite about its centre of mass in the form

$$(1 + e\cos\theta)\frac{d^2\delta}{d\theta^2} - 2e\sin\theta\frac{d\delta}{d\theta} + 3b^2\sin\delta = 4e\sin\theta, \tag{6.72}$$

where $\delta = 2\nu$ and $\nu$ is the angle between the radius-vector of the centre of mass and the axis of inertia, $C$ is the moment of inertia relative to this centre, $e$ is the eccentricity of the orbit; $\theta$ is the angular distance of the radius-vector from the perigee of the orbit; $b^2 = \dfrac{A-C}{B}$, where $A$ and $C$ are moments of inertia with respect to the first two principal central axes of inertia $A \geq C$; and $B$ is the moment of inertia with respect to the third principal central axis of inertia of the satellite; $b \leq 1$.

Equation (6.72) is a nonlinear differential equation of the second degree with periodic coefficients. The equation contains two parameters $b$ and $e$. When $e = 0$ (circular orbit) equation (6.27) reduces to the equation of the pendulum.

When $b = 0$ the equation (6.72) is integrable in elementary functions.

We consider three cases.

1) $e \ll 1$, the orbit is close to circular.

2) $b \ll 1$, the satellite is like a body with dynamic symmetry.

3) $\left|\dfrac{d\delta}{d\theta}\right| \gg 1$ the angular velocity of rotation of the satellite is much larger than that of the radius-vector of its centre of mass.

1) In the case $e \ll 1$, the equation (6.72) can be represented in the form:

$$\frac{d^2\delta}{d\theta^2} + 3b^2\sin\delta = e\Big(4\sin\theta + 2\sin\theta\frac{d\delta}{d\theta} + 3b^2\cos\theta\sin\delta\Big) + O(e^2). \tag{6.73}$$

When $e = 0$, one obtains an unexcited equation

$$\frac{d^2\delta}{d\theta^2} + 3b^2 \sin\delta = 0, \tag{6.74}$$

whose general solution describes the oscillatory, rotational or aperiodic motions and depends on two arbitrary constants $a_0$ and $\theta_0$. Aperiodic motion will not be considered here. Hence, the general solution of equation (6.74) has the form

$$\delta = z(\psi, a), \tag{6.75}$$

where

$$\psi = \omega(a)(\theta - \theta_0). \tag{6.76}$$

The function $z(\psi, a)$ satisfies the condition

$$z(\psi + 2\pi, a) = z(\psi, a) + \gamma \tag{6.77}$$

with every $a$; $\gamma = 0$ for oscillation and $\gamma = 2\pi$ for rotation.

Remember that for the equation (6.74) the integral of the action is

$$J(a) = \frac{\omega(a)}{2\pi} \int_0^{2\pi} z'^{2}_{\psi}(\psi, a) d\psi \tag{6.78}$$

and, according to the results obtained before

$$J'(a) = D(a). \tag{6.79}$$

By changing the variables in (6.73) with the help of the formulae:

$$\begin{aligned} \delta &= z(\psi, a), \\ \frac{d\delta}{d\theta} &= \omega(a) z'_{\psi}(\psi, a), \end{aligned} \tag{6.80}$$

we obtain

$$\begin{aligned} \frac{da}{d\theta} &= \frac{-e}{D(a)} F(\theta, \psi, a) z'_{\psi}(\psi, a), \\ \frac{d\psi}{d\theta} &= \omega(a) + \frac{e}{D(a)} F(\theta, \psi, a) z'_{a}(\psi, a), \end{aligned} \tag{6.81}$$

where $F(\theta, \psi, a)$ is the factor multiplying $e$ on the right hand side of equation (6.73), and the terms of degree $O(e^2)$ are neglected. The right-hand sides of (6.81) are periodic on $\psi$ and $\theta$ with period $2\pi$.

In the non-resonance case, when $\omega(a) \neq \frac{p}{q}$, where $p$ and $q$ are simple relative integers, the averaged equations for (6.81) become

$$\begin{aligned} \frac{da}{d\theta} &= 0, \\ \frac{d\psi}{d\theta} &= \omega(a). \end{aligned} \tag{6.82}$$

From here, we obtain: $a$ = const, $\psi = \omega(a)(\theta + \theta_0)$. Therefore, the exciting moment, causing the ellipticity of the orbit, does not affect the amplitude $a$ and the frequency $\omega(a)$ in the first approximation.

In the resonance case when

$$\omega(a) = \frac{p}{q} + O(e) \tag{6.83}$$

we introduce a new variable $\vartheta$, the phase lag, by means of the relationship

$$\psi = \frac{p}{q}(\theta + \vartheta). \tag{6.84}$$

Then, denoting by $a_0$ the value $a$ for which the exact resonance occurs:

$$\omega(a_0) = \frac{p}{q}, \tag{6.85}$$

we obtain the following equations with accuracy $O(e)$:

$$\begin{aligned} \frac{da}{d\theta} &= \frac{-e}{2\pi p D(a_0)} \int_0^{2\pi p} F\left(\frac{q}{p}\psi - \vartheta, \psi, a_0\right) z'_\psi(\psi, a_0)\, d\psi, \\ \frac{d\vartheta}{d\theta} &= \frac{q}{p}\omega'(a_0)(a - a_0) + \frac{eq}{2\pi p^2 D(a_0)} \int_0^{2\pi p} F\left(\frac{q}{p}\psi - \vartheta, \psi, a_0\right) z'_a(\psi, a_0)\, d\psi. \end{aligned} \tag{6.86}$$

The functions in the right-hand sides of (6.86) can be found by the aid of elliptic functions. The stationary solutions of (6.86) and their stability can be obtained as was done by Beletxki [4].

2) We now consider the second case $b \ll 1$. It is convenient to introduce a new independent variable $\tau$ in equation (6.72):

$$\begin{aligned} \tau &= 2\text{Arctan}\sqrt{\frac{1-e}{1+e}}\tan\frac{\theta}{2} - \frac{e\sqrt{1-e^2}\sin\theta}{1+e\cos\theta}, \\ \tau(\theta + 2\pi) &= \tau(\theta) + 2\pi, \end{aligned} \tag{6.87}$$

and a new unknown function $x$, the angle between the principal central axis of inertia and the radius-vector of the perigee:

$$x = \theta + \vartheta = \theta + \frac{\delta}{2} \cdot \tag{6.88}$$

We have the following equation of motion

$$\frac{d^2x}{d\tau^2} + \frac{3b^2}{2}\frac{(1+e\cos\theta)^3}{(1-e^2)^3}\sin 2(x-\theta) = 0, \tag{6.89}$$

where $\theta = \theta(\tau)$ is determined from (6.87).

When $b = 0$, the satellite rotates uniformly about its centre of mass: $x = C_1\tau + C_2$. With small $b$, the motion is close to uniform rotation.

We replace equation (6.89) by

$$\begin{aligned} \frac{dx}{d\tau} &= y, \\ \frac{dy}{d\tau} &= -\frac{3b^2}{2}\frac{(1-e\cos\theta)^3}{(1-e^2)^3}\sin 2(x-\theta). \end{aligned} \tag{6.90}$$

The solution of this equation is found in the form

$$x = \Omega\tau + \varphi, \quad y = \Omega + bz, \tag{6.91}$$

where $\Omega$ is a constant, and $\varphi$ and $z$ are new functions. We obtain the following equations in the standard form

$$\begin{aligned} \frac{d\varphi}{d\tau} &= bz, \\ \frac{dz}{d\tau} &= -\frac{3b^2}{2}\frac{(1+e\cos\theta)^3}{(1-e^2)^3}\sin 2(\Omega\tau + \varphi - \theta), \end{aligned} \tag{6.92}$$

where $b$ is a small parameter.

If $2\Omega$ is not an integer, then on averaging the right-hand sides of (6.92), we derive

$$\frac{d\varphi}{d\tau} = bz, \quad \frac{dz}{d\tau} = 0. \tag{6.93}$$

From here it follows that $z = \text{const}$, $\varphi = C\tau + C_0$ and therefore, as in the case of absence of excitation ($b = 0$), $x = C_1\tau + C_2$. We conclude that in the first approximation, the gravitational exciting moment does not affect the uniform rotation of the satellite.

Let $2\Omega = m$, where $m$ is an integer (when $m = 2$ there is the principal resonance. The period of rotation of the satellite is close to the period of its motion on the orbit). Averaging the equations (6.92), we derive

$$\frac{d\varphi}{d\tau} = bz, \qquad \frac{dz}{d\tau} = -\frac{3b^2}{2}\Phi_m(e)\sin 2\varphi, \tag{6.94}$$

where

$$\Phi_m(e) = \frac{1}{2\pi}\int_{-\pi}^{\pi} \frac{(1+e\cos\theta)^3}{(1-e^2)^3}\cos(m\tau - 2\theta)d\tau. \tag{6.95}$$

From (6.94), we can write

$$\frac{d^2(2\varphi)}{d\tau^2} + 3b^2\Phi_m(e)\sin 2\varphi = 0, \tag{6.96}$$

which coincides with the equation of motion of a pendulum.

Following (6.91) with $\Omega = m/2$ and $\varphi$ determined from (6.96), the function $x$ describes a rotation with angular velocity $\Omega$, on which the slow rotations or oscillations are superimposed. The position of equilibrium for the equation (6.96) $\varphi = \frac{n\pi}{2}$ $(n = 0, \pm1, \pm2, \ldots)$ corresponds to the uniform rotation with constant angular velocity. The stability of this rotation depends on the sign of the function $\Phi_m(e)$.

3) We now consider the rapid rotation of the satellite in the plane of the orbit, the period of rotation is much less than the period of motion of the satellite in its orbit. In this case, the equation (6.89) can be written in the form

$$\frac{d^2x}{d\tau^2} + g(x,\tau) = 0, \tag{6.97}$$

where the function $g(x,\tau)$ is periodic in $x$ with period $\pi$, and has a zero mean value for all $\tau$.

With the method presented above, for large values of $\Omega$: $\Omega \gg 1$, the asymptotic solution of (6.97) is found in the form

$$x = \Omega(\tau+\tau_0) + \frac{3b^2(1+e\cos\theta)^3}{8\Omega(1-e^2)^3}\Big\{\sin 2\big[\Omega(\tau+\tau_0) - \theta\big] - \sin 2\theta\Big\} + O\Big(\frac{1}{\Omega^3}\Big). \tag{6.98}$$

This solution consists of a rapid rotation with the angular frequency $\Omega$, on which the rapid periodic oscillation with frequency $2\Omega$ is superimposed, and slow oscillation with the phase $2\theta$.

We see that the averaging method can be used to solve many important problems of satellite dynamics.

## 7. Modified Averaging Methods

In nonlinear oscillation problems, there are three common kinds of equations of motion:

$$\begin{cases} \ddot{x} + \omega_1^2 x &= \varepsilon f(x, \dot{x}, y, \dot{y}, t), \\ \ddot{y} + \omega_2^2 y &= \varepsilon g(x, \dot{x}, y, \dot{y}, t). \end{cases} \tag{7.1}$$

$$\begin{cases} \ddot{x} + \omega_1^2 x &= \varepsilon f(x, \dot{x}, y, \dot{y}), \\ \ddot{y} + \omega_2^2 y &+ h\dot{y} = q(t) + \varepsilon g(x, \dot{x}, y, \dot{y}). \end{cases} \tag{7.2}$$

$$\begin{cases} \ddot{x} + \omega_1^2 x &= \varepsilon f(x, \dot{x}, y, \dot{y}), \\ \ddot{y} + \omega_2^2 y &+ h\dot{y} = g(x, \dot{x}). \end{cases} \tag{7.3}$$

Equations (7.1) describe the interaction between two non-linear oscillators with weak liaisons. The variables $x$ and $y$ have an identified role. For the equations (7.1), it is necessary to calculate the oscillations of the $x$ and $y$ - coordinates and their interaction.

Equations (7.2) appear in the problems of indirectly excited parametric oscillations (6, Chapter 4). The external force is applied to the $y$ - coordinate, but is in resonance with the $x$ - coordinate. The oscillation of the $y$ - coordinate is determined by external excitation. The influence of the $x$ - coordinate on the $y$ - coordinate is weak, but the influence of the $y$ - coordinate on the $x$ - coordinate is strong, hence, resonance occurs. To study the equations (7.2), one can first calculate the oscillation of the $y$ - coordinate, then calculate the oscillation of the $x$ - coordinate.

Equations (7.3) are an intermediate development of (7.1) and (7.2). The calculating process for them is as follows: First, find the general solution of the degenerate equation when $\varepsilon = 0$ corresponding to the $x$ - variable, this depends on two unknown constants. Then, substitute this solution into the $y$ - equation and find its undamped solution. Finally, substitute the last solution into the right-hand side of the $x$ - equation and study it by means of the averaging method.

We give below a more detailed investigation of the general equations of the type (7.3) [39,b].

Let us consider a system of $N + m$ differential equations of the form

$$\begin{aligned} &\sum_{j=1}^{N} (a_{ij}\ddot{x}_j + c_{ij}x_j) = \varepsilon f_i(x, \dot{x}, y, \dot{y}, t), \quad i = 1, 2, \dots N \\ &\sum_{r=1}^{n} (m_{kr}\ddot{y}_r + h_{kr}\dot{y} + n_{kr}y_r) = g_k(x, \dot{x}, t), \quad k = 1, 2, \dots, n \end{aligned} \tag{7.4}$$

where $(a_{ij})$, $(c_{ij})$, $(m_{kr})$, $(h_{kr})$ and $(n_{kr})$ are positive definite matrices; $\varepsilon$ is a small positive parameter. The $x$ - subsystem has natural frequencies $p_i$ $(i = 1, 2, ;N)$. The $x$ - variables will oscillate and are called oscillating variables. The $y$ - subsystem has the characteristic roots with negative real parts and therefore, the solution of the homogeneous part of the $y$ - subsystem $(g_k = 0)$ is damped. Hence, the $y$ - variables are called damped variables. It is seen that the oscillating variables strongly influence the damped variables. The influence of the damped variables on the oscillating variables is weak, although important, because the $x$ - subsystem is a quasilinear system.

First, we transform the $x$ - subsystem into the principal modes, and the $y$ - subsystem into a Jordan form:

$$\ddot{X}_i + p_i^2 X_i = \frac{\varepsilon}{M_i} F_i(X, \dot{X}, Y, \dot{Y}, t), \ (1 = 1, 2, \dots, N) \tag{7.5}$$

$$\begin{aligned}
\dot{Y}_1^{(s)} &= q_s Y_1^{(s)} + Y_2^{(s)} + G_1^{(s)}(X, \dot{X}, t), \\
\dot{Y}_2^{(s)} &= q_s Y_2^{(s)} + Y_3^{(s)} + G_2^{(s)}(X, \dot{X}, t), \\
&\dots\dots\dots\dots\dots\dots\dots\dots\dots \\
\dot{Y}_{ns-1}^{(s)} &= q_s Y_{ns-1}^{(s)} + Y_{ns}^{(s)} + G_{ns-1}^{(s)}(X, \dot{X}, t), \\
\dot{Y}_{ns}^{(s)} &= q_s Y_{ns}^{(s)} + G_{n_s}^{(s)}(X, \dot{X}, t), \\
&(s = 1, 2, \dots, w \leq 2n, \ n_1 + n_2 + \dots + n_w = 2n),
\end{aligned} \tag{7.6}$$

where

$$F_i(X, \dot{X}, Y, \dot{Y}, t) = \sum_{j=1}^{N} d_j^{(i)} f_j(X, \dot{X}, Y, \dot{Y}, t). \tag{7.7}$$

We now consider the resonance oscillations, letting

$$\begin{aligned}
x_i &= a_i \sin(p_i t + \varphi_i), \\
\dot{x}_i &= a_i p_i \cos(p_i t + \varphi_i). \\
&(i = 1, 2, \dots, N)
\end{aligned} \tag{7.8}$$

Substituting these expressions into the functions $G_j^{(s)}$, considering $a_i$ and $\varphi_i$ as constants, we obtain the general solution of the $y$ - subsystem:

$$\begin{aligned}
Y_1^{(s)} &= Y_1^{(s)o}(a, \varphi, t) + \left\{ \frac{t^{ns-1}}{(ns-1)!} P_{ns}^{(s)} + \frac{t^{ns-2}}{(ns-2)!} P_{ns-1}^{(s)} + \right. \\
&\quad \left. + \dots + \frac{t}{1!} P_2^{(s)} + P_1^{(s)} \right\} e^{q s t}, \\
Y_2^{(s)} &= Y_2^{(s)o}(a, \varphi, t) + \left\{ \frac{t^{ns-1}}{(ns-1)!} P_{ns}^{(s)} + \frac{t^{ns-2}}{(ns-2)!} P_{ns-1}^{(s)} + \right. \\
&\quad \left. + \dots + P_2^{(s)} \right\} e^{q s t}, \\
&\dots\dots\dots\dots\dots\dots\dots\dots\dots
\end{aligned} \tag{7.9}$$

$$Y^{(s)}_{ns-1} = Y^{(s)o}_{ns-1}(a,\varphi,t) + \left\{tP^{(s)}_{ns} + P^{(s)}_{ns-1}\right\}e^{q_s t},$$
$$Y^{(s)}_{ns} = Y^{(s)o}_{ns}(a,\varphi,t) + \left\{P^{(s)}_{ns}\right\}e^{q_s t},$$
$$(s = 1,2,\ldots,w \leq 2n,\ n_1 + n_2 + \cdots + n_w = 2n)$$

where $Y_1^{(s)o}$, $Y_2^{(s)o}$, ..., $Y_{ns}^{(s)o}$ are forced oscillation terms, and $P_1^{(s)}$, $P_2^{(s)}$, ..., $P_{ns}^{(s)}$ are arbitrary constants.

Considering (7.8) and (7.9) as the formulae of transformation of the variables from $(X, Y)$ to $(a, \varphi, Q)$, we have:

$$\begin{aligned}
\dot{a}_i &= \frac{\varepsilon}{M_i P_i} F_i \cos(p_i t + \varphi_i),\\
\dot{\varphi}_i &= \frac{-\varepsilon}{M_i P_i a_i} F_i \sin(p_i t + \varphi),\\
\dot{Q}_1^{(s)} &= q_s Q_1^{(s)} + Q_2^{(s)} + \varepsilon R_1^{(s)},\\
\dot{Q}_2^{(s)} &= q_s Q_2^{(s)} + Q_3^{(s)} + \varepsilon R_2^{(s)},\\
&\cdots\cdots\cdots\cdots\cdots\cdots\cdots\cdots\\
\dot{Q}^{(s)}_{ns-1} &= q_s Q^{(s)}_{ns-1} + Q^{(s)}_{ns} + \varepsilon R^{(s)}_{ns-1},\\
\dot{Q}^{(s)}_{ns} &= q_s Q^{(s)}_{ns} + \varepsilon R^{(s)}_{ns},\\
&(s = 1,2,\ldots,w \leq 2n,\ n_1 + n_2 + \cdots + n_w = 2n)
\end{aligned} \tag{7.10}$$

where $Q$ are the terms in the brackets in the right-hand sides of (7.9) and

$$R_u^{(s)} = \sum_{i=1}^{N} \frac{F_i}{M_i P_i a_i} \frac{\partial Y_u^{(s)o}}{\partial(p_i t + \varphi_i)} \sin(p_i t + \varphi_i) - \frac{\partial Y_u^{(s)o}}{\partial a_i} \cos(p_i t + \varphi_i). \tag{7.11}$$

In the first approximation, we have the following averaged equations:

$$\begin{aligned}
\dot{a}_i &= \frac{\varepsilon}{M_i P_i} \langle F_i \cos(p_i t + \varphi_i)\rangle,\\
\dot{\varphi}_i &= \frac{-\varepsilon}{M_i P_i a_i} \langle F_i \sin(p_i t + \varphi_i)\rangle,\\
\dot{Q}_1^{(s)} &= q_s Q_1^{(s)} + Q_2^{(s)},\\
\dot{Q}_2^{(s)} &= q_s Q_2^{(s)} + Q_3^{(s)},\\
&\cdots\cdots\cdots\cdots\cdots\cdots\\
\dot{Q}^{(s)}_{ns} &= q_s Q^{(s)}_{ns},
\end{aligned} \tag{7.12}$$

$(i = 1,2,\ldots,N,\ s = 1,2,\ldots,w \leq 2n,\ n_1 + n_2 + \cdots + n_w = 2n)$.

The subsystem $(Q)$ has only a trivial stationary zero solution, which is asymptotically stable, since $\mathrm{Re}(q_s) < 0$. Hence, the problem of studying the system

(7.12) leads to examining the subsystem $(a, \varphi)$, that is to considering the $x$ - subsystem.

## 8. Averaging Method and Stability of Motion in the Critical Case

This section deals with the stability of motion in that critical case, in which the characteristic equation of the system in the first approximation has a pair of imaginary roots [37,c]. The necessary and sufficient conditions for the stability of the zero solution are obtained by using the asymptotic method. Some examples explaining the method used are given. This method is effective in many practical problems.

Let

$$\begin{aligned} \dot{x} &= -\lambda y + \varepsilon X(x, y, \varepsilon), \\ \dot{y} &= \lambda x + \varepsilon Y(x, y, \varepsilon), \end{aligned} \tag{8.1}$$

be the differential equations of the perturbation motion, where $\varepsilon$ is a small positive parameter, $\lambda$ is a positive constant. $X$, $Y$ are the polynomial functions in $x$, $y$, whose expansions begin with terms of at least the second degree:

$$\begin{aligned} X(x, y, \varepsilon) &= X_2(x, y) + \varepsilon X_3(x, y) + \varepsilon^2 X_4(x, y) + \dots, \\ Y(x, y, \varepsilon) &= Y_2(x, y) + \varepsilon Y_3(x, y) + \dots \end{aligned}$$

Here $X_s(x, y)$, $Y_s(x, y)$ denote the terms in $x$, $y$ of the order $s$.

The problem of stability of the zero solution $x = y = 0$ of the system (8.1) can be resolved simply by the following discussion. It is noteworthy that, when the stability is lost, in the considered system lying on the boundary of the stability, increasing oscillations near to the harmonic occur. Therefore, we shall establish the relationship between the system (8.1) and a linear system with constant coefficients

$$\begin{aligned} \dot{x} &= ax + by, \\ \dot{y} &= cx + dy. \end{aligned} \tag{8.2}$$

The question of the stability of the zero solution $x = y = 0$ of the system (8.2) is answered by the sign of the real part of the roots of the characteristic equation

$$\sigma^2 - (a + d)\sigma + ad - bc = 0. \tag{8.3}$$

For reduction of the system (8.1) to the equivalent linear form (8.2), we note that the motion of the system (8.1) in the neighbourhood of the boundary of stability is near to the harmonic:

$$\begin{aligned} x &= r\cos\theta + \varepsilon u_1(r, \theta) + \varepsilon^2 u_2(r, \theta) + \dots, \\ y &= r\sin\theta + \varepsilon v_1(r, \theta) + \varepsilon^2 v_2(r, \theta) + \dots, \end{aligned} \tag{8.4}$$

where $u_i(r,\theta)$, $v_i(r,\theta)$ are periodic functions with the period $2\pi$ relative to $\theta$: $u_i(r,\theta) = u_i(r,\theta+2\pi)$, $v_i(r,\theta) = v_i(r,\theta+2\pi)$, not containing the harmonics $\sin\theta$, $\cos\theta$ and $r$ is a constant. The function $\theta$ satisfies the equation [6]:

$$\dot{\theta} = \lambda + \varepsilon B_1(r) + \varepsilon^2 B_2(r) + \dots \tag{8.5}$$

In order to find the unknown quantities $u_i$, $v_i$, $B_i$, we equate the coefficients of the corresponding powers of $\varepsilon$ in (8.1), using the expressions (8.4) for $x$, $y$. We have

$$\dot{x} = (\lambda + \varepsilon B_1 + \varepsilon^2 B_2 + \dots)\Big( - r\sin\theta + \varepsilon\frac{\partial u_1}{\partial\theta} + \varepsilon^2\frac{\partial u_2}{\partial\theta} + \dots\Big),$$
$$\dot{y} = (\lambda + \varepsilon B_1 + \varepsilon^2 B_2 + \dots)\Big(r\cos\theta + \varepsilon\frac{\partial v_1}{\partial\theta} + \varepsilon^2\frac{\partial v_2}{\partial\theta} + \dots\Big),$$

By equating coefficients of the first power of $\varepsilon$, we find that

$$\begin{aligned} -rB_1\sin\theta + \lambda\frac{\partial u_1}{\partial\theta} &= -\lambda v_1 + X(r\cos\theta, r\sin\theta, 0),\\ rB_1\cos\theta + \lambda\frac{\partial v_1}{\partial\theta} &= \lambda u_1 + Y(r\cos\theta, r\sin\theta, 0). \end{aligned} \tag{8.6}$$

Let

$$\begin{aligned} X(r\cos\theta, r\sin\theta, 0) &= g_0 + \sum_{n=1}^{\infty}(g_n\cos n\theta + h_n\sin n\theta),\\ Y(r\cos\theta, r\sin\theta, 0) &= c_0 + \sum_{n=1}^{\infty}(c_n\cos n\theta + d_n\sin n\theta), \end{aligned} \tag{8.7}$$

be the expansions of the functions $X$, $Y$, then we obtain:

$$g_1 = d_1 = 0, \quad rB_1 = -h_1 = c_1, \tag{8.8}$$

$$\begin{aligned} u_1 &= a_1 + \sum_{n=2}^{\infty}(\beta_{n1}\cos n\theta + \gamma_{n1}\sin n\theta),\\ v_1 &= a_2 + \sum_{n=2}^{\infty}(\beta_{n2}\cos n\theta + \gamma_{n2}\sin n\theta), \end{aligned} \tag{8.9}$$

where

$$\begin{aligned} &a_1 = -\frac{c_0}{\lambda}, \quad a_2 = \frac{g_0}{\lambda}, \quad \beta_{n1} = \frac{c_n - nh_n}{\lambda(n^2-1)},\\ &\beta_{n2} = \frac{g_n + nc_n}{\lambda(1-n^2)}, \quad \gamma_{n1} = \frac{c_n - ng_n}{\lambda(n^2-1)}, \quad \gamma_{n2} = \frac{h_n - nc_n}{\lambda(1-n^2)}. \end{aligned} \tag{8.10}$$

By linearization of the right-hand parts of the system (8.1), we can write:

$$\begin{aligned}\varepsilon X(x,y,\varepsilon) &= \varepsilon X(r\cos\theta+\varepsilon u_1+\ldots,r\sin\theta+\varepsilon v_1+\ldots,\varepsilon)\approx ax+by,\\ \varepsilon Y(x,y,\varepsilon) &= \varepsilon Y(r\cos\theta+\varepsilon u_1+\ldots,r\sin\theta+\varepsilon v_1+\ldots,\varepsilon)\approx cx+dy.\end{aligned}\tag{8.11}$$

Since in the expressions $u_i$, $v_i$, the harmonics $\cos\theta$, $\sin\theta$ are absent, then for determination of the constants $a$, $b$, $c$, $d$, we average both parts of (8.11) over the variable $\theta$, multiplied by $\frac{2}{r}\cos\theta$ and $\frac{2}{r}\sin\theta$. We have

$$\begin{aligned}
a &= \frac{\varepsilon}{\pi r}\int_0^{2\pi} X(r\cos\theta+\varepsilon u_1+\ldots,r\sin\theta+\varepsilon v_1+\ldots,\varepsilon)\cos\theta d\theta,\\
b &= \frac{\varepsilon}{\pi r}\int_0^{2\pi} X(r\cos\theta+\varepsilon u_1+\ldots,r\sin\theta+\varepsilon v_1+\ldots,\varepsilon)\sin\theta d\theta,\\
c &= \frac{\varepsilon}{\pi r}\int_0^{2\pi} Y(r\cos\theta+\varepsilon u_1+\ldots,r\sin\theta+\varepsilon v_1+\ldots,\varepsilon)\cos\theta d\theta,\\
d &= \frac{\varepsilon}{\pi r}\int_0^{2\pi} Y(r\cos\theta+\varepsilon u_1+\ldots,r\sin\theta+\varepsilon v_1+\ldots,\varepsilon)\sin\theta d\theta.
\end{aligned}\tag{8.12}$$

We can now write the system (8.1) in the following equivalent linear form:

$$\dot{x} = ax+(b-\lambda)y, \quad \dot{y} = (c+\lambda)x+dy. \tag{8.13}$$

The condition for the asymptotic stability of the zero solution of the system (8.13) is:

$$a+d<0, \tag{8.14}$$

$$ad-(b-\lambda)(c+\lambda) = \lambda^2+0(\varepsilon)>0. \tag{8.15}$$

For sufficiently small $\varepsilon$ the condition (8.15) is always satisfied. Consequently, the inequality (8.14) is the unique condition of asymptotic stability of the zero solution $x=y=0$. If $a+d>0$, then instability takes place.

The case $a+d=0$ is indeterminate. For the solution of the problem of stability in this case, it is necessary to write the terms with a higher order than in expansions (8.4). However, in practice it often requires the knowledge only of the form of solution in the first approximation $x=r\cos\theta$, $y=r\sin\theta$ or in the second approximation $x=r\cos\theta+\varepsilon u_1$, $y=r\sin\theta+\varepsilon v_1$.

We now consider some illustrative examples.

**Example 25**

We study the stability of the zero solution of the system

$$\begin{aligned} \dot{x} &= -y - \frac{a}{2}(x^2 + y^2)x, \\ \dot{y} &= x - \frac{a}{2}(x^2 + y^2)y. \end{aligned} \tag{8.16}$$

By letting

$$x = r\cos\theta, \quad y = r\sin\theta$$

we have

$$\begin{aligned} \dot{x} &= -\frac{a}{2}r^2 x - y, \\ \dot{y} &= x - \frac{a}{2}r^2 y. \end{aligned}$$

Therefore, $a+d = -ar^2$ and the zero solution of the system (8.16) is asymptotically stable for positive $a$ and unstable for negative $a$.

**Example 26**

In the equation:

$$\ddot{x} + x = a\dot{x}f(x, \dot{x}^2) + g(x, \dot{x}^2) \tag{8.17}$$

assuming that $x = r\sin\theta$, $\dot{x} = r\cos\theta$, we obtain:

$$a\dot{x}f(x, \dot{x}^2) + g(x, \dot{x}^2) \approx A\dot{x} + Bx,$$

where

$$\begin{aligned} A &= \frac{1}{r\pi}\int_0^{2\pi} \left[a\dot{x}f(x, \dot{x}^2) + g(x, \dot{x}^2)\right]\cos\theta d\theta = \\ &= \frac{a}{\pi}\int_0^{2\pi} f(r\sin\theta, r^2\cos^2\theta)\cos^2\theta d\theta. \end{aligned}$$

Consequently, the zero solution of the system (8.17) is asymptotically stable when $A < 0$ and unstable when $A > 0$.

**Example 27**

Consider the equation of perturbed motion of the form:

$$\ddot{y} + y = \varepsilon\psi(a\dot{y} - by)^2. \tag{8.18}$$

One can verify that the form of the solution of this equation in the first approximation does not solve the problem of stability. Therefore, we write its solution in the second approximation:

$$\begin{aligned} y &= r\sin\theta + \varepsilon\psi r^2(\alpha + \beta\cos 2\theta + \gamma\sin 2\theta), \\ \dot{y} &= r\cos\theta + 2\varepsilon\psi r^2(-\beta\sin 2\theta + \gamma\cos 2\theta), \end{aligned} \tag{8.19}$$

where

$$\alpha = \frac{a^2 + b^2}{2}, \quad \beta = \frac{b^2 - a^2}{6}, \quad \gamma = \frac{ab}{3}.$$

Substituting $y$, $\dot{y}$ from (8.19) into the right-hand side of equation (8.18) and linearizating it, we find:

$$\begin{aligned} \ddot{y} + y = \varepsilon\psi\Big\{a\big[r\cos\theta + 2\varepsilon\psi r^2(-\beta\sin 2\theta + \gamma\cos 2\theta)\big] - \\ - b\big[r\sin\theta + \varepsilon\psi r^2(\alpha + \beta\cos 2\theta + \gamma\sin 2\theta)\big]\Big\}^2 = A\dot{y} + By + \ldots, \end{aligned}$$

where

$$A = M_\theta\left\{\frac{2\varepsilon}{r}\psi(a\dot{y} - by)^2\cos\theta\right\} = -\frac{\varepsilon^2 r^2 ab}{2}(a^2 + b^2)\psi^2.$$

From here, it follows that if $ab > 0$ then the zero solution is asymptotically stable, and if $ab < 0$, it is unstable.

## APPENDIX 1

# PRINCIPAL COORDINATES

A system with $N$ degrees of freedom has $N$ principal modes of oscillations. It is possible to define generalized coordinates such that each of them can perform a harmonic oscillation only at one of the natural frequencies of the system. Such coordinates are called principal or normal coordinates.

* Consider first a system of two differential equations which is very often met in applied problems

$$a\ddot{x} + bx + cy = f_1,$$
$$d\ddot{y} + ey + hx = f_2, \tag{1}$$

where $f_1$ and $f_2$ are functions of coordinates and time, $a$, $b$, $c$, $d$, $e$ and $h$ are constants satisfied the conditions for the positiveness of $\Omega_1^2$, $\Omega_2^2$ (4).

Introduce now the principal coordinates $\xi_1, \xi_2, \ldots, \xi_n$ connected with the primary coordinates $x$, $y$ by the relations

$$x = \xi_1 + \xi_2, \quad y = \sigma_1\xi_1 + \sigma_2\xi_2, \tag{2}$$

where

$$\sigma_1 = \frac{h}{d\Omega_1^2 - e} = \frac{a\Omega_1^2 - b}{c},$$
$$\sigma_2 = \frac{h}{d\Omega_2^2 - e} = \frac{a\Omega_2^2 - b}{c}, \tag{3}$$

and $\Omega_1$, $\Omega_2$ are the natural frequencies - the roots of the characteristic equation:

$$(b - a\Omega^2)(e - d\Omega^2) - hc = 0,$$
$$\Omega_1^2 = \frac{1}{2ad}\left[ae + bd - \sqrt{(ae - bd)^2 + 4acdh}\right],$$
$$\Omega_2^2 = \frac{1}{2ad}\left[ae + bd + \sqrt{(ae - bd)^2 + 4acdh}\right]. \tag{4}$$

It is easy to verify that

$$\sigma_1.\sigma_2 = \frac{-ah}{dc}. \tag{5}$$

We have the following equations for the principal coordinates:

$$\ddot{\xi}_1 + \Omega_1^2\xi_1 = k_1(f_1 + \frac{c}{h}\sigma_1 f_2),$$
$$\ddot{\xi}_2 + \Omega_2^2\xi_2 = k_2(f_1 + \frac{c}{h}\sigma_2 f_2), \tag{6}$$

where

$$k_1 = \frac{\sigma_2}{a(\sigma_2 - \sigma_1)} = \frac{1}{a + \frac{dc}{h}\sigma_1^2}\,,$$
$$k_2 = \frac{\sigma_1}{a(\sigma_1 - \sigma_2)} = \frac{1}{a + \frac{dc}{h}\sigma_2^2}\,. \tag{7}$$

**Note :** The following inequalities are true for the case $ad > 0$, $hc > 0$:

$$\Omega_1^2 < \frac{e}{d}\,, \quad \frac{b}{a} < \Omega_2^2. \tag{8}$$

In fact, we have:

$$\Omega_{1,2}^2 = \frac{1}{2}\Big(\frac{e}{d} + \frac{b}{a} \mp \Delta^2\Big),$$

where

$$\Delta^2 = \sqrt{\Big(\frac{e}{d} - \frac{b}{a}\Big)^2 + 4\frac{hc}{ad}}\,.$$

It is obvious that:

$$\frac{e}{d} - \Omega_1^2 = \frac{1}{2}\Big(\frac{e}{d} - \frac{b}{a} + \Delta^2\Big), \tag{9}$$

$$\Omega_2^2 - \frac{e}{d} = \frac{1}{2}\Big(\frac{b}{a} - \frac{e}{d} + \Delta^2\Big), \tag{10}$$

$$\frac{b}{a} - \Omega_1^2 = \frac{1}{2}\Big(\frac{b}{a} - \frac{e}{d} + \Delta^2\Big), \tag{11}$$

$$\Omega_2^2 - \frac{b}{a} = \frac{1}{2}\Big(\frac{e}{d} - \frac{b}{a} + \Delta^2\Big). \tag{12}$$

Adding (9) and (10) we obtain:

$$\Big(\frac{e}{d} - \Omega_1^2\Big) + \Big(\Omega_2^2 - \frac{e}{d}\Big) = \Omega_2^2 - \Omega_1^2 = \Delta^2 > 0. \tag{13}$$

Multiplying (9) and (10) we have:

$$\Big(\frac{e}{d} - \Omega_1^2\Big)\Big(\Omega_2^2 - \frac{e}{d}\Big) = \frac{1}{4}\Big[\Delta^4 - \Big(\frac{e}{d} - \frac{b}{a}\Big)^2\Big] = \frac{hc}{ad} > 0. \tag{14}$$

From (13) and (14) it follows

$$\frac{e}{d} - \Omega_1^2 > 0, \quad \Omega_2^2 - \frac{e}{d} > 0. \tag{15}$$

Similarly, we have the inequalities (8):

$$\frac{b}{a} - \Omega_1^2 > 0, \quad \Omega_2^2 - \frac{b}{a} > 0. \tag{16}$$

**Example 28**

For equations:

$$
\begin{aligned}
m_1\ddot{x}_1 + c_1(x_1 - x_2) &= P\cos\omega t, \\
m_2\ddot{x}_2 + c_1(x_2 - x_1) + c_2x_2 &= -\alpha f(x_2),
\end{aligned}
\tag{17}
$$

we have

$$
a = m_1, \quad b = -c = c_1, \quad d = m_2, \quad e = c_1 + c_2, \quad h = -c_1.
$$

Hence,

$$
\begin{aligned}
\sigma_1 &= \frac{c_1}{c_1 + c_2 - m_2\Omega_1^2}, \quad \sigma_2 = \frac{c_1}{c_1 + c_2 - m_2\Omega_2^2}, \\
k_1 &= \frac{\sigma_2}{m_1(\sigma_2 - \sigma_1)}, \quad k_2 = \frac{\sigma_1}{m_1(\sigma_2 - \sigma_1)}, \\
\Omega_{1,2}^2 &= \frac{1}{2}\left[\frac{c_1}{m_1} + \frac{c_1 + c_2}{m_2} \mp \sqrt{\left(\frac{c_1}{m_1}\right)^2 + \left(\frac{c_1 + c_2}{m_2}\right)^2 + \frac{2c_1(c_1 - c_2)}{m_1m_2}}\right],
\end{aligned}
\tag{18}
$$

and equations in the principal coordinates are

$$
\begin{aligned}
x_1 &= \xi_1 + \xi_2, \quad x_2 = \sigma_1\xi_1 + \sigma_2\xi_2, \\
\ddot{\xi}_1 + \Omega_1^2\xi_1 &= \frac{P}{m_1}.\frac{\sigma_2}{\sigma_2 - \sigma_1}\cos\omega t + \frac{\alpha}{m_2}.\frac{1}{\sigma_2 - \sigma_1}f(\sigma_1\xi_1 + \sigma_2\xi_2), \\
\ddot{\xi}_2 + \Omega_2^2\xi_2 &= \frac{P}{m_1}.\frac{\sigma_1}{\sigma_1 - \sigma_2}\cos\omega t + \frac{\alpha}{m_2}.\frac{1}{\sigma_1 - \sigma_2}f(\sigma_1\xi_1 + \sigma_2\xi_2).
\end{aligned}
\tag{19}
$$

**Example 29**

For the system

$$
\begin{aligned}
\ddot{x} + (1 + p)x - py &= \varepsilon f_1, \\
n\ddot{y} + p(y - x) &= \varepsilon f_2,
\end{aligned}
\tag{20}
$$

we have

$$
\begin{aligned}
&a = 1, \quad b = 1 + p, \quad c = -p, \quad d = n, \quad e = -h = p, \\
&\sigma_1 = \frac{-p}{n\Omega_1^2 - p}, \quad \sigma_2 = \frac{-p}{n\Omega_2^2 - p}, \\
&k_1 = \frac{1}{1 + n\sigma_1^2}, \quad k_2 = \frac{1}{1 + n\sigma_2^2}, \\
&\Omega_{1,2}^2 = \frac{1}{2n}\left[p + n(1 + p) \mp \sqrt{[p - n(1 + p)]^2 + 4np^2}\right],
\end{aligned}
\tag{21}
$$

and the equations for $\xi_1$ and $\xi_2$ are

$$
\begin{aligned}
x &= \xi_1 + \xi_2, \quad y = \sigma_1\xi_1 + \sigma_2\xi_2, \\
\ddot{\xi}_1 + \Omega_1^2\xi_1 &= \frac{\varepsilon}{1 + n\sigma_1^2}(f_1 + \sigma_1 f_2), \\
\ddot{\xi}_2 + \Omega_2^2\xi_2 &= \frac{\varepsilon}{1 + n\sigma_2^2}(f_1 + \sigma_2 f_2).
\end{aligned}
\tag{22}
$$

* **We consider now a system of $n$ differential equations of second order**

$$\begin{aligned} m_1\ddot{x}_1 + c_1x_1 + \cdots + c_nx_n &= f_1, \\ m_2\ddot{x}_2 + d_1x_1 + \cdots + d_nx_n &= f_2, \\ \dots\dots\dots\dots\dots\dots\dots \\ m_n\ddot{x}_n + l_1x_1 + \cdots + l_nx_n &= f_n. \end{aligned} \tag{23}$$

**Suppose that, the characteristic determinant**

$$D(\Omega^2) = \begin{vmatrix} c_1 - m_1\Omega^2 & c_2 & \dots & c_n \\ d_1 & d_2 - m_2\Omega^2 & \dots & d_n \\ \vdots & \vdots & \ddots & \vdots \\ \ell_1 & \ell_2 & \dots & \ell_n - m_n\Omega^2 \end{vmatrix}$$

**has $n$ eigenvalues $\Omega_1^2$, $\Omega_2^2$, ..., $\Omega_n^2$ - the real positive roots of the equation $D(\Omega^2) = 0$. We have the following formulae, transforming (23) into the principal modes**

$$x_j = \sum_{k=1}^{n} d_j^{(k)}\xi_k, \quad j = 1, 2, \dots, n, \tag{24}$$

**where $d_j^{(k)} = \frac{\sigma_j^{(k)}}{\sigma_1^{(k)}}$ and $\sigma_j^{(k)} = D_j(\Omega_k^2)$ is the algebraic supplement of the element at j-th column and last row of $D(\Omega_k^2)$, i.e. $\sigma_j^{(k)} = (-1)^{j+k}D_{jk}$, where $D_{jk}$ is obtained from $D(\Omega^2)$ by striking out the last row and j-th column.**

**The equations for $\xi_k$ are:**

$$\ddot{\xi}_k + \Omega_k^2\xi_k = \frac{1}{M_k}\cdot\sum_{i=1}^{n} d_i^{(k)} f_i, \tag{25}$$

**where**

$$M_k = \sum_{i=1}^{n} m_i[d_i^{(k)}]^2, \quad k = 1, 2, \dots, n. \tag{26}$$

# APPENDIX 2

## SOME TRIGONOMETRIC FORMULAE OFTEN USED IN THE AVERAGING METHOD

1. $\sin^2\alpha = \frac{1}{2}(1 - \cos 2\alpha)$

2. $\cos^2\alpha = \frac{1}{2}(1 + \cos 2\alpha)$

3. $\sin\alpha\cos\alpha = \frac{1}{2}\sin 2\alpha$

4. $\sin^3\alpha = \frac{1}{4}(3\sin\alpha - \sin 3\alpha)$

5. $\cos^3\alpha = \frac{1}{4}(3\cos\alpha + \cos 3\alpha)$

6. $\sin\alpha\cos^2\alpha = \frac{1}{4}(\sin\alpha + \sin 3\alpha)$

7. $\cos\alpha\sin^2\alpha = \frac{1}{4}(\cos\alpha - \cos 3\alpha)$

8. $\sin^4\alpha = \frac{1}{8}(3 - 4\cos 2\alpha + \cos 4\alpha)$

9. $\cos^4\alpha = \frac{1}{8}(3 + 4\cos 2\alpha + \cos 4\alpha)$

10. $\sin^3\alpha\cos\alpha = \frac{1}{8}(2\sin 2\alpha - \sin 4\alpha)$

11. $\sin\alpha\cos^3\alpha = \frac{1}{8}(2\sin 2\alpha + \sin 4\alpha)$

12. $\sin^2\alpha\cos^2\alpha = \frac{1}{8}(1 - \cos 4\alpha)$

13. $\sin^5\alpha = \frac{1}{16}(\sin 5\alpha - 5\sin 3\alpha + 10\sin\alpha)$

14. $\cos^5\alpha = \frac{1}{16}(\cos 5\alpha + 5\cos 3\alpha + 10\cos\alpha)$

15. $\sin\alpha\sin\beta = \frac{1}{2}[\cos(\alpha - \beta) - \cos(\alpha + \beta)]$

16. $\sin\alpha\cos\beta = \frac{1}{2}[\sin(\alpha - \beta) + \sin(\alpha + \beta)]$

17. $\cos\alpha\cos\beta = \frac{1}{2}[\cos(\alpha - \beta) + \cos(\alpha + \beta)]$

# REFERENCES


1. Alifov A.A., Frolov K.V.: *Interaction of Nonlinear Oscillatory Systems with Power Supply* (in Russian), Nauka, Moscow, (1985), 328p.

2. Andronov A.A., Vitt A.A. and Khaikin S.E.: *The Theory of Oscillations* (in Russian), second edition, Fizmatgiz, Moscow, (1959), 915p.

3. Babakov I.M.: *The Theory of Oscillations* (in Russian), second edition, Nauka, Moscow, (1965), 625p.

4. Beletxkii V.V.: *Motion of an Artificial Satellite about its Centre of Mass* (in Russian), Moscow, Nauka, (1965), 416p.

5. Blaquiere A.: *Nonlinear System Analysis.* Academic Press, (1966), 392p.

6. Bogoliubov N.N., Mitropolskii Yu.A.: *Asymptotic Methods in the Theory of Nonlinear Oscillations.* (in Russian), fourth edition, Moscow, (1974), 410p.

7. Bolotin V.V.: *Dynamic Stability of Elastic Systems* (in Russian), Gostekhizdat, Moscow, (1956), 600p.

8. Bulgakov B.V.: *Oscillations* (in Russian), Gostekhizdat, Moscow, (1954), 890p.

9. Chernousko F.L.: *Resonance Phenomena in Motion of Satellite about Centre of Mass* (in Russian), *J. of computational mathematics and math. physics,* Moscow, 3, No.3, (1963).

10. Chetaev N.G.: *Stability of Motion* (in Russian), second edition, Gostekhizdat, Moscow, (1955), 207p.

11. Clendening W.R., Dubey R.N.: *An Analysis of Control Methods for Galloping System. Trans. ASME,* series B, vol.95, No.3, (1973).

12. Den Hartog J.P.: *Mechanical Vibrations,* Mc Graw-Hill, New York, (1947), 460p.

13. Frolov K.V.:

    a) *Decreasing the amplitude of oscillation of resonance systems by control of the parameters. Construction of Machines,* Moscow, No.3, (1965).

    b) *Some nonlinear effects in physical systems with variable parameters. Abh. DAW, Abt. Math. Naturwiss. Techn. Heft* 5, (1977).

14. Ganiev R.F. and Kononenko V.O.: *Oscillations of rigid bodies* (in Russian), Moscow, Nauka, (1976), 432p.

15. Hagedorn P.:

    a) *Nonlinear Oscillations.* Second edition, Oxford science publications, (1988), 311p.

    b) *Uber die Tilgung selbsterrergter Schwingungen. ZAMP.,* vol.29, (1978) pp. 815-821.

16. Hale J.K.: *Oscillations in Nonlinear Systems,* Mc Graw-Hill, New York, (1963), 229p.

17. Hayashi C.: *Nonlinear Oscillations in Physical Systems,* Mc Mc Graw-Hill, New York, (1964), 204p.

18. Inoue J. and Others: *On self-synchronization of Mechanical Vibrators. Bulletin of the JSME* vol.B, 2, (1970), No.56, pp. 258-263.

19. Kauderer H.: *Nichtlineare Mechanik.* Springer - Verlag Berlin (1958), 777p.

20. Kononenko V.O.: *Vibrating System with a Limited Power Supply,* London, (1969), 236p.

21. Kryloff N. and Bogoliuboff N.: *Introduction to Nonlinear Mechanics.* Pinceton, New York, (1949).

22. Lefschetz S.: *Differential Equations. Geometric Theory.* Second Edition, N.Y. Intersci Publ. (1957), 364p.

23. Leimanis E. and Minorsky N.: *Dynamics and Nonlinear Mechanics.* John wiley & sons, Inc. New York, (1958).

24. Magnus K.: *Vibrations.* Blackie & son Limited, London, (1965), 299p.

25. Malkin J.G.: *Some Problems in the Theory of Nonlinear Oscillations* (in Russian), Gostekhizdat, Moscow (1956), 491p.

26. Mansour W.M.: *Quenching of limit Cycles of a Van der Pol Oscillator. J. Sound Vibration* 25, (1972) pp.395-405.

27. Meirovitch L.: *Elements of Vibration Analysis.* Mc. Graw-Hill , Inc., (1975).

28. Minorsky N.: *Nonlinear Oscillations.* D. Van Nostrand Company, Inc., (1962), 714p.

29. Mitropolskii Yu.A.:

a) *Problems of the Asymptotic Theory of Nonstationary Oscillations* (in Russian), Nauka, Moscow, (1964), 431p.

b) *Averaging Method in Nonlinear Mechanics* (in Russian), Naukova Dumka, Kiev, (1971), 440p.

c) *Certains Aspects des Progres de la Methode de Centrage,* Edizioni cremonese, Roma (1973).

d) *Fundamental Trends in the Theory of Nonlinear Oscillations and their Development* (in Russian). *Proceedings of ICNO,* Kiev, Tom I, (1961), pp. 15-22.

e) The Development of Averaging Method. *Proceedings of ICNO IX.,* Kiev (1981).

30. Mitropolskii Yu.A. and Moseenkov B.I.: *Asymptotic Solutions of Equations with Partial Derivatives* (in Russian), Kiev, (1976), 590p.

31. Mitropolskii Yu.A. and Samoilenko A.M.: *On Asymptotic Expansions in Nonlinear Mechanics* (in Russian). *Ukrainian Math. Journal,* Kiev, No.1, (1979), pp.42-53.

32. Mitropolskii Yu.A., Homa G.P.: *The Mathematical Foundation of Asymptotic Methods of Nonlinear Mechanics* (in Russian). Naukova Dumka, Kiev, (1983), 215p.

33. Mitropolskii Yu.A., Lopatin A.K.: *Theoretico - Group Approach in the Asymptotic Methods of Nonlinear Mechanics* (in Ruissian). Naukova Dumka, Kiev, (1988), 272p.

34. Mitropolskii Yu.A., Nguyen Van Dao, Nguyen Dong Anh.: *Nonlinear Oscillations in the Systems of Arbitrary Order* (in Russian). Naukova Dumka, Kiev, (1992), 329p.

35. Moiseev N.N.: *Asymptotic Methods of Nonlinear Mechanics* (in Russian). Nauka, Moscow, (1969), 380p.

36. Nayfeh A.H. and Mook D.T.: *Nonlinear Oscillations.* Wiley, (1979).

37. Nguyen Van Dao:

a) *Fundamental methods of the theory of nonlinear oscillations.* Education Publishers, Hanoi (1971), 193p.

b) *Nonlinear Oscillations of Higher Order Systems.* NCSR Vietnam, Hanoi (1979), 64p.

c) *An Approximate Method for Investigation of the Stability of Motion in the Critical Case. Bulletin de L'Academie Polonaise des Sciences,* No.2, (1971), pp. 17-21.

d) *On the Phenomenon of Parametric Resonance of a Nonlinear Vibrator Under the Action of Electromagnetic Force. Proc. of Vibration Problems,* Warsaw, Poland, No. 3, (1972), pp. 281-291.

e) *Interaction between Parametric and Forced Oscillations in Multidimensional Systems. J. of Technical Physics,* Warsaw, Poland, No.3, (1975).

f) *Interaction of Subharmonic Oscillations. J. of Technical Physics,* Warsaw, Poland, No.2, (1975), pp. 227-237.

g) *Some Properties of the Generalized Van der Pol Equation. Journal of Technical Physics,* Warsaw, Poland No.2, (1976), pp. 183-190.

h) *Parametric Oscillations of Dynamical Systems Under the Influence of Nonlinear Frictions. Acta Mathematica Vietnam.* Tom 3, No.2, (1978).

i) *Asymptotic Method for Study of Quasilinear Differential Equations of Higher Order. Asymptotic Methods of Mathematical Physics,* Kiev, Naukova Dumka, (1988), pp. 191-203.

j) *Lanchester Damper and Voigt Dynamic Absorber for Quenching Self-excited Vibrations. Proccedings of VIII International Congress on Theory of Mechanisms & Machines,* Prague, vol.6, August (1991).

k) *Dynamic Absorber Effect for Quenching Self-excited Vibration of Mechanical systems with Limit Energy Resource. Nonlinear Problems of Machine Dynamics,* Moscow, Nauka, (1992), pp. 63-69.

38. Nguyen Van Dao, Nguyen Van Dinh:

a) *Dynamic Absorber Effect for Self-excited Systems. Advances in Mechanics,* Warsaw, vol. 14, No.1, (1991).

b) *Dynamic Absorber for Systems with Distributed Parameters. Proccedings of NCSR Vietnam,* vol.2, (1990), pp. 3-23.

c) *Asymptotic Method for Investigation of Multifrequency Oscillations of Quasilinear System of Second Order. Proceedings of NCSR Vietnam,* Vol. 1, No 1, (1987), pp. 10-15.

39. Nguyen Van Dinh:

a) *The Tuned Absorber in Self-excited System. Journal of Mechanics,* Hanoi, No. 3-4, (1979), pp. 21-30.

b) *The Dynamics Absorber in Quasilinear Systems.* Dissertation, Hanoi, (1980), 62p.

40. Osinski Z.:

a) *Vibration Theory.* Warsaw, Poland (1978).

b) *Damping of Mechanical Vibrations.* Warsaw, Poland (1979).

41. Panovko Ya.G.: *Introduction to the Theory of Mechanical Oscillations* (in Russian). Nauka, Moscow, (1977), 233p.

42. Pisarenko G.S.: *Energy Dissipation in Mechanical Vibrations* (in Russian). Kiev (1962), 436p.

43. Pontryagin L.S.: *Ordinary Differential Equations* (in Russian). Fizmatgiz, Moscow, (1961), 311p.

44. Roseau M.: *Vibrations in Mechanical Systems.* Springer Verlag, (1989), 515p.

45. Rubanik V.P. *Oscillations in Quasilinear Systems with Time Lag.* (in Russian) Nauka, Moscow, (1969), 287p.

46. Schmidt G.: *Parametererregte Schwingungen.* Berlin, (1975), 313p.

47. Schmidt G. and Tondl A.: *Nonlinear Vibrations.* Cambridge Univ. Press. and Akademic-Verlag (1986).

48. Starzhinskii V.M.: *Applied Methods in the Theory of Nonlinear Oscillations.* Mir Publishers, Moscow, (1980), 264p.

49. Stoker J.J.: *Nonlinear Vibrations.* Interscience Publishers LTD, London, (1957).

50. Tondl A.:

a) *Quenching of Self-Excited Vibrations. Equilibrium Aspects. Journal Sound and Vibration 42* (1975) pp. 251-260.

b) Quenching of Self-Excited Vibration: *One and two frequency vibration. Journal Sound and Vibration* 42 (1975) pp. 261-271.

c) Quenching of Self-excited Vibrations : *Effect of Dry Friction. Journal Sound and Vibration* 45, (1976) pp. 285-294.

51. Tran Kim Chi:

a) *Asymptotic Solutions of Quasilinear Differential Equations of Third Order with Delay. Ucrainian Math. Journal* (in Russian), No.3, (1983), pp. 392-397.

b) *Asymptotic Method for Construction of Differential Equations of $N$ - Order with Slowly Varying Parameters* (autonomous case) (in Russian), *Ucrainian Math. Journal,* T. 33, No 3, (1981), pp. 427-429.

c) *Asymptotic Method for Construction of Differential Equations of $N$ - Order with Slowly Varying Parameters in Non-autonomous Case* (in Russian), *Ucrainian Math. Journal,* T. 33, No 4, (1981), pp. 567-570.

d) *Parametric Oscillation with High Degree Term at Modulation Depth. Proc. of 11th International Conference on Nonlinear Oscillations,* Budapest, (1987), pp. 498-501.

52. Tran Kim Chi, Nguyen Van Dao:

a) *Construction of the Solutions of Nonlinear Higher Order Differential Equations. Acta Technica CSAV,* Prague, No.3, (1980), pp. 358-368.

b) *Nonlinear Differential Equation with Self-excited and Parametric Excitations. The Colloquium on Differential Equations and Applications.* Budapest, Hungary, (1991), pp. 65-73.

c) *Asymptotic Method for Studying of the Quasilinear Oscillations of Dynamic Systems of Higher Order. Advances in Mechanics,* Warsaw, vol. 3, No 4, (1980), pp. 3-21.

53. Volosov V.M. and Morgunov V.I.: *The Method of Averaging in the Theory of Nonlinear Systems.* (in Russian), Izd-vo, MGU, Moscow, (1971).

# INDEX

## J



## L



## M



## O

P



Q



R

# Mechanics

## *SOLID* MECHANICS AND ITS APPLICATIONS

*Series Editor:* G.M.L. Gladwell

44. D.A. Hills, P.A. Kelly, D.N. Dai and A.M. Korsunsky: *Solution of Crack Problems.* The Distributed Dislocation Technique. 1996 ISBN 0-7923-3848-0
45. V.A. Squire, R.J. Hosking, A.D. Kerr and P.J. Langhorne: *Moving Loads on Ice Plates.* 1996 ISBN 0-7923-3953-3
46. A. Pineau and A. Zaoui (eds.): *IUTAM Symposium on Micromechanics of Plasticity and Damage of Multiphase Materials.* Proceedings of the IUTAM Symposium held in Sèvres, Paris, France. 1996 ISBN 0-7923-4188-0
47. A. Naess and S. Krenk (eds.): *IUTAM Symposium on Advances in Nonlinear Stochastic Mechanics.* Proceedings of the IUTAM Symposium held in Trondheim, Norway. 1996 ISBN 0-7923-4193-7
48. D. Ieşan and A. Scalia: *Thermoelastic Deformations.* 1996 ISBN 0-7923-4230-5
49. J. R. Willis (ed.): *IUTAM Symposium on Nonlinear Analysis of Fracture.* Proceedings of the IUTAM Symposium held in Cambridge, U.K. 1997 ISBN 0-7923-4378-6
50. A. Preumont: *Vibration Control of Active Structures.* An Introduction. 1997 ISBN 0-7923-4392-1
51. G.P. Cherepanov: *Methods of Fracture Mechanics: Solid Matter Physics.* 1997 ISBN 0-7923-4408-1
52. D.H. van Campen (ed.): *IUTAM Symposium on Interaction between Dynamics and Control in Advanced Mechanical Systems.* Proceedings of the IUTAM Symposium held in Eindhoven, The Netherlands. 1997 ISBN 0-7923-4429-4
53. N.A. Fleck and A.C.F. Cocks (eds.): *IUTAM Symposium on Mechanics of Granular and Porous Materials.* Proceedings of the IUTAM Symposium held in Cambridge, U.K. 1997 ISBN 0-7923-4553-3
54. J. Roorda and N.K. Srivastava (eds.): *Trends in Structural Mechanics.* Theory, Practice, Education. 1997 ISBN 0-7923-4603-3
55. Yu. A. Mitropolskii and N. Van Dao: *Applied Asymptotic Methods in Nonlinear Oscillations.* 1997 ISBN 0-7923-4605-X

Kluwer Academic Publishers – Dordrecht / Boston / London

MIX
Papier aus verantwortungsvollen Quellen
Paper from responsible sources
FSC® C105338

If you have any concerns about our products,
you can contact us on
**ProductSafety@springernature.com**

In case Publisher is established outside the EU,
the EU authorized representative is:
**Springer Nature Customer Service Center GmbH**
**Europaplatz 3, 69115 Heidelberg, Germany**

Printed by Libri Plureos GmbH
in Hamburg, Germany